Group Theory and Numerical Analysis

Volume 39

CRM PROCEEDINGS & LECTURE NOTES

Centre de Recherches Mathématiques
Université de Montréal

Group Theory and Numerical Analysis

P. Winternitz
D. Gomez-Ullate
A. Iserles
D. Levi
P. J. Olver
R. Quispel
P. Tempesta
Editors

The Centre de Recherches Mathématiques (CRM) of the Université de Montréal was created in 1968 to promote research in pure and applied mathematics and related disciplines. Among its activities are special theme years, summer schools, workshops, postdoctoral programs, and publishing. The CRM is supported by the Université de Montréal, the Province of Québec (FCAR), and the Natural Sciences and Engineering Research Council of Canada. It is affiliated with the Institut des Sciences Mathématiques (ISM) of Montréal, whose constituent members are Concordia University, McGill University, the Université de Montréal, the Université du Québec à Montréal, and the Ecole Polytechnique. The CRM may be reached on the Web at www.crm.umontreal.ca.

American Mathematical Society
Providence, Rhode Island USA

2000 *Mathematics Subject Classification.* Primary 34–06, 35–06; Secondary 20–06, 39–06.

Library of Congress Cataloging-in-Publication Data

Workshop on Group Theory and Numerical Analysis (2003 : Montréal, Québec)
Group theory and numerical analysis / P. Winternitz...[et al.], editors.
p. cm. — (CRM proceedings & lecture notes, ISSN 1065-8580; v. 39)
Includes bibliographical references.
ISBN 0-8218-3565-3 (alk. paper)
1. Group theory–Congresses. 2. Numerical analysis–Congresses. I. Winternitz, Pavel. II. Title III. Series.

QA174.W67 2003
512′.2—dc22 2005048328

♾ The paper used in this book is acid-free and falls within the guidelines
established to ensure permanence and durability.
This volume was submitted to the American Mathematical Society
in camera ready form by the Centre de Recherches Mathématiques.
Visit the AMS home page at `http://www.ams.org/`

10 9 8 7 6 5 4 3 2 1 10 09 08 07 06 05

Contents

Preface

The Workshop on Group Theory and Numerical Analysis, organized under the auspices of the Centre de Recherches Mathématiques, Université de Montréal, took place in Montréal, Québec, May 26–31, 2003.

This meeting brought together scientists working in several different but mutually related areas. The unifying theme was the application of group theory and geometrical methods to the solution of differential and difference equations. The emphasis was on the combination of analytical and numerical methods and also the use of symbolic computation.

The workshop attracted 51 participants from 9 countries: Canada (24), USA (12), UK (5), Norway (3), Mexico (2), Russia (2), Australia (1), Italy (1), and Ukraine (1). A total of 34 lectures was presented.

The 20 contributions presented in this volume cover several (mutually overlapping) lines of research:

- *Lie group integrators, geometric integrators, exponential integrators* (Talks presented by A. M. Bloch/A. Iserles, E. Celledoni, D. Lewis, E. Mansfield/ G. R. W. Quispel, and B. Owren).

- *Symbolic computation and solutions of ordinary differential equations, partial differential equations and differential-difference equations* (Talks presented by D. Baldwin, W. Hereman, E. S. Cheb-Terrab, and G. Reid).

- *Symmetry preserving discretization of ordinary and partial differential equations* (Talks presented by C. Cyr-Gagnon, R. Kozlov).

- *Discrete and finite Fourier transforms and data processing* (Talks presented by A. M. Atoyan, J. Patera and K. B. Wolf).

- *Boundary layer perturbation theory and its symmetry respecting discretization* (Talk by C. M. Bender).

- *Discrete symmetries of difference and differential difference equations* (Talk by D. Levi).

- *Numerical methods for treating rapid oscillations* (Talk by A. Iserles).

- *Orthogonal polynomials related to Padé approximations* (Talk by A. Zhedanov).

- *Applications in biophysics and physics* (Talks by J. Tuszynski and W. Zakrzewski).

This volume has the character of a monograph and should represent a useful reference book for scientists working in this highly topical field.

Among the review talks presented at the Workshop, but not included in these Proceedings, we mention that at least two are available elsewhere, namely P. J. Olver's talk corresponds to the publication of P. Kim and P. J. Olver, *Geometric integration via multi-space*, Regular and Chaotic Dynamics **9** (2004), no. 3,

213–226 and P. Winternitz's talk is included in D. Levi, P. Winternitz, Continuous Symmetries of Difference Equations, in nlin.SI/0502004.

The Workshop Organizing Committee consisted of

- P. Winternitz (Chair, CRM, Université de Montréal, Canada).
- D. Gomez-Ullate (Universitat Politècnica de Catalunya, Barcelona, Spain).
- A. Iserles (Cambridge University, United Kingdom).
- D. Levi (Universita' degli Studi Roma Tre, Italy).
- P. J. Olver (University of Minnesota, Minneapolis, USA).
- R. Quispel (La Trobe University, Melbourne, Australia).
- P. Tempesta (SISSA, Trieste, Italy).

The organizers thank the CRM and NSF for financial support making the Workshop possible.

P. Winternitz
Montréal, April 2005

List of Participants

Ana Inés Ansaldo
Institut universitaire de
Gériatrie de Montréal
ana.ines.ansaldo@umontreal.ca

Armen M. Atoyan
Université de Montréal
atoyan@CRM.UMontreal.CA

Douglas Baldwin
Colorado School of Mines
dbaldwin@mines.edu

Carl Bender
Washington University
cmb@wuphys.wustl.edu

Mhenni Benghorbal
London, Ontario
mbenghor@uwo.ca

Anthony Bloch
University of Michigan
abloch@umich.edu

Anne Bourlioux
Université de Montréal
bourliou@dms.umontreal.ca

Elena Celledoni
NTNU
elenac@math.ntnu.no

Edgardo S. Cheb-Terrab
Waterloo, Ontario
ecterrab@maplesoft.com

Jeongoo Cheh
University of Minnesota
cheh@math.umn.edu

Peter A. Clarkson
University of Kent
p.a.clarkson@ukc.ac.uk

Catherine Cyr-Gagnon
Université de Montréal
cyr@dms.umontreal.ca

Vladimir Dorodnitsyn
Russian Academy of Sciences
dorod@spp.keldysh.ru

David Gomez-Ullate
Università di Bologna
david.gomez-ullate@@upc.edu

Alfred Michel Grundland
Univ. du Québec à Trois-Rivières
grundlan@CRM.UMontreal.CA

Willy Hereman
Colorado School of Mines
whereman@mines.edu

Véronique Hussin
Université de Montréal
hussin@dms.umontreal.ca

Arieh Iserles
University of Cambridge
a.iserles@damtp.cam.ac.uk

Pilwon Kim
University of Minnesota
pwkim@math.umn.edu

Roman Kozlov
University of Oslo
kozlov@ifi.uio.no

Frédéric Laliberté
McGill University
flalib@po-box.mcgill.ca

François Lemaire
University of Western Ontario
lemaire@orcca.on.ca

Decio Levi
Universita' di Roma Tre University
levi@fis.uniroma3.it

Debra Lewis
University of California
lewis@math.ucsc.edu

Simon J. Malham
Heriot-Watt University
simonm@ma.hw.ac.uk

Elizabeth L. Mansfield
University of Kent at Canterbury
E.L.mansfield@ukc.ac.uk

Peter J. Olver
University of Minnesota
olver@math.umn.edu

Brynjulf Owren
NTNU
brynjulf.owren@math.ntnu.no

Jiri Patera
Université de Montréal
patera@CRM.UMontreal.CA

Alexei V. Penskoi
Université de Montréal

G. Reinout W. Quispel
La Trobe University
r.quispel@latrobe.edu.au

Greg Reid
University of Western Ontario
reid@uwo.ca

David Richter
William Paterson university
richterd@wpunj.edu

Nicolas Robidoux
Laurentian University
nrobidoux@pims.math.ca

Robin Scott
University of Western Ontario
rscott2@uwo.ca

Vadim Shapiro
University of Wisconsin (Madison)
vshapiro@engr.wisc.edu

Neil Stewart
Université de Montréal
stewart@iro.UMontreal.CA

Piergiulio Tempesta
SISSA
tempesta@fm.sissa.it

Zora Thomova
SUNY-Institute of Technology
thomovz@sunyit.edu

Paul Tupper
McGill University
tupper@math.mcgill.ca

Alexander Turbiner
UNAM
turbiner@nuclecu.unam.mx

Jacek Tuszynski
University of Alberta
jtus@phys.ualberta.ca

Francis Valiquette
Université de Montréal
francis.valiquette@umontreal.ca

Pavel Winternitz
Université de Montréal
wintern@CRM.UMontreal.CA

Kurt Bernardo Wolf
Universidad Nacional Autonoma
de Mexico
bwolf@fis.unam.mx

Thomas Wolf
Brock University
twolf@brocku.ca

Ravil Yamilov
Russian Academy of Sciences
yamilov@imat.rb.ru

Wojciech J. M. Zakrzewski
University of Durham
w.j.Zakrzewski@durham.ac.uk

Alexander Zhalij
Institute of Mathematics, Kiev, Ukraine
zhaliy@imath.kiev.ua

Alexei Zhedanov
Donetsk Institute for Physics and Technology
zhedanov@kinetic.ac.donetsk.ua

Wenqin Zhou
University of Western Ontario
wzhou7@uwo.ca

Centre de Recherches Mathématiques
CRM Proceedings and Lecture Notes
Volume **39**, 2005

Continuous Extension of the Discrete Cosine Transform, and Its Applications to Data Processing

A. M. Atoyan and J. Patera

ABSTRACT. We are developing methods for practical calculation of discrete transforms on compact semisimple Lie groups, abbreviated hereafter as DGT for *Discrete Group transform*. DGT provides a pair of exact Fourier transforms, direct and inverse, of a discrete function g_j produced from a continuous function $g(t)$ at the points $\{t_j \mid j = 0, 1, \dots, N\}$ of a uniform grid F_N of elements of finite adjoint order N in the fundamental region F of the relevant Lie group. For the simplest implementation of this method, corresponding to the group SU(2), the fundamental region F_N is reduced to a one-dimensional segment, and the DGT happens to result in the *discrete cosine transform* (DCT). It is often considered to be basically a specific case of the *standard* discrete Fourier transform, abbreviated typically as DFT. Here we show, however, that the DCT is in fact very different from the DFT when the properties of their *continuous extension* (CE) from the discrete $\{t_j \in F_N\}$ to all $t \in \mathsf{F}$ are studied. Unlike the continuous extension of DFT:

(a) the CEDCT function $f_N(t)$ uniformly converges to $g(t)$ at all $t \in F$ with increasing N;
(b) the derivative of $f_N(t)$ converges to the derivative of $g(t)$; and
(c) the *localization principle* holds for CEDCT.

Similar are the properties of the multi-dimensional DCTs, which are separable transforms, formally corresponding to the Lie groups SU(2) × SU(2). Here we show, on the example of 2-dimensional DCT, that these properties become very useful for a number of practical applications of CEDCT, such as interpolation and zooming of images, image & data compression, etc.

1. DCT as a discrete Fourier transform on SU(2) group

The Discrete Cosine Transform (DCT), which is known since 1974 [**1**], and is often considered as a specific case of the standard Discrete Fourier Transform (DFT), has proven to be very effective for a number of practical applications, with general performance characteristics exceeding the ones of the DFT (see [**10**] for a review). As we have recently shown in [**2**], the DCT happens to implement the

---

2000 *Mathematics Subject Classification.* Primary: 65T40, 65T50; Secondary: 94A08.

The authors acknowledge partial support of the National Science and Engineering Research Council of Canada, of NATEQ, of Lockheed Martin Canada, and of NATO.

This is the final form of the paper.

one-dimensional case of a broad class of generally multi-dimensional discrete transforms on the compact semisimple Lie groups [**4**]. The simplest possible case, which corresponds to the group SU(2). Generally, these Fourier transforms make 'full use' of the symmetry properties intrinsic to the group. That provides qualitative explanation for the good properties of such transforms.

In general, there are two ingredients and a basic property on which our method is based. These are the *elements of finite order* (EFO's) [**3,4**] and the *orbit functions* [**4, 9**] of the group, and the property of orthogonality of a subset of orbit functions [**4**] on a discrete grid of points in $\mathbb{R}^n$.

Recall that an EFO is an element X of the group G (generally, of a rank n), such that $X^N = I$ for some natural number N, and that it is described by a vector in the fundamental region $\mathsf{F} \subset \mathbb{R}^n$ of the group G. There is a one-to-one correspondence between elements of F and conjugacy classes of elements of G.

The action of the Weyl group W of G transforms each vector $\lambda \in \mathsf{F}$ into a finite set $W(\lambda) \subset \mathbb{R}^n$ of vectors equidistant from the origin, which makes the Weyl group orbit of λ. An orbit function $\Phi_\lambda(r)$ is defined [**4,5**] as the finite sum of exponentials through this set,

$$\Phi_\lambda(r) = \sum_{\mu \in W(\lambda)} e^{2\pi i (\mu | r)}, \quad r \in \mathbb{R}^n. \tag{1.1}$$

Here $(\mu \mid r)$ is the scalar product in $\mathbb{R}^n$. Although $\Phi_\lambda(r)$ is clearly defined (and well behaved) in the entire real Euclidean space $\mathbb{R}^n$, we are particularly interested in its values when $r \in \mathsf{F}$, and furthermore, when the coordinates of r are rational in a suitable basis naturally defined by the geometry of F (basis of fundamental weights).

The simplest example of the general problem of discrete Fourier transforms on compact Lie groups (hereafter abbreviated DGT for Discrete Group Transform), is the case of rank-1 compact Lie group SU(2).

Referring for details to [**2**], we remind that the Lie group SU(2) can be realized as a set of all unitary 2×2 matrices A, with determinant equal to one. Any unitary matrix can be diagonalized by a unitary transformation. Therefore every element of SU(2) is conjugate to a diagonal matrix X in the defining 2-dimensional representation. All elements which can be simultaneously diagonalized form a maximal torus T of SU(2):

$$\mathsf{T} = \left\{ X(\theta) = \begin{pmatrix} e^{2\pi i\theta} & 0 \\ 0 & e^{-2\pi i\theta} \end{pmatrix} \;\middle|\; -\frac{1}{2} < \theta \le \frac{1}{2} \right\}. \tag{1.2}$$

Using $B = \left(\begin{smallmatrix} 0 & 1 \\ -1 & 0 \end{smallmatrix}\right) \in \mathrm{SU}(2)$, we have $BX(\theta)B^{-1} = X(-\theta)$. Therefore every element of SU(2) is conjugate to just one element in the subset $F \subset T$, where

$$F = \left\{ X(\theta) = \begin{pmatrix} e^{2\pi i\theta} & 0 \\ 0 & e^{-2\pi i\theta} \end{pmatrix} \;\middle|\; 0 \le \theta \le \frac{1}{2} \right\}. \tag{1.3}$$

The fundamental region F of SU(2) is thus the closed segment $\mathsf{F} = [0, \frac{1}{2}]$. The Weyl group orbit $W(\lambda)$ of an element λ consists of $\{\lambda, -\lambda\}$, since in case of SU(2) group the action of the Weyl group is reduced to reflection of λ with respect to the point of origin. In case of the element $\lambda = 0$ the Weyl group consists of only one element.

Consider now only natural $\lambda = k \in \mathbb{Z}^{\ge 0}$. Orbit functions (1.1) in case of SU(2) are then reduced to cosine functions of harmonic order $k/2$, which can be therefore both integer and half-integer:

$$\Phi_k(\theta) = \begin{cases} e^{2\pi i k\theta} + e^{-2\pi i k\theta} = 2\cos(2\pi k\theta) & \text{for } k > 0, \\ 1 & \text{for } k = 0. \end{cases} \tag{1.4}$$

For any $N \in \mathbb{Z}^{>0}$, the point set $S = \{\theta_j = \frac{1}{2}jN \mid j = 0, 1, \ldots, N\}$ determines the conjugacy classes of EFO's whose adjoint order divides N. That is, $X(\theta_j)^N = I$, where I is the identity matrix, $I = \left(\begin{smallmatrix} 1 & 0 \\ 0 & 1 \end{smallmatrix}\right)$.

Consider the orbit functions $\Phi_k(\theta)$ on the set S,

$$\tfrac{1}{2}\Phi_k(\theta_j) = \psi_k(\theta_j) = \cos(\pi kj/N)\,.$$

Note that this is exactly the basis for the well-known *discrete cosine transform* (DCT), or, according to [**1, 10**], the DCT of type 1. The fundamental *discrete orthogonality* property of orbit functions on a suitable grid, in the SU(2) case is reduced to a known equality

$$\langle \psi_k, \psi_m \rangle_N = \sum_{j=0}^{N} C_{N,j} \cos\frac{\pi jk}{N} \cos\frac{\pi jm}{N} = \frac{2N}{C_{N,k}}\delta_{km}, \quad 0 \le k, m \le N, \tag{1.5}$$

where $C_{N,k} = 1$ for $k = 0, N$ and $C_{N,k} = 2$ otherwise, and δ_{km} is the Kroneker symbol.

The orthogonality equation (1.5) is used for decomposition of a discrete function $\{g_k \mid k = 0, 1, \ldots, N\}$, given at $(N+1)$ points of a uniform grid $\{t_k = k(T_0/N) \mid k = 0, 1, \ldots, N\}$ into the series of $N+1$ cosine functions of discrete argument.

Let $\{g_k = g(t_k)\}$ be the values of some continuous function $g(t) \mid t \in [0, T_0])$ at the grid points $\{t_k \mid k = 0, 1, \ldots, N\}$. The "discrete transform of $g(t)$" onto the basis $\{\psi_m \mid 0, 1, \ldots, N\}$ corresponds to finding the coefficients $\{a_m \mid m = 0, 1, \ldots, N\}$ in the set of $(N+1)$ linear transformation equations

$$g_k = \sum_{m=0}^{N} a_m \psi_m(\theta_k) = \sum_{m=0}^{N} a_m \cos\frac{\pi mk}{N}, \quad 0 \le k \le N. \tag{1.6}$$

Using the orthogonality (1.5), one can invert (1.6):

$$a_j = \sum_{k=0}^{N} D_{jk}^{[N]} g_k \quad \text{where} \quad D_{jk}^{[N]} = \frac{C_{N,j}C_{N,k}}{2N} \cos\frac{\pi jk}{N}. \tag{1.7}$$

Here $D_{jk}^{[N]}$ are the elements of $(N+1) \times (N+1)$ matrix $D^{[N]}$ of the DGT on SU(2), which is easily reduced to the transform matrix of the DCT-1 in its common form [**10**] after renormalization by a factor $\sqrt{2C_{N,j}C_{N,k}/N}$.

In the current literature the DCT is often being considered as a particular case of another discrete transform, namely the standard *discrete Fourier transform* abbreviated generally as DFT. The DFT is given by (e.g. see [**7, 8**])

$$u_j = \frac{1}{N} \sum_{k=0}^{N-1} g_k e^{-i/2\pi Nkj}, \tag{1.8}$$

for the direct transform, and the inverse transform is:

$$g_k = \sum_{j=0}^{N-1} u_j e^{i/2\pi Nkj}. \tag{1.9}$$

Thus, it is based on the integer modes of the exponential function $e^{2\pi i x}$ for the normalized variable $x = t/T_0$. Note that the DFT includes only the first N values of the N-interval grid function, but not the last $(N+1)$th one, because the full-order exponential harmonics are periodic functions with a period T_0, so the DFT forces a "recovery of the periodicity" of $g(t)$ at the last, $(N+1)$th point. Meanwhile, the DCT is formally based on the harmonic modes $\{2\pi kx\}$ up to order $k \leq N/2$, but also includes the modes of half-integer order $(k+1/2)$, which allow $g(0) \neq g(T_0)$.

The reason why DCT is considered as a particular case of DFT, albeit with a better performance characteristics (see [**10**]), is because the N-interval DCT can be formally calculated through the 2N-interval DFT coefficients. Consider an even extension of the function $\{g_k\}$ with $k = 0, \ldots, N$ onto the interval $[T_0, 2T_0]$, such that $g_{2N-k} = g_k$ at all $(2N+1)$ sampling points t_k of the $(2N)$-interval grid on the segment $[0, 2T_0]$. The $2N$-point DFT coefficients of this new function can be represented as

$$
\begin{aligned}
(1.10)\qquad u_j^{(2N)} = \frac{1}{2N}\Bigg(g_0 + \sum_{k=1}^{N-1} g_k e^{-2\pi i k j/2N} + g_N e^{-\pi i j} \\
+ \sum_{k_1=1}^{N-1} g_{k_1} e^{-2\pi i k_1 j/2N(2N-k_1)}\Bigg) \\
= \sum_{k=0}^{N} \frac{C_{N,k}}{2N} \cos\left(\frac{\pi j k}{N}\right) g_k = \frac{1}{C_{N,j}} a_j^{(N)}.
\end{aligned}
$$

Thus, the first $\{j = 0, 1, \ldots, N\}$ coefficients of such $2N$-point DFT represent a linear rescaling of the coefficients of the $(N+1)$-point DCT with a factor $\frac{1}{2}$, except for the cases $j = 0$ and $j = N$ when $C_{N,J} = 1$. Despite this fact, these 2 discrete transforms appear as qualitatively very different, when we consider the properties of the *continuous extensions* of the (*inverse*) DCTs and DFTs.

2. Continuous extension of the DCT

Recall that the inverse transforms of both DCT and DFT, given by (1.6) and (1.9), correspond to "lossless" discrete transforms, as far as they return the initial sets $\{g_k\}$ at the given (equidistant) grid points $\{t_k\}$. It seems then natural also to ask how well it is possible to recover the original continuous function $g(t)$ by a Fourier series also on the entire continuous segment $[0, T_0]$. Putting for simplicity $T_0 = 1$, in which case k/N is simply equal to t_k, let us define the *continuous extension of* DCT (hereafter CEDCT) between the grid points as follows.

Proposition 1. *Let $g(t)$ be a complex valued function of $t \in [0,1]$, taking values $g_k = g(t_k)$ on an equidistant point grid $\{t_k = k/N \mid k = 0, 1, \ldots, N\}$. The function*

$$
(2.1)\qquad f_N(t) = \sum_{j=0}^{N} a_j \cos(\pi j t), \quad t \in [0,1],
$$

with the discrete transform coefficients

$$
(2.2)\qquad a_j = \sum_{k=0}^{N} \frac{C_{N,j} C_{N,k}}{2N} g_k \cos\frac{\pi j k}{N},
$$

represents a continuous Fourier extension of the (inverse) DCT *of $\{g_k\}$, which is exact at all $N+1$ points of the grid, $f(t_k) = g_k$.*

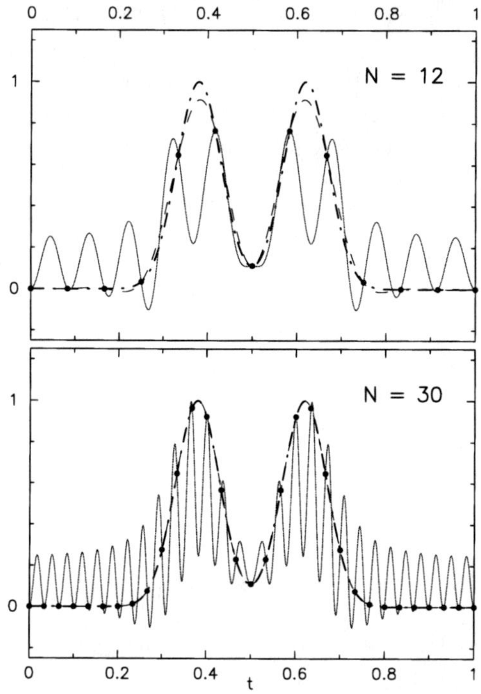

FIGURE 1. Approximation of an analytic function $g(t)$ (dot-dashed curves), composed of 2 Gaussian functions with dispersions $\sigma = 0.05$ each, by the CEDCT function $f_N(t)$ (dashed curves) for $N = 12$ and $N = 30$. The solid curves show continuous extensions of DFT. The big dots show the values of the grid function $\{g_k \equiv g(t_k)\}$ for $\{k = 0, \ldots, N\}$.

Generally, similar to DCT, one also can continuously extend to all points of the segment $[0, 1]$ the discrete inverse transforms of any other types which would be presented in the form $g_k = \sum_m a_m \Psi_m(t_k)$, where $\{\Psi_m(t)\}$ is the basis of the transform. Because $g_k \equiv g(t_k)$, substitution of the discrete argument t_k (which may not necessarily correspond to an *equidistant* grid in $[0, 1]$ in general case) with a continuous variable t will naturally result in a *continuous extension* (CE) of that transform to the points between the discrete grid. In particular, such extension is possible for the standard DFT, which is based on the harmonics $\Psi_m(t) = e^{2\pi imt}$.

In Fig. 2 we show by dashed and solid curves, respectively, the continuous extensions of DCT and DFT for the discrete functions originated from a function $g(t)$ composed of the sum of 2 Gaussians with the dispersion $\sigma = 0.05$ and located symmetrically with respect to the median point $t = 0.5$ of the interval $[0, 1]$. The difference between CEDCT and CEDFT is striking. The reason for the profound oscillations in CEDFT could be qualitatively understood if we note that for the transformation of the N-point vector the DFT and formally uses the trigonometric harmonics $e^{2\pi imt_k}$ of the full order up to $m \leq N - 1$, whereas the DCT is based on the harmonics of both integer and half-integer orders, but only up to full order $m \leq N/2$.

Besides the property of fast convergence of $f_N(t)$ given by (1.6) to the continuous original function $g(t)$, in [**2**] we have also proven some other important properties of CEDCT, which are analogous to the properties of the canonical *continuous* Fourier transform polynomials. These properties include the so called *localization principle* and the *differentiability* of the CEDCT polynomials.

Localization principle. Let a sequence of N-interval grid functions $\{g_k\}_N$, with various N, be originated from a smooth function $g(t)$ with a bounded derivative $g'(t)$ on the interval $[0,1]$. Then at any given $t_0 \in [0,1]$ and for any small $\Delta > 0$ and $\varepsilon > 0$ there exists $N_{\varepsilon,\Delta}$ such that for all $N > N_{\varepsilon,\Delta}$ the value of the continuous extension of the discrete group transform $f_N(t_0)$ is defined within the accuracy ε only by the values of $g(t)$ in the Δ-neighborhood of t_0, i.e.: for any $\varepsilon, \Delta > 0 \exists N_{\varepsilon,\Delta}$ such that for all $N > N_{\varepsilon,\Delta} \Longrightarrow$

$$\left| f_N(t_0) - \sum_{j=0}^{N} a_j^* \cos(\pi j t_0) \right| < \varepsilon,$$

where coefficients a_j^* are calculated according to the equation (2.2), but where the sum is taken only over a small subset of the grid points $\{k'\}$ satisfying the condition $t - \Delta < k'/N < t + \Delta$.

In other words, the *localization* property of the transform means that for large N an interpolation which is provided by $f_N(t)$ at any chosen t^0 depends *only* on the values of g_k in the near vicinity of t_0. This property ensures, for example, that the variations or errors in g_k at some distant segment would not have a significant impact on the result of calculations at the point t_0.

The property of differentiability of CEDCT can be formulated in the following proposition (see [**2**] for a proof):

> Let $g(t)$ be a smooth function with bounded second derivative on the interval $t \in [0,1]$, which produces the N-interval grid function $\{g_k \equiv g(t_k) \mid k = 0, 1, \ldots, N\}$. Then the function $f'_N(t) \equiv df_N/dt$ produced by term-by-term differentiation of the continuous extension of the discrete Fourier transform on SU(2), $f_N(t)$, converges with increasing N to $g'(t)$ at any $t \in (0,1)$.

The property of differentiability of the CEDCT is illustrated in Fig. 2, where on the two panels on the left we show the convergence of the CEDCT function $f_N(t)$, shown by dashed curve, to the original function $g(t) = \exp[-(t/0.17)^4] + 0.7\exp[-(t/0.1)^2]$ (solid curve) with the increase of N. The dot-dashed curves correspond to the trigonometric polynomials $s_K(t)$ in the form corresponding to a canonical Fourier expansion of a continuous function (e.g. see [**12**]), but where we use the DFT coefficients $\{u_j\}$ given by (1.8), instead of the Fourier coefficients found by integration of $g(t)$. Namely,

$$s_K(t) = u_0 + 2\mathrm{Re}\sum_{j=1}^{K} u_j e^{i2\pi jt/T}, \tag{2.3}$$

where the coefficients u_j are given by (1.8). Note the difference in the multiplication factor 2 in this series at $j \geq 1$ as compared with the standard DFT extension of (1.9). We chose $K = N/2$ in order to have a polynomial of the same maximum harmonic order as $f_N(t)$. The discarding of the harmonics of the order exceeding

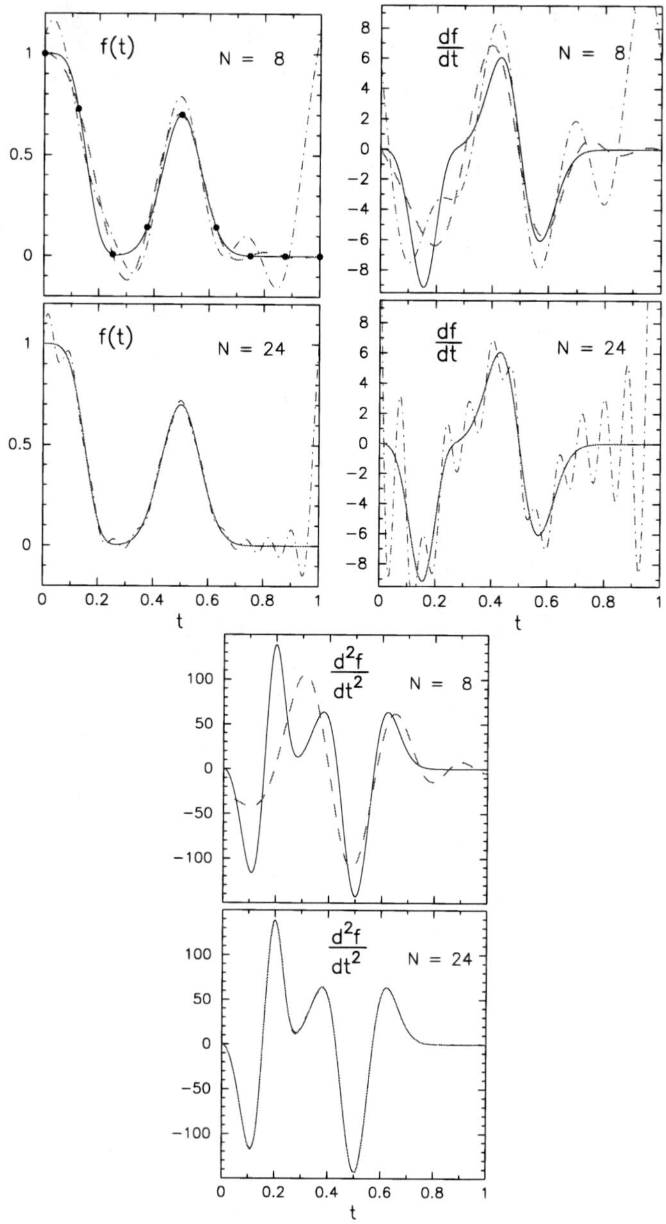

FIGURE 2. Approximations for the continuous function $g(t)$ and its first and second derivatives dg/dt and d^2g/dt^2 (all plotted by solid lines), which are provided by the CEDCT function $f_N(t)$ and its first and second derivatives (all plotted by dashed lines), respectively, in case of $N = 8$ and $N = 24$. The dashed-dotted curves on the left-side and the central panels correspond to the trigonometric polynomials of the order $K = N/2$ with the coefficients calculated in DFT approximation (see text). Big dots on the upper-left panel show the values of the grid function $\{g_k\}$.

$N/2$, as compared with the DFT series, results in the recovery of the convergence property in $s_K(t)$ polynomials with increasing N. This is reached, however, on the expence of loosing the property of *exactness* of that transform at the grid points, which is apparent on the upper-left panel in Fig. 2.

The central panels in Fig. 2 demonstrate convergence of the first derivative of the CEDCT (dashed curves) to the derivative of dg/dt. Meanwhile, the derivative ds_K/dt does not converge, and remains strongly oscillating with further increase of N. The two panels on the right in Fig. 2 show that for this function $g(t)$ also the second derivative $d^2 f_N/dt^2$ converges to $d^2 g/dt^2$. It is worth noticing that although we could not prove it strictly (*unlike* with the case of the first derivative), our numerical study of the problem of differentiability of the second derivative of CEDCT has shown that such property does hold, if the first derivatives of the original function $g(t)$ are equal to zero at both edges $t = 0$ and $t = 1$, as it is the case in Fig. 2. However, unlike the case with the first derivative, we could not prove this property mathematically.

3. Possible applications of CEDCT

Our study of the properties of CEDCTs shows that (unlike continuous extensions of standard DFTs) these continuous trigonometric polynomials produced from discrete functions have properties similar to the properties of canonical continuous Fourier transforms. It is also important that generalization of this transform to any multi-dimensional space is straightforward, as far as multi-dimensional DCT is a separable transform. [1]. The later implies that all good properties of DCT also hold for discrete functions defined on multidimensional rectangular grids. This then suggests a number of practical applications for DCT/CEDCT.

3.1. Continuous interpolation of images. The first possibility would be to use the CEDCT for interpolation of the discrete data, such as two- (or multi-) dimensional images, between the points of the greed. In the 2-dimensional case the CEDCT is reduced to the following proposition [**2**]:

Let $G_{jk} = G(x_j, y_k)$ be values of a bounded function $G(x,y)$ on the rectangular grid points $x_j = jX_0/M$, $y_k = kY_0/N$; $j \in \{0, 1, \ldots, M\}$, $k \in \{0, 1, \ldots, N\}$. A trigonometric function given by finite Fourier series

$$F_{MN}(x,y) = \sum_{m=0}^{M}\sum_{n=0}^{N} A_{mn} \cos\frac{\pi m x}{X_0} \cos\frac{\pi n y}{Y_0}, \tag{3.1}$$

with

$$A_{mn} = \sum_{j=0}^{M}\sum_{k=0}^{N} \frac{C_{M,m}C_{M,j}C_{N,n}C_{N,k}}{4MN} G_{jk} \cos\frac{\pi m j}{M} \cos\frac{\pi n k}{N}, \tag{3.2}$$

continuously extends the (inverse) 2-dimensional DCT of the grid function $\{G_{jk}\}$ onto the rectangular area ($x \in [0, X_0], y \in [0, Y_0]$), and satisfies the equality

$$F_{MN}(x_j, y_k) = G_{jk} \quad \text{for all } j \in \{0, 1, \ldots, M\},\ k \in \{0, 1, \ldots, N\}.$$

[1]Note that from the group-theoretical point of view, multi-dimensional DCT corresponds to the discrete transform on the semisimple Lie group $SU(2) \times \cdots \times SU(2)$

In Fig. 3.1 the upper left panel shows, in terms of brightness distribution, the original grid function produced by 2 Gaussian ellipsoids with dispersions $\sigma_{1,\perp} = \sigma_{2,\perp}$ in transverse directions to the respective major axes equal to the distance $\Delta = 1/N$ between the nearest points of a square grid, $M = N$. The major axes of the ellipsoids are inclined at a small angle 20° to each other, and the distance between the ellipsoids is not very large. It is obvious from that panel, that for the chosen ratio between Δ and $\sigma_\perp$ the granularity of the discrete pattern becomes very damaging for the discrete image. On the upper-right panel of Fig. 3.1 we show the reconstruction of the continuous image in the form of 2-dimensional CEDCT. For comparison, on the bottom left panel we show the contour plot of the exact (i.e. the original) function $G(x, y)$, and the bottom right panel shows its approximation by 2-dimensional CEDGT. It is obvious that the reconstructed CEDGT image not only recovers the directions of the ellipsoids and their maxima, but it practically coincides with the exact image. The difference between these two contour plots can be observed only in the lowest contours on the right plotted by dashed lines corresponding to the level slightly below zero, $F_{MN} = -0.001$.

In order to understand the applicability of CEDCT approach for interpolation of real images (which, unlike the analytic functions, typically contain noise at a certain level), in Fig. 3.1 we use the CEDCT approach for interpolation of the test image "Lena". On the upper panel a fragment of the original (256×256)-pixel image is shown which is zoomed to a large size, such that the granularity of the discrete image becomes apparent. On the panel (b) we show the "continuous" image of that fragment, where the surface density of the points has been increased by a factor $3 \times 3 = 9$ using CEDCT interpolation. For comparison, on the lower panel we show the same fragment, but which is now taken from another, (512×512)-pixel version of the test image "Lena", i.e. also an original image with the resolution 2×2 higher. Comparing Fig. 3.1(b) with Fig. 3.1(c) shows that these 2 images look very similar (perhaps, except for some difference in the brightness levels of the 2 original images used).

Note that the spatial density of dots (or pixels) in Fig. 3.1b is higher than in the Fig. 3.1(c) by a factor $9/4 = 2.25$, and for the CEDCT image it can be easily increased further, allowing "zooming" formally to any high level. However, further zooming of the fragment on the panel (c) by another factor of 2 will make the discrete structure of that image visible. It is also noteworthy that the size of the data file containing all DCT coefficients $\{A_{mn}\}$ needed for calculations of the CEDCT image on Fig. 3.1(b) is by a factor of 4 *smaller* than the data size of the higher resolution image on Fig. 3.1(c).

3.2. Data compression by CEDCT. The latter notion suggests that perhaps the DCT/CEDCT approach could be used more effectively than presently for purposes of image compression. We remind in this respect that in 90-s the DCT has been successfully implemented as the basic transform in the image compression standard JPEG (which stands for "Joint Photographic Experts Group"). In the standard JPEG approach (see [**11**] for detailed description) the image is first divided into adjacent blocks of size 8×8, and then each block of data $\{G_{jk}\}_i$, or a sub-image, is transformed by DCT into the wave-number space, resulting in 64 coefficients $\{A_{jk} \mid j, k = 0, 1 \ldots, 7\}$. Then JPEG uses the property that most of the DCT coefficients corresponding to higher order harmonics are typically much smaller than the coefficients of the lower-order modes (i.e. in a definite sense, the

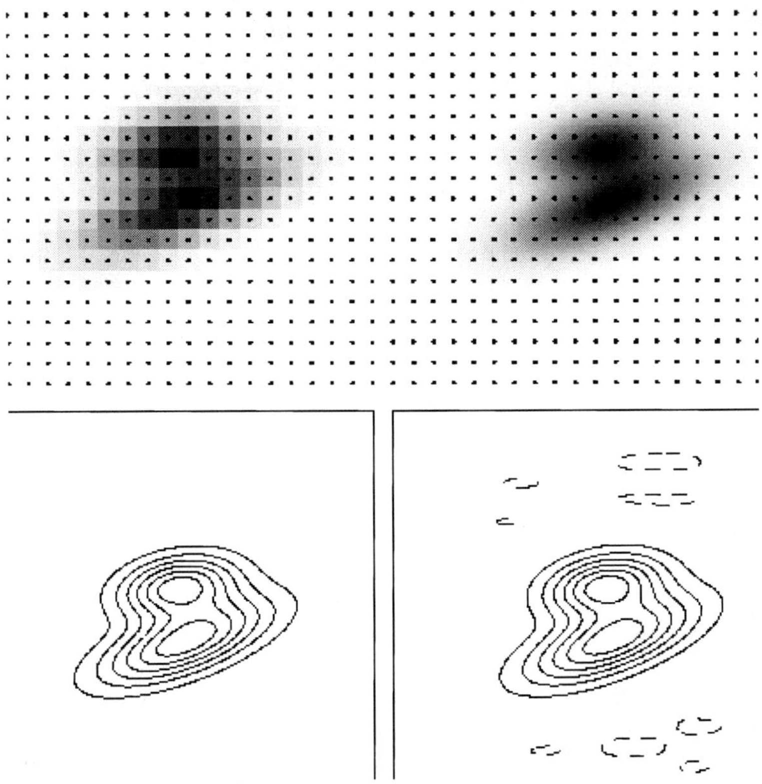

FIGURE 3. **U**pper panels: The images, in terms of brightness distributions, corresponding to the grid function $\{G_{j,k}\}$ shown on the left panel, and its reconstruction by 2-dimensional CEDCT series $F_{M,N}(x,y)$ on the right. **L**ower panels: Contour plots of the analytic function $G(x,y)$ which has produced the discrete function $\{G_{j,k}\}$ (on the left), and of its approximation by 2-dimensional CEDCT function $F_{M,N}(x,y)$ (on the right).

basic information about the image is contained in the "upper-left corner" of the DCT matrix $\{A_{jk}\}$). This allows to discard a very significant amount of the coefficients A_{jk} with absolute values below a certain level (usually to be related with the maximum amplitude in the sub-image), keeping only those coefficients which pass through such "low-pass" filtering procedure. Denoting the mean number of such coefficients averaged over all sub-blocks of the entire image as $C_{\rm pass}$, such procedure by its own would effectively correspond to *compression* of the original image by a factor $R_{\rm DCT} = 64/C_{\rm pass}$ [2]. Note that such a compression is apparently a "lossy compression", i.e. the initial values of $\{G_{jk}\}$ cannot be recovered exactly when the inverse DCT with the truncated set of coefficients $\{A_{jk}\}'$ is applied. However, an attractive feature of DCT is that in this way the JPEG compression could

[2]note that the final JPEG compression value is typically by a factor of 2 higher than $R_{\rm DCT}$, which is reached by an additional (*lossless*) compression at the level of coding, see [11]

(a)

(b)

(c)

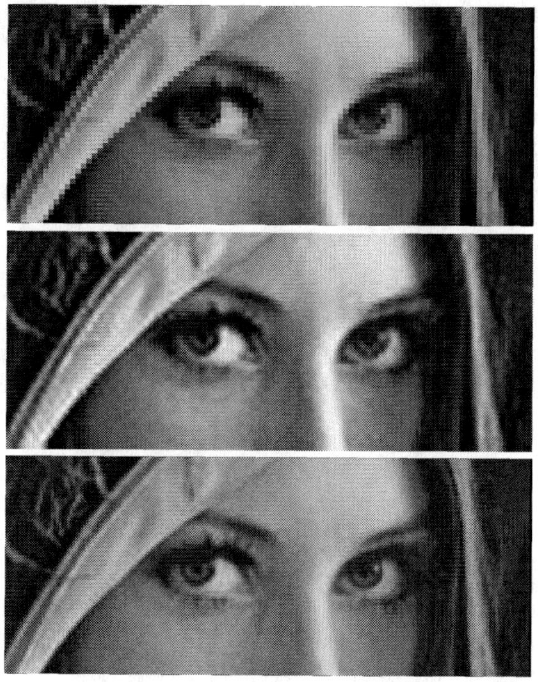

FIGURE 4. A fragment of the original (256×256)-pixel version of the test image "Lena" which is highly "zoomed" (enlarged) to expose the resolution limits of the image (panel (a)), and the CEDCT representation of that fragment (panel (b)) plotted with the spatial density of points by $3 \times 3 = 9$ times higher than the initial grid. For comparison, on panel (c) we show the same fragment, but taken from the other original, (512×512)-pixel version of the image "Lena".

reach very significant levels, while preserving good similarity of the reconstructed compressed image with the original.

The latter should not be very surprising in the view of good convergence properties of CEDCT series discussed above. Because nullification of low-amplitude coefficients A_{jk} corresponds mostly to removal of higher-frequency modes, effectively the process of JPEG compression results in the use of cosine polynomials truncated to some low orders. And if the CEDCT series of order N are really analogs of canonical continuous Fourier transform polynomials of harmonic order $N/2$, then the inverse DCT should provide a reasonable approximation $\{\widetilde{G}_{jk}\}$ for the initial $\{G_{jk}\}$ after truncation to lower harmonic orders as well. Good performance of the standard JPEG is actually a proof of this suggestion.

If so, we can also then suppose that perhaps the use of blocks with sizes larger than 8×8 would allow to reach image compression levels significantly higher than it is possible with the standard JPEG. Indeed, in the latter case there are only 64 coefficients A_{jk} in the image, therefore the DCT compression factor has an absolute maximum $R_{\mathrm{DCT}} \leq 64$ as far as at least one of those coefficients must be kept (e.g. the coefficient A_{00}, which defines the mean brightness in the block). The increase

FIGURE 5. A 240×240 fragment of the original (256 × 256)-pixel version of the test image "Lena" compressed to the levels of $R_{\mathrm{DCT}} \approx 16$ (2 top panels) and $R_{\mathrm{DCT}} \approx 35$ (2 panels in the middle raw), using the block sizes 8×8 (2 panels on the left) and 30×30 (2 panels on the right). The single image in the bottom corresponds to $R_{\mathrm{DCT}} \approx 110$ for the block size 60×60. Note that this would effectively correspond to the total image compression factor > 200 if one takes into account the possibility for further ("lossless") compression by another factor of $\simeq 2$ by coding procedures.

of the block size should then allow a much larger number of coefficients to work with, and application of the truncated CEDCT series would allow even a continuous interpolation and zooming of the compressed image.

In order to check the validity of this hypothesis, in Fig. 3.2 we demonstrate the practicability of this idea by compressing the test image "Lena" using blocks of the size $N_{\rm sub} \times N_{\rm sub}$ with $N_{\rm sub} = 30$, compared with the standard JPEG procedure with $N_{\rm sub} = 8$. Comparison between the images on the left panels with the images on the right corresponding to the same values of the compression parameter $R_{\rm DCT}$ clearly shows advantage of the use of larger sub-image blocks. It is noteworthy that not only the general view of the compressed image is improved, but also the so called "block-effect", which is distortion of the image caused by mismatch of the image intensities on the edges of adjacent blocks due to "lossy" compression, and which is considered as a common weakness of the JPEG compression, becomes significantly less profound with the increase of $N_{\rm sub}$.

On the single panel on the bottom of Fig. 3.2 we show that in case of further increase of $N_{\rm sub}$ to $N_{\rm sub} = 60$, a compression even to levels $R_{\rm DCT} \geq 100$, absolutely excluded in case of $N_{\rm sub} = 8$, would still keep significant similarity with the original image. In that case on average less than 36 coefficients $A_{jk} \neq 0$ from the nominal 3600 $\{A_{jk}\}$ in the CEDCT series per each block is used. Note also that the compression method implemented in Fig. 3.2 corresponds to one of the simplest possible "low-pass" filtering procedures. In particular, we expect that further improvements in the DCT compression would be possible if we make use of not only the convergence property of CEDCT function itself (which actually is in the basis of good performance characteristics of DCT compression), but also of the convergence of the first derivatives of 2-dimensional CEDCT. The latter property would allow compression schemes invoking difference maps of nearest neighbors on the grid. The discussion of such possibilities is, however, outside of the scope of this paper.

4. Summary

Summarizing the results presented above, a Fourier transform of a one-dimensional discrete function $\{g_k \mid k = 0, 1, \ldots, N\}$, which is originated from a continuous function $g(t)$ on the equidistant grid $\{t_k \in [0, T_0]\}$, onto the orbit functions of compact Lie groups, in the case of SU(2) group is reduced to the well known discrete cosine transform, or more exactly to DCT-I. The principal difference between the DCT and the standard DFT consists in the fact that DCT is based on the functions $\cos(k\pi t/T_0)$ with $k \leq N$, which correspond to trigonometric harmonics of both integer and half-integer orders $n = k/2 \leq N/2$, whereas the DFT utilizes trigonometric functions of integer n only, but extending to orders $n \leq N$. This results in vital subsequent differences in the properties of DFT and DCT transforms. Since both DFT and DCT belong to the class of exact discrete transforms, these differences become apparent only when we consider the properties of *continuous extensions* of these transforms as continuous trigonometric polynomials onto the points between the grid [**2**].

If the function $g(t)$ is a continuous function, then the CEDCT sequence results in the function $f_N(t)$ which is converging to the original $g(t)$ at all $t \in [0, T_0]$ with increasing N. This property does not hold for the continuous extension of the standard DFT sequence, which shows profound oscillations between the grid points

t_k. This feature also implies a significantly smaller vulnerability of DCT to the truncations (or any approximation errors) in the process of filtering as compared with the standard DFT.

Similar to canonical *continuous* Fourier transform polynomials with coefficients calculated by exact integrations, the CEDCT series satisfies the principle of *locality*. This property insures, in particular, that the computation errors connected, e.g., with noise or uncertainties in one segment of the data will not significantly affect the reconstructed data on the distant segment of data.

Similar to the canonical continuous Fourier transform polynomials, the derivative of CEDCT series $f_N(t)$ converges to the derivative of $g(t)$ in the limit $N \to \infty$ at all $t \in (0, T_0)$, provided that the second derivative of $g(t)$ is an integrable function on the interval $[0, T_0]$. For CEDCT this property is valid both when $g(0) = g(T_0)$ and $g(0) \neq g(T_0)$. It does not necessarily hold for other types of discrete Fourier transforms which might themselves be converging, like the series given by (2.3), but which produce nonetheless a non-converging derivative series, as demonstrated in Fig. 2.

All these good properties of CEDCT, which are valid also in case of multi-dimensional CEDCTs, can be effectively exploited in a number of practical applications. In this paper we have considered some possible applications of DCT/CEDCT in the 2-dimensional case. Application of CEDCT to the discrete images has demonstrated (see Fig. 3.1) that at least in case of images without significant noise the continuous representation of the image and its zooming by CEDCT has a potential of revealing (or separation) of structures of the spatial size of order of, and even smaller than, the distance between the grid points. Application of CEDCT to test images has also proved the practicability of the "zooming" of the real images by CEDCT method. Important also is that a good quality of the zoomed image in that case is reached without any additional increase of the actual size (i.e. in Bytes) of the image file.

At last, we have suggested above that the potential of DCT for the purposes of data compression, in the lines of the JPEG image compression ideology, can be increased very significantly simply if one uses sub-images with block sizes significantly larger than $N_{\rm sub} = 8$ as implemented currently in the standard JPEG.

In this paper we did not discuss number of other applications of CEDCT. In particular, our recent studies show that there are interesting possibilities for effective and fast suppression of noise in the data using DCT and CEDCT. The close resemblance between the properties of DCT and of the *continuous* Fourier transform, as well the fact that for DCT one does not require the periodicity of the originating functions at the boundaries of the definition region of the function, all contain promises for effectiveness of the application of CEDCT in the area of numerical solution of differential equations. All these discussions we will. however, defer to our forthcoming papers

References

1. N. Ahmed, T. Natarajan, and K. R. Rao, *Discrete cosine transform*, IEEE Trans. Comput. **23** (1974), 90–93.
2. A. Atoyan and J. Patera, *Properties of continuous Fourier extension of the discrete cosine transform and its multidimensional generalization*, J. Math. Phys. (2003) (submitted); `math-ph/0309039`.

3. V. G. Kac, *Automorphisms of finite order of semisimple Lie algebras*, J. Funct. Anal. Appl. **3** (1969), 252–254.
4. R. V. Moody and J. Patera, *Characters of elements of finite order in simple Lie groups*, SIAM J. Algebraic Discrete Methods **5** (1984), 359–383.
5. ———, *Computation of character decompositions of class functions on compact semisimple Lie groups*, Math. Comp. **48** (1987), 799–827.
6. ———, *Elements of finite order in simple Lie groups and their applications*, XIIIth International Colloquium on Group Theoretical Methods in Physics (College Park, 1984) (W. Zachary, ed.), World Sci. Publishing, Singapore, 1984, pp. 308–316.
7. H. J. Nussbaumer, *Fast Fourier transform and convolution algorithms*, Springer Ser. Inform. Sci., vol. 2, Springer-Verlag, Berlin-New York, 1981.
8. A. V. Oppenheim and R. W. Schafer, *Digital signal processing*, Prentice-Hall, Inc., Englewood Cliffs, NJ, 1975.
9. J. Patera, *Algebraic solutions of the Neumann boundary value problems on fundamental region of a compact semisimple Lie group*, Proceedings of the Workshop on group theory and numerical methods, (Montreal, 2003) (to appear).
10. K. R. Rao and P. Yip, *Discrete cosine transform*, Algorithms, Advantages, Applications, Academic Press, Inc., Boston, MA, 1990.
11. D. S. Taubman and M. W. Marcellin, JPEG2000: *image compression fundamentals, standards, and practice*, Kluwer, Boston, 2002.
12. G. P. Tolstov, *Fourier series*, 2nd English transl., Dover Publications, Inc., New York, 1976 (Russian).

Centre de Recherches Mathématiques, Université de Montréal, C.P. 6128, succ. Centre-ville, Montreal, Qc H3C 3J7, Canada.

E-mail address, A. Atoyan: atoyan@crm.umontreal.ca

E-mail address, J. Patera: patera@crm.umontreal.ca

Centre de Recherches Mathématiques
CRM Proceedings and Lecture Notes
Volume **39**, 2005

Symbolic Algorithms for the Painlevé Test, Special Solutions, and Recursion Operators for Nonlinear PDEs.

D. Baldwin, W. Hereman, and J. Sayers

ABSTRACT. This paper discusses the algorithms and implementations of three MATHEMATICA packages for the study of integrability and the computation of closed-form solutions of nonlinear polynomial PDEs.

The first package, `PainleveTest.m`, symbolically performs the Painlevé integrability test. The second package, `PDESpecialSolutions.m`, computes exact solutions expressible in hyperbolic or elliptic functions. The third package, `PDERecursionOperator.m`, generates and tests recursion operators.

1. Introduction

The investigation of complete integrability of nonlinear partial differential equations (PDEs) is a nontrivial matter [**12**]. Likewise, finding the explicit form of solitary wave and soliton solutions requires tedious, unwieldy computations which are best performed using computer algebra systems. For example, the symbolic computation of solitons with Hirota's direct method and the homogenization method are covered in [**13, 16**].

Recently, progress has been made using MATHEMATICA and MAPLE in applying the inverse scattering transform (IST) method to compute solitons for the Camassa-Holm equation [**17**]. Before applying the IST method (a nontrivial exercise in analysis!), one would like to know if the PDE is completely integrable or what elementary travelling wave solutions exist. This is where the symbolic algorithms and packages presented in this paper come into play.

In this paper we introduce three algorithms and related MATHEMATICA packages [**4**] which may greatly aid the investigation of integrability and the search for exact solutions. In Section 2 we present the algorithm for the well-known Painlevé integrability test [**1, 5, 6**], which was recently implemented as `PainleveTest.m`. Section 3 outlines the algorithm behind `PDESpecialSolutions.m`, which allows one

2000 *Mathematics Subject Classification.* Primary 35Q53, 37K10; Secondary 35K40, 37K05.

This material is based upon work supported by the National Science Foundation (NSF) under Grants Nos. DMS-9732069, DMS-9912293 and CCR-9901929.

This is the final form of the paper.

to automatically compute exact solutions expressible in hyperbolic or elliptic functions; full details are presented in [**3**]. In Section 4 we give an algorithm for computing and testing recursion operators [**11**]; the package `PDERecursionOperator.m` automates the steps. The latter package builds on the code `InvariantsSymmetries.m` [**9**], which computes conserved densities, fluxes, and generalized symmetries for polynomial equations. Our codes are publicly available, see [**4**].

In Sections 2 and 3 we consider systems of M polynomial differential equations,

$$\boldsymbol{\Delta}\big(\mathbf{u}(\mathbf{x}), \mathbf{u}'(\mathbf{x}), \mathbf{u}''(\mathbf{x}), \ldots, \mathbf{u}^{(m)}(\mathbf{x})\big) = \mathbf{0}, \tag{1.1}$$

where the dependent variable $\mathbf{u}$ has M components u_i, the independent variable $\mathbf{x}$ has N components x_j, and $\mathbf{u}^{(m)}(\mathbf{x})$ denotes the collection of mixed derivative terms of order m. We assume that any arbitrary coefficients parameterizing the system are strictly positive and denoted by lower-case Greek letters. However, the algorithm in Section 4 only applies to a special case of (1.1), namely evolution equations in $(1 + 1)$ dimensions.

Two carefully selected examples will illustrate the algorithms. The first example is the Kaup–Kupershmidt (KK) equation (see e.g. [**8, 22**]),

$$u_t = 5u^2u_x + \frac{25}{2}u_xu_{2x} + 5uu_{3x} + u_{5x}. \tag{1.2}$$

The second example is the Hirota–Satsuma system of coupled Korteweg-de Vries (KdV) equations (see e.g. [**1**]),

$$\begin{aligned} u_t &= \beta(6uu_x + u_{3x}) - 2vv_x, \quad \beta > 0, \\ v_t &= -3uv_x - v_{3x}. \end{aligned} \tag{1.3}$$

The computations for both examples will be done by the software.

2. The Painlevé test

The Painlevé test verifies whether a system of ODEs or PDEs satisfies the necessary conditions for having the Painlevé property. There is some variation in what is meant by the Painlevé property [**18**]. As defined in [**2**], for a PDE to have the Painlevé property, all ODE reductions of the PDE must have the Painlevé property. While [**2**] requires that all movable singularities of all solutions are poles, the more general definition used by Painlevé himself requires that all solutions of the ODE are single-valued around all movable singularities. A later version [**23**] allows testing of PDEs directly without having to reduce them to ODEs. For a thorough discussion of the Painlevé property, see [**5, 6**].

Definition 2.1. A PDE has the Painlevé property if its solutions in the complex plane are single-valued in the neighborhood of all its movable singularities.

2.1. Algorithm and implementation. Following [**23**], we assume a Laurent expansion for the solution

$$u_i(\mathbf{x}) = g^{\alpha_i}(\mathbf{x}) \sum_{k=0}^{\infty} u_{i,k}(\mathbf{x}) g^k(\mathbf{x}), \qquad u_{i,0}(\mathbf{x}) \neq 0 \quad \text{and} \quad \alpha_i \in \mathbb{Z}, \tag{2.1}$$

where at least one of the leading exponents α_i is a negative integer and $u_{i,k}(\mathbf{x})$ is an analytic function in the neighborhood of $g(\mathbf{x})$. The solution should be single-valued in the neighborhood of the non-characteristic, movable singular manifold $g(\mathbf{x}) = 0$, which can be viewed as the surface of the movable poles in the complex plane.

The algorithm for the Painlevé test is composed of the following three steps:

Step 1 (Determine the dominant behavior). It suffices to substitute

$$u_i(\mathbf{x}) = u_{i,0}(\mathbf{x}) g^{\alpha_i}(\mathbf{x}) \tag{2.2}$$

into (1.1) to determine the leading exponents α_i (one of which must be a negative integer) and the function $u_{i,0}(\mathbf{x})$. In the resulting polynomial system, equating every two possible lowest exponents of $g(\mathbf{x})$ in each equation gives a linear system to determine α_i. The linear system is then solved for α_i.

If one or more exponents α_i remain undetermined, we assign integer values to the free α_i (if applicable, up to a user-specified non-negative boundary, `DominantBehaviorMax`, with default value -1) so that every equation in (1.1) has at least two different terms with equal lowest exponents. Once α_i is known, we substitute (2.2) back into (1.1) and solve for $u_{i,0}(\mathbf{x})$.

Step 2 (Determine the resonances). For each α_i and $u_{i,0}(\mathbf{x})$, we calculate the integers r for which $u_{i,r}(\mathbf{x})$ is an arbitrary function in (2.1). We substitute

$$u_i(\mathbf{x}) = u_{i,0}(\mathbf{x}) g^{\alpha_i}(\mathbf{x}) + u_{i,r}(\mathbf{x}) g^{\alpha_i + r}(\mathbf{x}) \tag{2.3}$$

into (1.1), keeping only the lowest order terms in $g(\mathbf{x})$, and requiring that the coefficients of $u_{i,r}(\mathbf{x})$ equate to zero. This is done by computing the roots for r of $\det Q = 0$, where the $M \times M$ matrix Q satisfies

$$Q \cdot \mathbf{u}_r = \mathbf{0}, \quad \mathbf{u}_r = (u_{1,r} \;\; u_{2,r} \;\; \cdots \;\; u_{M,r})^{\mathrm{T}}. \tag{2.4}$$

Step 3 (Find the constants of integration and check compatibility conditions). For the system to possess the Painlevé property, the arbitrariness of $u_{i,r}(\mathbf{x})$ must be verified up to the highest resonance level. That is, all compatibility conditions must be trivially satisfied. This is done by substituting

$$u_i(\mathbf{x}) = g^{\alpha_i}(\mathbf{x}) \sum_{k=0}^{r_{\mathrm{Max}}} u_{i,k}(\mathbf{x}) g^k(\mathbf{x}) \tag{2.5}$$

into (1.1), where r_{Max} is the highest positive integer resonance. For the system to have the Painlevé property, there must be as many arbitrary constants of integration at resonance levels as resonances at that level. Furthermore, all constants of integration $u_{i,k}(\mathbf{x})$ at non-resonance levels must be unambiguously determined.

2.2. Examples of the Painlevé test.

Example 2.1 (Kaup–Kupershmidt). To determine the dominant behavior, we substitute (2.2) into (1.2) and pull off the exponents of $g(\mathbf{x})$ (see Table 1).

Removing duplicates and non-dominant exponents, we are left with

$$\{\alpha_1 - 5, 2\alpha_1 - 3, 3\alpha_1 - 1\}. \tag{2.6}$$

Considering all possible balances of two or more exponents leads to $\alpha_1 = -2$.

Substituting $u(\mathbf{x}) = u_{1,0}(\mathbf{x}) g^{-2}(\mathbf{x})$ into (1.2) and solving for $u_{1,0}(\mathbf{x})$ gives us

$$u_{1,0}(\mathbf{x}) = -24 g_x^2(\mathbf{x}) \quad \text{and} \quad u_{1,0}(\mathbf{x}) = -3 g_x^2(\mathbf{x}). \tag{2.7}$$

For the first branch, substituting $u(\mathbf{x}) = -24 g_x^2(\mathbf{x}) g^{-2}(\mathbf{x}) + u_{1,r}(\mathbf{x}) g^{r-2}(\mathbf{x})$ into (1.2), keeping the lowest order terms, and taking the coefficient of $u_{1,r}(\mathbf{x})$, gives

$$-2(r+7)(r+1)(r-6)(r-10)(r-12) g_x^5(\mathbf{x}) = 0. \tag{2.8}$$

TABLE 1. The exponents of $g(\mathbf{x})$ for (1.2).

Term	Exponents of $g(\mathbf{x})$ with duplicates removed
u_t	$\alpha_1 - 1$
u_{5x}	$\alpha_1 - 5, \alpha_1 - 4, \alpha_1 - 3, \alpha_1 - 2, \alpha_1 - 1$
$5uu_{3x}$	$2\alpha_1 - 3, 2\alpha_1 - 2, 2\alpha_1 - 1$
$\frac{25}{2}u_x u_{2x}$	$2\alpha_1 - 3, 2\alpha_1 - 2$
$5u^2u_x$	$3\alpha_1 - 1$

Hence, $r = -7, -1, 6, 10, 12$. The first branch is not a principal branch due to the presence of a negative resonance other than $r = -1$. The latter is called the universal resonance and corresponds to the arbitrariness of the manifold function $g(\mathbf{x})$. In [**7**], a perturbative Painlevé approach is presented which properly treats negative integer resonances. As is common practice, our algorithm ignores them.

The constants of integration at level j are found by substituting (2.5) into (1.2), where $r_{\text{Max}} = 12$, and pulling off the coefficients of $g^j(\mathbf{x})$. The first few are

$$\text{at } j = 1\colon\ u_{1,1}(\mathbf{x}) = 24 g_{2x}(\mathbf{x}), \tag{2.9}$$

$$\text{at } j = 2\colon\ u_{1,2}(\mathbf{x}) = \frac{6g_{2x}^2(\mathbf{x}) - 8g_x(\mathbf{x})g_{3x}(\mathbf{x})}{g_x^2(\mathbf{x})}, \tag{2.10}$$

$$\text{at } j = 3\colon\ u_{1,3}(\mathbf{x}) = \frac{6g_{2x}^3(\mathbf{x}) - 8g_x(\mathbf{x})g_{2x}(\mathbf{x})g_{3x}(\mathbf{x}) + 2g_x^2(\mathbf{x})g_{4x}(\mathbf{x})}{g_x^4(\mathbf{x})}. \tag{2.11}$$

The compatibility conditions are satisfied at resonance levels 6, 10, and 12. The remaining constants of integration $u_{1,j}(\mathbf{x})$ were computed but not shown here.

Likewise, in the second branch, substitute $u(\mathbf{x}) = -3g_x^2(\mathbf{x})g^{-2}(\mathbf{x}) + u_{1,r}(\mathbf{x})g^{r-2}(\mathbf{x})$ into (1.2), and proceed as before to get $r = -1, 3, 5, 6, 7$. Thus, the second branch is a principal branch. The constants of integration at $j = 1, 2$ and 4 are again found by substituting (2.5) into (1.2) and pulling off the coefficients of $g^j(\mathbf{x})$. This gives,

$$\text{at } j = 1\colon\ u_{1,1}(\mathbf{x}) = 3g_{2x}(\mathbf{x}), \tag{2.12}$$

$$\text{at } j = 2\colon\ u_{1,2}(\mathbf{x}) = \frac{3g_{2x}^2(\mathbf{x}) - 4g_x(\mathbf{x})g_{3x}(\mathbf{x})}{4g_x^2(\mathbf{x})}. \tag{2.13}$$

Coefficient $u_{1,4}$ at level $j = 4$ is not shown here due to length. At the resonance levels $r = 3, 5, 6, 7$, the compatibility conditions happen to be satisfied. So, (1.2) passes the Painlevé test. It is well-known (see e.g. [**8, 22**]) that (1.2) is completely integrable.

Example 2.2 (Hirota–Satsuma). The Hirota–Satsuma system illustrates the subtleties of determining the dominant behavior.

As in Example 2.1, we substitute (2.2) into (1.3) and pull off the exponents of $g(\mathbf{x})$ (listed in Table 2).

Removing non-dominant exponents and duplicates by term, we get $\{\alpha_1 - 3, 2\alpha_1 - 1, 2\alpha_2 - 1\}$ from Δ_1 and $\{\alpha_2 - 3, \alpha_1 + \alpha_2 - 1\}$ from Δ_2.

Equating the possible dominant exponents from Δ_2 gives us $\alpha_2 - 3 = \alpha_1 + \alpha_2 - 1$ or $\alpha_1 = -2$. Unexpectedly, $\alpha_1 = -2$ balances two of the possible dominant terms

TABLE 2. The exponents of $g(\mathbf{x})$ for (1.3).

Term	Exponents of $g(\mathbf{x})$
u_t	$\alpha_1 - 1$
$-\beta u_{3x}$	$\alpha_1 - 3, \alpha_1 - 3, \alpha_1 - 3, \alpha_1 - 2, \alpha_1 - 2, \alpha_1 - 1$
$-6\beta u u_x$	$2\alpha_1 - 1$
$2vv_x$	$2\alpha_2 - 1$
v_t	$\alpha_2 - 1$
v_{3x}	$\alpha_2 - 3, \alpha_2 - 3, \alpha_2 - 3, \alpha_2 - 2, \alpha_2 - 2, \alpha_2 - 1$
$3uv_x$	$\alpha_1 + \alpha_2 - 1$

in Δ_1, and we are free to choose α_2 such that

$$2\alpha_1 - 1 \leq 2\alpha_2 - 1 \quad \text{or} \quad -2 \leq \alpha_2. \tag{2.14}$$

Using the default value -1 for `DominantBehaviorMax`, we have $-2 \leq \alpha_2 \leq -1$. Hence, $\alpha_2 = -1$ or $\alpha_2 = -2$.

Using the two solutions for α_i and solving for $u_{i,0}$ results in

$$\begin{cases} \alpha_1 = -2, & u_{1,0}(\mathbf{x}) = -4g_x^2(\mathbf{x}), \\ \alpha_2 = -2, & u_{2,0}(\mathbf{x}) = \pm 2\sqrt{6\beta}\, g_x^2(\mathbf{x}), \end{cases} \tag{2.15}$$

$$\begin{cases} \alpha_1 = -2, & u_{1,0}(\mathbf{x}) = -2g_x^2(\mathbf{x}), \\ \alpha_2 = -1, & u_{2,0}(\mathbf{x}) \text{ arbitrary}. \end{cases} \tag{2.16}$$

We substitute (2.3) into (1.3) while using the results for α_i and $u_{i,0}(\mathbf{x})$. For (2.15), substituting $u(\mathbf{x}) = -4g_x^2(\mathbf{x})g^{-2}(\mathbf{x}) + u_{1,r}(\mathbf{x})g^{r-2}(\mathbf{x})$ and $v(\mathbf{x}) = \pm 2\sqrt{6\beta}\, g_x^2(\mathbf{x})g^{-2}(\mathbf{x}) + u_{2,r}(\mathbf{x})g^{r-2}(\mathbf{x})$ into (1.3) and keeping the most singular terms gives

$$Q \cdot \mathbf{u}_r = \begin{pmatrix} -(r-4)(r^2-5r-18)\beta g_x^3(\mathbf{x}) & \pm 12\sqrt{6\beta}\, g_x^3(\mathbf{x}) \\ \mp 4(r-4)\sqrt{6\beta}\, g_x^3(\mathbf{x}) & (r-2)(r-7)r g_x^3(\mathbf{x}) \end{pmatrix} \begin{pmatrix} u_{1,r}(\mathbf{x}) \\ u_{2,r}(\mathbf{x}) \end{pmatrix} = \mathbf{0}.$$

Setting

$$\det Q = -\beta(r+2)(r+1)(r-3)(r-4)(r-6)(r-8)g_x^6(\mathbf{x}) = 0$$

yields $r = -2, -1, 3, 4, 6, 8$. This is not a principal branch.

As in the previous example, the constants of integration at level j are found by substituting (2.5) into (1.3) and pulling off the coefficients of $g^j(\mathbf{x})$. At $j = 1$,

$$\begin{pmatrix} -66\beta g_x^3(\mathbf{x}) & \mp 12\sqrt{6\beta}\, g_x^3(\mathbf{x}) \\ \mp 12\sqrt{6\beta}\, g_x^3(\mathbf{x}) & 6g_x^3(\mathbf{x}) \end{pmatrix} \begin{pmatrix} u_{1,1}(\mathbf{x}) \\ u_{2,1}(\mathbf{x}) \end{pmatrix} = \begin{pmatrix} -120\beta g_x^3(\mathbf{x})g_{2x}(\mathbf{x}) \\ \mp 60\sqrt{6\beta}\, g_x^3(\mathbf{x})g_{2x}(\mathbf{x}) \end{pmatrix}. \tag{2.17}$$

Thus,

$$u_{1,1}(\mathbf{x}) = 4g_{2x}(\mathbf{x}), \quad u_{2,1}(\mathbf{x}) = \pm 2\sqrt{6\beta}\, g_{2x}(\mathbf{x}). \tag{2.18}$$

At $j = 2$,

$$(2.19)\qquad \begin{aligned} u_{1,2}(\mathbf{x}) &= \frac{3g_{2x}^2(\mathbf{x}) - g_x(\mathbf{x})\big(g_t(\mathbf{x}) + 4g_{3x}(\mathbf{x})\big)}{3g_x^2(\mathbf{x})},\\ u_{2,2}(\mathbf{x}) &= \pm\frac{3\beta g_{2x}^2(\mathbf{x}) - 4\beta g_x(\mathbf{x})g_{3x}(\mathbf{x}) - (1+2\beta)g_t(\mathbf{x})g_x(\mathbf{x})}{\sqrt{6\beta}g_x^2(\mathbf{x})}. \end{aligned}$$

The remaining constants of integration are omitted due to length. The compatibility conditions at $r = 3$ and 4 are satisfied. At $r = 6$ and $r = 8$, the compatibility conditions require $\beta = \frac{1}{2}$.

Likewise, for (2.16), the resonances following from the substitution of $u(\mathbf{x}) = -2g_x^2(\mathbf{x})g^{-2}(\mathbf{x}) + u_{1,r}(\mathbf{x})g^{r-2}(\mathbf{x})$ and $v(\mathbf{x}) = u_{2,0}(\mathbf{x})g^{-2}(\mathbf{x}) + u_{2,r}(\mathbf{x})g^{r-1}(\mathbf{x})$ into (1.3) are $r = -1, 0, 1, 4, 5, 6$. This is a principal branch. The zero resonance explains the arbitrariness of $u_{2,0}(\mathbf{x})$. Similarly, we computed all constants of integration, but ran into a compatibility conditions at $r = 5$ and $r = 6$, which requires $\beta = \frac{1}{2}$. Therefore, (1.3) passes the Painlevé test if $\beta = \frac{1}{2}$, a fact confirmed by other analyses of integrability [**1**].

3. Travelling wave solutions in hyperbolic or elliptic functions

The traveling wave solutions of many nonlinear ODEs and PDEs from soliton theory (and elsewhere) can be expressed as polynomials of hyperbolic or elliptic functions. For instance, the bell shaped sech-solutions and kink shaped tanh-solutions model wave phenomena in fluid dynamics, plasmas, elastic media, electrical circuits, optical fibers, chemical reactions, bio-genetics, etc. A discussion is given in [**15**], while a multitude of references to tanh-based techniques and applications can be found in [**3, 19**].

In this section we discuss the tanh-, sech-, cn- and sn-methods as they apply to nonlinear polynomial systems of ODEs and PDEs in multi-dimensions.

3.1. Algorithm and implementation. All four flavors of the algorithm share the same five basic steps.

Step 1 (Transform the PDE into a nonlinear ODE). We seek solutions in the traveling frame of reference,

$$(3.1)\qquad \xi = \sum_{j=1}^{N} c_j x_j + \delta,$$

where the components c_j of the wave vector and the phase δ are constants.

In the tanh method, we seek polynomial solutions expressible in hyperbolic tangent, $T = \tanh\xi$. Based on the identity $\cosh^2\xi - \sinh^2\xi = 1$,

$$(3.2)\qquad \tanh'\xi = \operatorname{sech}^2\xi = 1 - \tanh^2\xi,$$

$$(3.3)\qquad \tanh''\xi = -2\tanh\xi + 2\tanh^3\xi, \text{ etc.}$$

Therefore, the first and all higher-order derivatives of $T = \tanh\xi$ are polynomial in T. Consequently, repeatedly applying the chain rule,

$$(3.4)\qquad T = \tanh(\xi)\colon \frac{\partial\bullet}{\partial x_j} = \frac{\partial\xi}{\partial x_j}\frac{dT}{d\xi}\frac{d\bullet}{dT} = c_j(1 - T^2)\frac{d\bullet}{dT}$$

transforms (1.1) into a coupled system of nonlinear ODEs.

TABLE 3. Values for $R(F)$ in (3.11).

F	$R(F)$
T	0
S	$1 - S^2$
CN	$(1 - \mathrm{CN}^2)(1 - m + m\,\mathrm{CN}^2)$
SN	$(1 - \mathrm{SN}^2)(1 - m\,\mathrm{SN}^2)$

Similarly, using the identity $\tanh^2\xi + \operatorname{sech}^2\xi = 1$, we get

$$\operatorname{sech}'\xi = -\operatorname{sech}\xi\tanh\xi = -\operatorname{sech}\xi\sqrt{1-\operatorname{sech}^2\xi}. \tag{3.5}$$

Likewise, for Jacobi's elliptic functions with modulus m, we use the identities

$$\mathrm{sn}^2(\xi;m) = 1 - \mathrm{cn}^2(\xi;m), \quad \text{and} \quad \mathrm{dn}^2(\xi;m) = 1 - m + m\,\mathrm{cn}^2(\xi;m), \tag{3.6}$$

to write, for example, cn' in terms of cn :

$$\mathrm{cn}'(\xi;m) = -\,\mathrm{sn}(\xi;m)\,\mathrm{dn}(\xi;m) = -\sqrt{\big(1-\mathrm{cn}^2(\xi;m)\big)\big(1-m+m\,\mathrm{cn}^2(\xi;m)\big)}. \tag{3.7}$$

Then, repeatedly applying the chain rules,

$$S = \operatorname{sech}(\xi)\colon \frac{\partial\bullet}{\partial x_j} = -c_j S\sqrt{1-S^2}\frac{d\bullet}{dS}, \tag{3.8}$$

$$\mathrm{CN} = \mathrm{cn}(\xi;m)\colon \frac{\partial\bullet}{\partial x_j} = -c_j\sqrt{(1-\mathrm{CN}^2)(1-m+m\,\mathrm{CN}^2)}\frac{d\bullet}{d\mathrm{CN}}, \tag{3.9}$$

$$\mathrm{SN} = \mathrm{sn}(\xi;m)\colon \frac{\partial\bullet}{\partial x_j} = c_j\sqrt{(1-\mathrm{SN}^2)(1-m\mathrm{SN}^2)}\frac{d\bullet}{d\mathrm{SN}}, \tag{3.10}$$

we transform (1.1) into a coupled system of nonlinear ODEs of the form

$$\mathbf{\Gamma}(F, \mathbf{u}(F), \mathbf{u}'(F), \ldots) + \sqrt{R(F)}\,\mathbf{\Pi}(F, \mathbf{u}(F), \mathbf{u}'(F), \ldots) = \mathbf{0}, \tag{3.11}$$

where F is either T, S, CN, or SN, and $R(F)$ is defined in Table 3.

Step 2 (Determine the degree of the polynomial solutions). Since we seek polynomial solutions

$$U_i(F) = \sum_{j=0}^{M_i} a_{ij}F^j, \tag{3.12}$$

the leading exponents M_i must be determined before the a_{ij} can be computed. The process for determining M_i is quite similar to the one for finding α_i in Section 2.

Substituting $U_i(F)$ into (3.11), the coefficients of every power of F in every equation must vanish. In particular, the highest degree terms must vanish. Since the highest degree terms only depend on F^{M_i} in (3.12), it suffices to substitute $U_i(F) = F^{M_i}$ into (3.11). Doing so, we get

$$\mathbf{P}(F) + \sqrt{R(F)}\,\mathbf{Q}(F) = 0, \tag{3.13}$$

where $\mathbf{P}$ and $\mathbf{Q}$ are polynomials in F. Equating every two possible highest exponents in each P_i and Q_i gives a linear system for determining M_i. The linear system is then solved for M_i.

If one or more exponents M_i remain undetermined, we assign a strictly positive integer value to the free M_i, so that every equation in (3.11) has at least two different terms with equal highest exponents in F.

Step 3 (Derive the algebraic system for the coefficients a_{ij}). To generate the system for the unknown coefficients a_{ij} and wave parameters c_j, substitute (3.12) into (3.11) and set the coefficients of F^j to zero. The resulting nonlinear algebraic system for the unknown a_{ij} is parameterized by the wave parameters c_j and the parameters in (1.1), if any.

Step 4 (Solve the nonlinear parameterized algebraic system). The most difficult aspect of the method is solving the nonlinear algebraic system. To solve the system we designed a customized, yet powerful, nonlinear solver.

The nonlinear algebraic system is solved under the following assumptions:

(1) all parameters (the lower-case Greek letters) in (1.1) are strictly positive. (Vanishing parameters may change the exponents M_i in Step 2). To compute solutions corresponding to negative parameters, reverse the signs of the parameters in (1.1).
(2) the coefficients of the highest power terms ($a_{i\,M_i}$, $i = 1, \ldots, M_i$) in (3.12) are all nonzero (for consistency with Step 2).
(3) all c_j are nonzero (demanded by the physical nature of the solutions).

Step 5 (Build and test solutions). Substitute the solutions from Step 4 into (3.12) and reverse Step 1 to obtain the explicit solutions in the original variables. It is prudent to test the solutions by substituting them into (1.1), which we do.

3.2. Examples of travelling wave solutions.

Example 3.1 (Kaup–Kupershmidt). While the tanh-, sech-, cn- and sn-methods find solutions for (1.2), we demonstrate the steps of the algorithm using the tanh-method. After which, we summarize the results for the other methods.

First, transform (1.2) into a nonlinear ODE by repeatedly applying chain rule (3.4). The resulting ODE is

$$
\begin{aligned}
(3.14)\quad & 2c_2U' \\
&+ c_1\Big[10U^2U' + c_1^2\Big[25(T^2-1)U'[2TU' + (T^2-1)U''] \\
&\quad + 10U\big[(6T^2-2)U' + 6T(T^2-1)U'' + (T^2-1)^2U''' \\
&\quad\quad + c_1^2\big(16(15T^4 - 15T^2 + 2)U' + 40(T^2-1)\big(6T(2T^2-1)U'' \\
&\quad\quad + (6T^4 - 7T^2 + 1)U''' + T(T^2-1)^2U^{(4)}\big) + (T^2-1)^4U^{(5)}\big)\big]\Big]\Big] = 0,
\end{aligned}
$$

where $T = \tanh(\xi)$ and $U = U(T)$.

Next, to compute the degree of the polynomial solution(s), substitute $U(T) = T^{M_1}$ into (3.14) and pull off the exponents of T (see Table 4). Remove duplicates and non-dominant exponents, to get

$$(3.15)\qquad \{3M_1 - 1, 2M_1 + 1, M_1 + 3\}.$$

TABLE 4. The exponents of T after substituting $U(T) = T^{M_1}$.

Term	Exponents of T with duplicates removed
u_t	$M_1 - 1$
$u^2 u_x$	$3M_1 - 1$
$\frac{25}{2} u_x u_{2x}$	$2M_1 + 1, 2M_1 - 1, 2M_1 - 3$
$5uu_{3x}$	$2M_1 + 1, 2M_1 - 1, 2M_1 - 3$
u_{5x}	$M_1 + 3, M_1 + 1, M_1 - 1, M_1 - 3, M_1 - 5$

Consider all possible balances of two or more exponents to find $M_1 = 2$.

Substitute

$$U(T) = a_{10} + a_{11}T + a_{12}T^2 \tag{3.16}$$

into (3.14) and equate the coefficients of T^j to zero (where $i = 0, 1, \dots, 5$) to get

$$\begin{aligned} &(a_{12} + 3c_1^2)(a_{12} + 24c_1^2) = 0,\\ &a_{11}(5a_{12}^2 + 55a_{12}c_1^2 + 24c_1^4) = 0,\\ &a_{11}(5a_{10}^2 c_1 - 10a_{10}c_1^3 + 25a_{12}c_1^3 + 16c_1^5 + c_2) = 0,\\ &a_{11}(a_{11}^2 + 6a_{10}a_{12} + 6a_{10}c_1^2 - 48a_{12}c_1^2 - 24c_1^4) = 0,\\ &4a_{11}^2 a_{12} + 4a_{10}a_{12}^2 + 11a_{11}^2 c_1^2 + 24a_{10}a_{12}c_1^2 - 56a_{12}^2 c_1^2 - 192a_{12}c_1^4 = 0,\\ &10a_{10}a_{11}^2 c_1 + 10a_{10}^2 a_{12}c_1 - 35a_{11}^2 c_1^3\\ &\qquad - 80a_{10}a_{12}c_1^3 + 50a_{12}^2 c_1^3 + 272a_{12}c_1^5 + 2a_{12}c_2 = 0. \end{aligned} \tag{3.17}$$

Solve the nonlinear algebraic system with the assumption that a_{12}, c_1, and c_2 are all nonzero. Two solutions are obtained:

$$\begin{cases} a_{10} = 16c_1^2, & a_{11} = 0,\\ a_{12} = -24c_1^2, & c_2 = -176c_1^5, \end{cases} \quad \text{and} \quad \begin{cases} a_{10} = 2c_1^2, & a_{11} = 0,\\ a_{12} = -3c_1^2, & c_2 = -c_1^5, \end{cases} \tag{3.18}$$

where c_1 is arbitrary.

Substitute the solutions into (3.16) and return to $u(x,t)$ to get

$$u(x,t) = 16c_1^2 - 24c_1^2 \tanh^2(c_1 x - 176c_1^5 t + \delta), \tag{3.19}$$

$$u(x,t) = 2c_1^2 - 3c_1^2 \tanh^2(c_1 x - c_1^5 t + \delta). \tag{3.20}$$

Using the sech-method, one finds

$$u(x,t) = -8c_1^2 + 24c_1^2 \operatorname{sech}^2(c_1 x - 176c_1^5 t + \delta), \tag{3.21}$$

$$u(x,t) = -c_1^2 + 3c_1^2 \operatorname{sech}^2(c_1 x - c_1^5 t + \delta). \tag{3.22}$$

Alternatively, the latter solutions can be found directly from the tanh-solutions by using the identity $\tanh^2 \xi = 1 - \operatorname{sech}^2 \xi$.

Using the sn-method, one gets

$$u(x,t) = 8c_1^2\left[1 + m - 3m \operatorname{sn}^2\left(c_1 x - 176c_1^5(m^2 - m + 1)t + \delta; m\right)\right], \tag{3.23}$$

$$u(x,t) = c_1^2\left[1 + m - 3m \operatorname{sn}^2(c_1 x - c_1^5(m^2 - m + 1)t + \delta; m)\right]. \tag{3.24}$$

The cn-solutions computed with the cn-method are not explicitly shown since they can be obtained from the sn-solutions using the identity $\operatorname{sn}^2\xi = 1 - \operatorname{cn}^2\xi$.

Example 3.2 (Hirota–Satsuma). As in the previous example, the tanh-, sech-, cn- and sn-methods all find solutions for (1.3). In this example, however, we will illustrate the steps using the sech-method.

Transform (1.3) into a coupled system of ODEs, apply the chain rule (3.8) and cancel the common $S\sqrt{1-S^2}$ factors to get

$$\begin{aligned} (3.25)\quad & c_2U_1' - 6\beta c_1U_1U_1' - \beta c_1^3[(1-6S^2)U_1' \\ &\qquad + 3S(1-2S^2)U_1'' + S^2(1-S^2)U_1'''] + 2c_1U_2U_2' = 0, \\ & c_2U_2' + 3c_1U_1U_2' + c_1^3[(1-6S^2)U_2' + 3S(1-2S^2)U_2'' + S^2(1-S^2)U_2'''] = 0. \end{aligned}$$

To find the degree of the polynomials, substitute $U_1(S) = S^{M_1}$, $U_2(S) = S^{M_2}$ into (3.25) and equate the highest exponents from Δ_2 to get

$$(3.26)\qquad M_1 + M_2 - 1 = 1 + M_2 \quad \text{or} \quad M_1 = 2.$$

The maximal exponents coming from Δ_1 are $2M_1 - 1$ (from the U_1U_1' term), M_1+1 (from U_1'''), and $2M_2 - 1$ (from U_2U_2').

Since $M_1 = 2$ balances at least two of the possible dominant exponents in Δ_1, namely $2M_1 - 1$ and $M_1 + 1$, one is again left with $1 \le M_2 \le M_1 = 2$, or

$$(3.27)\qquad \begin{cases} M_1 = 2, & U_1(S) = a_{10} + a_{11}S + a_{12}S^2, \\ M_2 = 1, & U_2(S) = a_{20} + a_{21}S, \end{cases}$$

$$(3.28)\qquad \begin{cases} M_1 = 2, & U_1(S) = a_{10} + a_{11}S + a_{12}S^2, \\ M_2 = 2, & U_2(S) = a_{20} + a_{21}S + a_{22}S^2. \end{cases}$$

To derive the algebraic system for a_{ij}, substitute (3.27) into (3.25), cancel common numerical factors, and organize the equations (according to complexity):

$$(3.29)\qquad \begin{aligned} a_{11}a_{21}c_1 &= 0, \\ \beta a_{11}c_1(3a_{12} - c_1^2) &= 0, \\ \beta a_{12}c_1(a_{12} - 2c_1^2) &= 0, \\ a_{21}c_1(a_{12} - 2c_1^2) &= 0, \\ a_{21}(3a_{10}c_1 + c_1^3 + c_2) &= 0, \\ 6\beta a_{10}a_{11}c_1 - 2a_{20}a_{21}c_1 + \beta a_{11}c_1^3 - a_{12}c_2 &= 0, \\ 3\beta a_{11}^2c_1 + 6\beta a_{10}a_{12}c_1 - a_{21}^2c_1 + 4\beta a_{12}c_1^3 - a_{12}c_2 &= 0. \end{aligned}$$

Similarly, after substitution of (3.28) into (3.25), one gets

$$
\begin{aligned}
&a_{22}c_1(a_{12} - 4c_1^2) = 0,\\
&a_{21}(3a_{10}c_1 + c_1^3 + c_2) = 0,\\
&c_1(a_{12}a_{21} + 2a_{11}a_{22} - 2a_{21}c_1^2) = 0,\\
&c_1(3\beta a_{11}a_{12} - a_{21}a_{22} - \beta a_{11}c_1^2) = 0,\\
&c_1(3\beta a_{12}^2 - a_{22}^2 - 6\beta a_{12}c_1^2) = 0,\\
&6\beta a_{10}a_{11}c_1 - 2a_{20}a_{21}c_1 + \beta a_{11}c_1^3 - a_{11}c_2 = 0,\\
&3a_{11}a_{21}c_1 + 6a_{10}a_{22}c_1 + 8a_{22}c_1^3 + 2a_{22}c_2 = 0,\\
&3\beta a_{11}^2c_1 + 6\beta a_{10}^2a_{12}c_1 - a_{21}^2c_1 - 2a_{20}a_{22}c_1 + 4\beta a_{12}c_1^3 - a_{12}c_2 = 0.
\end{aligned}
\tag{3.30}
$$

Since a_{12}, a_{21}, β, c_1, and c_2, are nonzero, the solution of (3.29) is

$$
\begin{cases}
a_{10} = -(c_1^3 + c_2)/(3c_1), & a_{20} = 0,\\
a_{11} = 0, & a_{21} = \pm\sqrt{4\beta c_1^4 - 2(1+2\beta)c_1c_2},\\
a_{12} = 2c_1^2.
\end{cases}
\tag{3.31}
$$

For a_{12}, a_{22}, β, c_1, and c_2 nonzero, the solution of (3.30) is

$$
\begin{cases}
a_{10} = -(4c_1^3 + c_2)/(3c_1), & a_{20} = \pm[4\beta c_1^3 + (1+2\beta)c_2]/(c_1\sqrt{6\beta}),\\
a_{11} = 0, & a_{21} = 0,\\
a_{12} = 4c_1^2, & a_{22} = \mp 2c_1^2\sqrt{6\beta}.
\end{cases}
\tag{3.32}
$$

The solutions of (1.3) involving sech are then

$$
\begin{aligned}
u(x,t) &= -\frac{c_1^3 + c_2}{3c_1} + 2c_1^2\,\mathrm{sech}^2(c_1x + c_2t + \delta),\\
v(x,t) &= \pm\sqrt{4\beta c_1^4 - 2(1+2\beta)c_1c_2}\;\mathrm{sech}(c_1x + c_2t + \delta),
\end{aligned}
\tag{3.33}
$$

and

$$
\begin{aligned}
u(x,t) &= -\frac{4c_1^3 + c_2}{3c_1} + 4c_1^2\,\mathrm{sech}^2(c_1x + c_2t + \delta),\\
v(x,t) &= \pm\frac{4\beta c_1^3 + (1+2\beta)c_2}{c_1\sqrt{6\beta}} \mp 2c_1^2\sqrt{6\beta}\,\mathrm{sech}^2(c_1x + c_2t + \delta).
\end{aligned}
\tag{3.34}
$$

In both sets of solutions, c_1, c_2, β, and δ are arbitrary.

4. Recursion Operators

The previous two algorithms work for (1.1). The algorithm for recursion operators presented in Section 4.2 uses the concepts of dilation invariance, densities, and symmetries. Thus, in contrast to more general approaches [**20**], our method only works for polynomial systems of evolution equations in $(1+1)$ dimensions,

$$
\mathbf{u}_t = \mathbf{F}(\mathbf{u}, \mathbf{u}_x, \mathbf{u}_{2x}, \ldots, \mathbf{u}_{mx}),
\tag{4.1}
$$

where $\mathbf{u}(x,t)$ has M components u_i and $\mathbf{u}_{mx} = \partial^m\mathbf{u}/\partial x^m$. For brevity, we write $\mathbf{F}(\mathbf{u})$, although $\mathbf{F}$ typically depends on $\mathbf{u}$ and its x-derivatives up to order m. If present, any parameters in (4.1) are strictly positive and denoted by lower-case Greek letters.

The algorithm in Section 4.2 will use the concepts of dilation invariance, densities, and symmetries.

4.1. Scaling invariance, densities, symmetries. A PDE is *dilation invariant* if it is invariant under a dilation symmetry.

Example 4.1 (Kaup–Kupershmidt). As an example, (1.2) is invariant under the dilation (scaling) symmetry

$$(t, x, u) \to (\lambda^{-5}t, \lambda^{-1}x, \lambda^{2}u), \tag{4.2}$$

where λ is an arbitrary parameter, leaving λ^7 as a common factor upon scaling.

To find the dilation symmetry, set the *weight* of the x-derivative to one, $w(D_x) = 1$, and require that all terms in (4.1) have the same weight. For (1.2), we have

$$w(u) + w(D_t) = 3w(u) + 1 = 2w(u) + 3 = 2w(u) + 3 = w(u) + 5. \tag{4.3}$$

So, $w(u) = 2$ and $w(D_t) = 5$, Consequently, in (1.2) the sum of the weights or *rank* of each term is 7.

A *generalized symmetry*, $\mathbf{G}(\mathbf{u})$, leaves (4.1) invariant under the replacement $\mathbf{u} \to \mathbf{u} + \epsilon\mathbf{G}$ within order ϵ. Hence, $\mathbf{G}$ must satisfy the linearized equation [**21**]

$$D_t\mathbf{G} = \mathbf{F}'(\mathbf{u})[\mathbf{G}], \tag{4.4}$$

on solutions of (4.1). $\mathbf{F}'(\mathbf{u})[\mathbf{G}]$ is the Fréchet derivative of $\mathbf{F}$ in the direction of $\mathbf{G}$.

A *recursion operator*, $\mathcal{R}$, is a linear integro-differential operator which links generalized symmetries [**21**]

$$\mathbf{G}^{(j+s)} = \mathcal{R}\mathbf{G}^{(j)}, \quad j \in \mathbb{N}, \tag{4.5}$$

where s is the seed and $\mathbf{G}^{(j)}$ is the j-th symmetry. The symmetries are linked consecutively if $s = 1$. This happens in most, but not all, cases.

A *conservation law* [**21**],

$$D_t\rho(x,t) + D_xJ(x,t) = 0, \tag{4.6}$$

valid for solutions of (4.1), links a conserved density $\rho(x,t)$ with the associated flux $J(x,t)$.

If (4.1) is scaling invariant, then its conserved densities, fluxes, generalized symmetries, and recursion operators are also dilation invariant. One could say they 'inherit' the scaling symmetry of the original PDE. The existence of an infinite number of symmetries and an infinite number of conservation laws are good indicators for complete integrability [**21**]. Such predictors, albeit useful, do not provide information about the actual analytic behavior of solutions of PDEs which is also at the core of complete integrability [**1**].

4.2. Algorithm and implementation.

Step 1 (Compute conserved densities and generalized symmetries). For details about the algorithms and symbolic computation of conserved densities and generalized symmetries, see [**8–10**].

Step 2 (Determine the rank of the recursion operator). The rank of the recursion operator is determined by the difference in ranks of the generalized symmetries it links,

$$\text{rank}\,\mathcal{R}_{ij} = \text{rank}\,\mathbf{G}_i^{(k+s)} - \text{rank}\,\mathbf{G}_j^{(k)}, \tag{4.7}$$

where $\mathcal{R}$ is an $M \times M$ matrix and $\mathbf{G}$ has M components.

Step 3 (Determine the form of the recursion operator). The recursion operator naturally splits into two pieces [**11**],

$$\mathcal{R} = \mathcal{R}_0 + \mathcal{R}_1, \tag{4.8}$$

where $\mathcal{R}_0$ is a differential operator and $\mathcal{R}_1$ is an integral operator. The differential operator, $\mathcal{R}_0$, is a linear combination (with constant coefficients) of terms of type $D_x^i u^j$ $(i, j \in \mathbb{Z}^*)$, which must be of the correct rank. To standardize the form of $\mathcal{R}_0$, propagate D_x to the right, for example, $D_x^2 u = u_{2x} I + u_x D_x + u D_x^2$.

The integral operator, $\mathcal{R}_1$, is composed of the terms

$$\sum_i \sum_j \tilde{c}_{ij} \mathbf{G}^{(i)} D_x^{-1} \otimes \rho^{(j)\prime} \tag{4.9}$$

of the correct rank, where $\otimes$ is the matrix outer product and $\rho^{(j)\prime}$ is the covariant (Fréchet derivative of $\rho^{(j)}$). To standardize the form of $\mathcal{R}_1$, propagate D_x to the left, for example, $D_x^{-1} u_x D_x = u_x I - D_x^{-1} u_{2x} I$.

Step 4 (Determine the coefficients). To determine the coefficients in the form of the recursion operator, we substitute $\mathcal{R}$ into the defining equation [**11, 22**],

$$\frac{\partial \mathcal{R}}{\partial t} + \mathcal{R}'[\mathbf{F}] + \mathcal{R} \circ \mathbf{F}'(\mathbf{u}) - \mathbf{F}'(\mathbf{u}) \circ \mathcal{R} = 0, \tag{4.10}$$

where $\circ$ denotes a composition of operators, $\mathcal{R}'[\mathbf{F}]$ is the Fréchet derivative of $\mathcal{R}$ in the direction of $\mathbf{F}$, and $\mathbf{F}'(\mathbf{u})$ is the Fréchet derivative with entries

$$\mathbf{F}'_{ij}(\mathbf{u}) = \sum_{k=0}^{m} \left(\frac{\partial F_i}{\partial (u_j)_{kx}} \right) D_x^k. \tag{4.11}$$

4.3. Examples of scalar and matrix recursion operators.

Example 4.2 (Kaup–Kupershmidt). Using the weights $w(u) = 2$, $w(D_x) = 1$, and $w(D_t) = 5$, we find

$$\begin{aligned} &\operatorname{rank} G^{(1)} = 3, && \operatorname{rank} G^{(2)} = 7, && \operatorname{rank} G^{(3)} = 9, \\ &\operatorname{rank} G^{(4)} = 13, && \operatorname{rank} G^{(5)} = 15, && \operatorname{rank} G^{(6)} = 19. \end{aligned} \tag{4 12}$$

We guess that $\operatorname{rank} \mathcal{R} = 6$ and $s = 2$, since $\operatorname{rank} G^{(2)} - \operatorname{rank} G^{(1)} \neq \operatorname{rank} G^{(3)} - \operatorname{rank} G^{(2)}$ but $\operatorname{rank} G^{(3)} - \operatorname{rank} G^{(1)} = \operatorname{rank} G^{(4)} - \operatorname{rank} G^{(2)} = 6$.

Thus, taking all $D_x^i u^j$ $(i, j \in \mathbb{Z}^*)$ such that $\operatorname{rank} D_x^i u^j = 6$ gives

$$\begin{aligned} \mathcal{R}_0 = {} & c_1 D_x^6 + c_2 u D_x^4 + c_3 u_x D_x^3 + c_4 u^2 D_x^2 + c_5 u_{2x} D_x^2 \\ & + c_6 u u_x D_x + c_7 u_{3x} D_x + c_8 u^3 I + c_9 u_x^2 I + c_{10} u u_{2x} I + c_{11} u_{4x} I. \end{aligned} \tag{4.13}$$

Using the densities $\rho^{(1)} = u$ and $\rho^{(2)} = 3u_x^2 - 4u^3$, and the symmetries $G^{(1)} = u_x$, and $G^{(2)} = F(u) = 5u^2 u_x + \frac{25}{2} u_x u_{2x} + 5 u u_{3x} + u_{5x}$ from (1.2), we compute

$$\begin{aligned} \mathcal{R}_1 &= \tilde{c}_{12} G^{(1)} D_x^{-1} \rho^{(2)\prime} + \tilde{c}_{21} G^{(2)} D_x^{-1} \rho^{(1)\prime} \\ &= \tilde{c}_{12} u_x D_x^{-1} (6 u_x D_x - 12 u^2 I) + \tilde{c}_{21} G^{(2)} D_x^{-1} I \\ &= c_{12} u_x [D_x^{-1} (u_{2x} I + 2u^2 I) - u_x I] + c_{13} G^{(2)} D_x^{-1}. \end{aligned} \tag{4.14}$$

Substituting $\mathcal{R} = \mathcal{R}_0 + \mathcal{R}_1$ and $G^{(2)} = F$ into (4.10) gives us 49 linear equations for c_i. Solving, we find

$$\begin{aligned} c_1 &= \frac{4c_9}{69}, & c_2 &= \frac{8c_9}{23}, & c_3 &= \frac{24c_9}{23}, & c_4 &= \frac{12c_9}{23}, & c_5 &= \frac{98c_9}{69}, & c_6 &= \frac{40c_9}{23}, \\ c_7 &= \frac{70c_9}{69}, & c_8 &= \frac{16c_9}{69}, & c_{10} &= \frac{82c_9}{69}, & c_{11} &= \frac{26c_9}{69}, & c_{12} &= \frac{2c_9}{69}, & c_{13} &= \frac{4c_9}{69}, \end{aligned} \tag{4.15}$$

where c_9 is arbitrary. Taking $c_9 = 69/4$, we find the recursion operator in [**22**]:

$$\begin{aligned} \mathcal{R} = {} & D_x^6 + 6uD_x^4 + 18u_xD_x^3 + 9u^2D_x^2 \\ & + \frac{49}{2}u_{2x}D_x^2 + 30uu_xD_x + \frac{35}{2}u_{3x}D_x + 4u^3I + \frac{69}{4}u_x^2I \\ & + \frac{41}{2}uu_{2x}I + \frac{13}{2}u_{4x}I + \frac{1}{2}u_xD_x^{-1}(u_{2x} + 2u^2)I + G^{(2)}D_x^{-1}. \end{aligned} \tag{4.16}$$

Example 4.3 (Hirota–Satsuma). Only when $\beta = \frac{1}{2}$ does (1.3) have infinitely many densities and symmetries. The first few are

$$\begin{gathered} \rho^{(1)} = u, \quad \rho^{(2)} = 3u^2 - 2v^2, \\ \mathbf{G}^{(1)} = \begin{pmatrix} u_x \\ v_x \end{pmatrix}, \quad \mathbf{G}^{(2)} = \begin{pmatrix} F_1 \\ F_2 \end{pmatrix} = \begin{pmatrix} \beta(6uu_x + u_{3x}) - 2vv_x \\ -(3uv_x + v_{3x}) \end{pmatrix}. \end{gathered} \tag{4.17}$$

We also computed the $\mathbf{G}^{(3)}$ and $\mathbf{G}^{(4)}$, but they are not shown due to length. Solving the equations for the weights,

$$\begin{cases} w(u) + w(D_t) = 2w(u) + 1 = w(u) + 3 = 2w(v) + 1, \\ w(v) + w(D_t) = w(u) + w(v) + 1 = w(v) + 3, \end{cases} \tag{4.18}$$

yields $w(u) = w(v) = 2$ and $w(D_t) = 3$. Based on these weights, $\operatorname{rank} \rho^{(1)} = 2$, $\operatorname{rank} \rho^{(2)} = 4$, and

$$\begin{aligned} \operatorname{rank} \mathbf{G}^{(1)} &= \begin{pmatrix} 3 \\ 3 \end{pmatrix}, & \operatorname{rank} \mathbf{G}^{(2)} &= \begin{pmatrix} 5 \\ 5 \end{pmatrix}, \\ \operatorname{rank} \mathbf{G}^{(3)} &= \begin{pmatrix} 7 \\ 7 \end{pmatrix}, & \operatorname{rank} \mathbf{G}^{(4)} &= \begin{pmatrix} 9 \\ 9 \end{pmatrix}. \end{aligned} \tag{4.19}$$

We would first guess that $\operatorname{rank} \mathcal{R}_{ij} = 2$ and $s = 1$. If indeed the symmetries were linked consecutively, then

$$\mathcal{R}_0 = \begin{pmatrix} c_1D_x^2 + c_2uI + c_3vI & c_4D_x^2 + c_5uI + c_6vI \\ c_7D_x^2 + c_8uI + c_9vI & c_{10}D_x^2 + c_{11}uI + c_{12}vI \end{pmatrix}. \tag{4.20}$$

Using (4.17), we have

$$\mathcal{R}_1 = \tilde{c}_{11}\mathbf{G}^{(1)}D_x^{-1} \otimes \rho^{(1)\prime} = \tilde{c}_{11} \begin{pmatrix} u_x \\ v_x \end{pmatrix} D_x^{-1} \otimes \begin{pmatrix} I & 0 \end{pmatrix} = c_{13} \begin{pmatrix} u_xD_x^{-1} & 0 \\ v_xD_x^{-1} & 0 \end{pmatrix}.$$

Substituting $\mathcal{R} = \mathcal{R}_0 + \mathcal{R}_1$ into (4.10), we find $c_1 = \cdots = c_{13} = 0$. Therefore, either the form of $\mathcal{R}$ is incorrect or the system does not have a recursion operator. Let us now repeat the steps taking $s = 2$, so $\operatorname{rank} \mathcal{R}_{ij} = 4$. Then,

$$\mathcal{R} = \begin{pmatrix} (\mathcal{R}_0)_{11} & (\mathcal{R}_0)_{12} \\ (\mathcal{R}_0)_{21} & (\mathcal{R}_0)_{22} \end{pmatrix} + \tilde{c}_{12}\mathbf{G}^{(1)}D_x^{-1} \otimes \rho^{(2)\prime} + \tilde{c}_{21}\mathbf{G}^{(2)}D_x^{-1} \otimes \rho^{(1)\prime}, \tag{4.21}$$

with $(\mathcal{R}_0)_{ij}$ a linear combination of $\{D_x^4, uD_x^2, vD_x^2, u_xD_x, v_xD_x, u^2, uv, v^2, u_{2x}, v_{2x}\}$. For instance,

$$(\mathcal{R}_0)_{12} = c_{11}D_x^4 + c_{12}uD_x^2 + c_{13}vD_x^2 + c_{14}u_xD_x + c_{15}v_xD_x + c_{16}u^2I + c_{17}uvI + c_{18}v^2I + c_{19}u_{2x}I + c_{20}v_{2x}I. \tag{4.22}$$

Using (4.17), the first term of $\mathcal{R}_1$ in (4.21) is

$$\mathcal{R}_1^{(1)} = \tilde{c}_{12}\mathbf{G}^{(1)}D_x^{-1} \otimes \rho^{(2)'} = \tilde{c}_{12}\begin{pmatrix} u_x \\ v_x \end{pmatrix} D_x^{-1} \otimes \begin{pmatrix} 6uI & -4vI \end{pmatrix} = c_{41}\begin{pmatrix} 3u_xD_x^{-1}uI & -2u_xD_x^{-1}vI \\ 3v_xD_x^{-1}uI & -2v_xD_x^{-1}vI \end{pmatrix}.$$

The second term of $\mathcal{R}_1$ in (4.21) is

$$\mathcal{R}_1^{(2)} = \tilde{c}_{21}\mathbf{G}^{(2)}D_x^{-1} \otimes \rho^{(1)'} = \tilde{c}_{21}\begin{pmatrix} F_1(\mathbf{u}) \\ F_2(\mathbf{u}) \end{pmatrix} D_x^{-1} \otimes \begin{pmatrix} I & 0 \end{pmatrix} = c_{42}\begin{pmatrix} F_1(\mathbf{u})D_x^{-1} & 0 \\ F_2(\mathbf{u})D_x^{-1} & 0 \end{pmatrix}.$$

Substituting the form of $\mathcal{R} = \mathcal{R}_0 + \mathcal{R}_1 = \mathcal{R}_0 + \mathcal{R}_1^{(1)} + \mathcal{R}_1^{(2)}$ into (4.10), the linear system for c_i has a non-trivial solution. Solving the linear system, we finally obtain

$$\mathcal{R} = \begin{pmatrix} \mathcal{R}_{11} & \mathcal{R}_{12} \\ \mathcal{R}_{21} & \mathcal{R}_{22} \end{pmatrix}, \tag{4.23}$$

where

$$\begin{aligned}
\mathcal{R}_{11} &= D_x^4 + 8uD_x^2 + 12u_xD_x + 16u^2I + 8u_{2x}I - \frac{16}{3}v^2I \\
&\quad + 4u_xD_x^{-1}uI + 12uu_xD_x^{-1} + 2u_{3x}D_x^{-1} - 8vv_xD_x^{-1}, \\
\mathcal{R}_{12} &= -\frac{20}{3}vD_x^2 - \frac{16}{3}v_xD_x^1 - \frac{16}{3}uvI - \frac{4}{3}v_{2x}I - \frac{8}{3}u_xD_x^{-1}vI \\
\mathcal{R}_{21} &= -10v_xD_x^1 - 12v_{2x}I + 4v_xD_x^{-1}uI - 12uv_xD_x^{-1} - 4v_{3x}D_x^{-1}, \\
\mathcal{R}_{22} &= -4D_x^4 - 16uD_x^2 - 8u_xD_x^1 - \frac{16}{3}v^2I - \frac{8}{3}v_xD_x^{-1}vI.
\end{aligned}$$

A similar algorithm for the symbolic computation of recursion operators of systems of differential-difference equations (DDEs) is given elsewhere in these proceedings [**14**].

References

1. M. J. Ablowitz, P. A. Clarkson, *Solitons, nonlinear evolution equations and inverse scattering*, London Math. Soc. Lect. Note Ser. **149** (1991), Cambridge University Press, Cambridge, U.K.
2. M. J. Ablowitz and A. Ramani and H. Segur, *A connection between nonlinear evolution equations and ordinary differential equations of P-type. I–II*, J. Math. Phys. **21** (1980), 715–721 and 1006–1015.
3. D. Baldwin, Ü. Göktaş, W. Hereman, L. Hong, R. S. Martino, and J. C. Miller, *Symbolic computation of exact solutions expressible in hyperbolic and elliptic functions for nonlinear PDEs*, J. Symbolic Comput. **37** (2004) 669-705.
4. D. Baldwin and W. Hereman, *The three Mathematica packages* `PainleveTest.m`, `PDESpecialSolutions.m`, *and* `PDERecursionOperator.m` *are available under software at* `http://www.mines.edu/fs_home/whereman/`.
5. A. R. Chowdhury, *Painlevé Analysis and its Applications*, Monogr. and Surveys in Pure Appl. Math. vol. 105, Chapman & Hall/CRC, Boca Raton, FL, 2000.
6. R. Conte (ed.), *The Painlevé property: One century later*, CRM Ser. in Math. Phys., Springer-Verlag, New York, 1999.

7. R. Conte, A. P. Fordy, and A. Pickering, *A perturbative Painlevé approach to nonlinear differential equations*, Phys. D **69** (1993), 33–58.
8. Ü. Göktaş and W. Hereman, *Symbolic computation of conserved densities for systems of nonlinear evolution equations*, J. Symbolic Comput. **24** (1997), 591–621.
9. ———, *The Mathematica package* `InvariantSymmetries.m` *is available since* 1997 *from* `http://www.mathsource.com/cgi-bin/msitem?0208-932`.
10. ———, *Algorithmic computation of higher-order symmetries for nonlinear evolution and lattice equations*, Adv. Comput. Math. **11** (1999), 55–80.
11. W. Hereman and Ü. Göktaş, *Integrability tests for nonlinear evolution equations*, Computer Algebra Systems: A Practical Guide (M. Wester, ed.), Wiley, New York, 1999, pp. 211–232.
12. W. Hereman, Ü. Göktaş, M. D. Colagrosso, and A. J. Miller, *Algorithmic integrability tests for nonlinear differential and lattice equations*, Comput. Phys. Comm. **115** (1998), 428–446.
13. W. Hereman and A. Nuseir, *Symbolic methods to construct exact solutions of nonlinear partial differential equations*, Math. Comput. Simulation **43** (1997), 13–27.
14. W. Hereman, J. A. Sanders, J. Sayers, and J. P. Wang, *Symbolic computation of conserved densities, generalized symmetries, and recursion operators for nonlinear differential-difference equations*, this volume.
15. W. Hereman and M. Takaoka, *Solitary wave solutions of nonlinear evolution and wave equations using Macsyma*, J. Phys. A **23** (1990), 4805–4822.
16. W. Hereman and W. Zhang, *Symbolic software for soliton theory*, Acta Appl. Math. **39** (1995), 316–378.
17. R. S. Johnson, *On solutions of the Camassa–Holm equation*, Proc. Roy. Soc. London Ser. A **459** (2003), 1687–1708.
18. M. D. Kruskal, N. Joshi, and R. Halburd, *Analytic and asymptotic methods for nonlinear singularity analysis: A review and extensions of tests for the Painlevé property*, Integrability and Nonlinear Systems (Pondicherry, 1996), Lecture Notes in Phys., vol. 495, Springer, Berlin, 1997, pp. 171–205.
19. W. Malfliet and W. Hereman, *The Tanh method. I. Exact solutions of nonlinear evolution and wave equations*, Phys. Scripta **54** (1996), 563–568.
20. A. V. Mikhailov and V. S. Novikov, *Perturbative symmetry approach*, J. Phys. A **35** (2002), 4775–4790.
21. P. J. Olver, *Applications of Lie groups to differential equations*, 2nd ed. Grad. Texts in Math. **107** (1993), Springer-Verlag, New York.
22. J. P. Wang, *Symmetries and conservation laws of evolution equations*, Ph.D. Thesis, Thomas Stieltjes Institute for Mathematics, Amsterdam, 1998.
23. J. Weiss, M. Tabor, and G. Carnevale, *The Painlevé property for partial differential equations*, J. Math. Phys. **24** (1983), 522–526.

Department of Mathematical and Computer Sciences, Colorado School of Mines, Golden, CO 80401-1887, USA.

E-mail address: `dbaldwin@Mines.EDU`

Department of Mathematical and Computer Sciences, Colorado School of Mines, Golden, CO 80401-1887, USA.

E-mail address: `whereman@Mines.EDU`

Physics Department 59-33, California Institute of Technology, Pasadena, CA 91125, USA.

E-mail address: `jack@its.caltech.edu`

Centre de Recherches Mathématiques
CRM Proceedings and Lecture Notes
Volume **39**, 2005

Continuum Limit of Lattice Approximation Schemes

C. M. Bender

ABSTRACT. Boundary-layer perturbation theory problems are inherently singular. However, it is known that discretizing the problem by introducing a lattice may convert such problems into regular perturbation problems. The singular nature of boundary-layer problems is then relegated to and hidden in the continuum limit, the subtle limit in which the lattice spacing tends to zero. If the lattice is introduced cavalierly, then extrapolating to zero lattice spacing gives a sequence of extrapolants that at first approach the correct limit and then veer off, thereby revealing the asymptotic nature of such problems. However, discretizing the problem following the procedures described here yields lattice approximations that have a smooth and regular continuum limit. These ideas are illustrated by three nonlinear ordinary differential equations: the cubic equation that describes instantons, an oscillator equation having a quadratic nonlinearity, and the Blasius equation.

1. Introduction

It is useful to discretize a quantum field theory by introducing a lattice. The advantage of this approach is that the lattice serves as a regulator; divergent quantities in the continuum theory become finite in the corresponding lattice theory. However, the lattice spacing must be removed at the end of the computation to obtain the solution to the original continuum theory. Removing the lattice spacing (taking the continuum limit of the lattice theory) is a highly nontrivial procedure.

In this talk we examine the problem of reconstructing the original continuum theory from a discrete lattice approximation in a simpler and more concrete context, namely, boundary-layer theory. A boundary-layer problem is a differential equation and an associated set of boundary conditions in which the highest derivative in the differential equation equation is multiplied by a small parameter δ. The solution to a boundary-layer problem consists of two regions: an *inner* region, called a *boundary layer*, in which the solution is rapidly varying, and an *outer* region in which the solution is slowly varying. The width of the inner region vanishes as $\delta \to 0$. Boundary-layer problems are *singular* perturbation problems [**19**]. This means that the solution to the boundary-layer problem does not have a convergent Taylor series representation in powers of δ.

2000 *Mathematics Subject Classification.* Primary: 65L12, 39A05; Secondary: 34B15.

Supported in part by the US Department of Energy.

This is the final form of the paper.

When a boundary-layer problem is *discretized* by introducing a lattice with lattice spacing a, then the resulting problem often becomes a *regular* perturbation problem (one whose solution has the form of a Taylor series in powers of δ [**2, 9, 21**]). This transformation from a singular to a regular perturbation problem is the differential-equation analog of the lattice regulation that occurs in the context of quantum field theory. The singular nature of the problem then resurfaces when the continuum limit of the discretized problem is taken. This singular limit is nontrivial. One cannot just set the parameter a to zero; rather, one must perform the delicate limit $a \to 0$ using numerical extrapolation techniques. Extrapolation techniques used in the past have been based on Padé approximation methods.

Unfortunately, the continuum limit $a \to 0$ is so singular that, not surprisingly, the accuracy of numerical results that have been obtained in the past is not unimpressive. In fact, in very high order approximations the numerical extrapolants typically approach the exact answer for a while and then veer off. Thus, there appears to be a maximally obtainable accuracy that cannot be improved by going to higher order in powers of δ.

In this talk we propose a resolution to this dilemma. There are *many* ways to discretize a differential equation problem. We enumerate two criteria that enable us to choose uniquely the optimal discretization scheme for the differential equation. The optimal scheme has the advantage that the extrapolants to the continuum limit approach the $a = 0$ value *smoothly*. Thus, as the order of the approximation is increased, the numerical error in these extrapolants continues to decrease.

Our organization is straightforward. In Section 2 we examine the instanton solution to a cubic nonlinear differential equation:

$$\delta^2 y''(x) = [y(x)]^3 - y(x). \tag{1.1}$$

The boundary conditions $y(0) = 0$ and $y(\infty) = 1$ give rise to a boundary layer at $x = 0$; the objective is to calculate the value of $y'(0)$. This problem has the virtue that the exact solution, called an instanton, is known analytically. We examine various discretization schemes for this problem, and we formulate a set of criteria for selecting the best of these schemes.

In Section 3 we study an oscillator having a quadratic nonlinearity:

$$\delta^2 y''(x) = [y(x)]^2 - 2y(x). \tag{1.2}$$

We seek soliton solutions satisfying the boundary conditions $y(\pm\infty) = 2$. There are two *even-parity* solutions to this problem, a trivial constant solution $y \equiv 2$ and a nontrivial solution. Our objective here is to find the value of $y(0)$ for the nontrivial solution. Again, the exact solution to this problem is known analytically.

Finally, in Section 4 we look at the very difficult problem of a boundary-layer solution to the Blasius equation

$$\delta y'''(x) + y(x)y''(x) = 0, \tag{1.3}$$

where $y(0) = 0$, $y'(0) = 0$, and $y'(\infty) = 1$. The goal is to calculate the value of $y''(0)$. The solution to this third problem is not known analytically; it can only be obtained by using numerical methods. We hope that the success of the lattice techniques discussed here will inspire a re-examination of the strong-coupling lattice techniques that have been used in the past to study quantum field theory [**3–8, 10–13, 16–18**]. We feel that these improved discretization schemes will lead to dramatically improved numerical results in quantum field theory.

2. Boundary-layer approximation to an instanton

As our first example we consider the boundary-value problem

$$\delta^2 y''(x) = [y(x)]^3 - y(x), \quad y(0) = 0, \; y(\infty) = 1. \tag{2.1}$$

The solution to this equation is called an *instanton.* This instanton arises in the context of a semiclassical approximation to the functional integral representing a ϕ^4 Euclidean quantum field theory.

The exact closed-form solution to the problem in (2.1) is known:

$$y(x) = \tanh[x/(\delta\sqrt{2})]. \tag{2.2}$$

Observe that when $\delta \ll 1$ the solution exhibits a *boundary layer* (a narrow region of rapid variation) at the origin $x = 0$; as $\delta \to 0$, the solution becomes discontinuous at $x = 0$. The solution is slowly varying elsewhere.

Our objective here is to use a lattice approximation to determine the value of $y'(0)$. Of course, from (2.2) we already know the exact answer:

$$y'(0) = 1/(\delta\sqrt{2}). \tag{2.3}$$

There are many ways to discretize the differential equation in (2.1). For example, an apparently natural way to introduce a lattice is to replace this differential equation by the difference equation

$$\varepsilon(y_{n+1} - 2y_n + y_{n-1}) = y_n^3 - y_n, \tag{2.4}$$

where $\varepsilon \equiv \delta^2/a^2$.

The underlying reason for introducing the variable ε is that it is *dimensionless.* We emphasize that the limit of zero lattice spacing does not make sense because one cannot take a dimensional quantity such as a to be "small." This is because one can always redefine the units so that the dimensional quantity has exactly the same numerical size. One can only take the lattice limit *relative to another quantity* in the theory having the same dimensions as a; this other quantity is δ. Thus, the continuum limit of the theory is achieved by performing the limit $\varepsilon \to \infty$.

There is another limit in the theory, namely, $\varepsilon \to 0$. We borrow some terminology from quantum field theory [**23–26**] and refer to this limit as the *ultralocal* limit. In the ultralocal limit the kinematic (derivative) terms on the left side of (2.1) vanish and there is a balance of the local terms on the right side. In the language of boundary-layer theory, this limit is called the *outer limit.*

The calculational procedure is to treat the lattice spacing a as being *fixed* and δ as small, or equivalently, to treat the parameter ε as small. Then, at each lattice point n we expand y_n as a series in powers of ε. Finally, from the power series for y_1 and y_0, we try to recover the derivative $y'(0)$.

We do not describe the details of this calculation here because they are given in Refs. [**2,9**]. However, in brief, we incorporate the boundary conditions by requiring that $y_0 = 0$ and that $y_\infty = 1$. Second, we observe that at $\varepsilon = 0$ (the ultralocal limit of the lattice problem) there is a simple solution $y_0 = 0$ and $y_n = 1$ $(n > 0)$. Next, we expand about this unperturbed solution and obtain a series expansion for

y_n for each value of n:

$$\begin{aligned}
y_0 &= 0,\\
y_1 &= 1 - \tfrac{1}{2}\varepsilon + \tfrac{1}{8}\varepsilon^2 - 0\varepsilon^3 + \tfrac{11}{28}\varepsilon^4 + \cdots,\\
y_2 &= 1 - \tfrac{1}{4}\varepsilon^2 + \tfrac{5}{16}\varepsilon^3 - \tfrac{15}{32}\varepsilon^4 + \cdots,\\
y_3 &= 1 - \tfrac{1}{8}\varepsilon^3 + \tfrac{9}{32}\varepsilon^4 + \cdots,\\
y_4 &= 1 - \tfrac{1}{16}\varepsilon^4 + \cdots,
\end{aligned}$$

and so on.

The boundary-layer structure at $x = 0$ is incorporated in this discrete problem in an interesting fashion: The first dependence on ε in the power series for y_n occurs at the ε^n term. Thus, as n increases, the ε dependence becomes weaker; this is the lattice version of the *outer region* where the solution to the boundary-layer problem is a slowly varying function of x. As n tends to ∞, y_n has the degenerate power series 1 and the boundary condition at $x = \infty$ is correctly incorporated.

We must now calculate $y'(0)$ in the continuum limit. Since the lattice representation of $y'(x)$ is $\lim_{a\to 0}(y_{n+1} - y_n)/a$, we have from (2.5)

$$\begin{aligned}
y'(0) &= \lim_{\varepsilon\to\infty} \tfrac{1}{a}(y_1 - y_0)\\
&= \lim_{\varepsilon\to\infty} \tfrac{1}{a}\left(1 - \tfrac{1}{2}\varepsilon + \tfrac{1}{8}\varepsilon^2 - 0\varepsilon^3 + \tfrac{11}{28}\varepsilon^4 + \cdots\right)\\
&= \tfrac{1}{\delta} \lim_{\varepsilon\to\infty} \sqrt{\varepsilon}\left(1 - \tfrac{1}{2}\varepsilon + \tfrac{1}{8}\varepsilon^2 - 0\varepsilon^3 + \tfrac{11}{28}\varepsilon^4 + \cdots\right).
\end{aligned} \tag{2.5}$$

If we compare with the exact answer in (2.3), we see that the factor of $1/\delta$ is correct. However, it is not obvious how to obtain the numerical result

$$\tfrac{1}{\sqrt{2}} = \lim_{\varepsilon\to\infty} \sqrt{\varepsilon}\left(1 - \tfrac{1}{2}\varepsilon + \tfrac{1}{8}\varepsilon^2 - 0\varepsilon^3 + \tfrac{11}{28}\varepsilon^4 + \cdots\right). \tag{2.6}$$

We have now encountered a very difficult problem associated with the interchange of limits. In principle, we must first obtain the perturbation expansion to all orders in powers of ε and sum the series. Second, we must take the limit $\varepsilon \to \infty$. Unfortunately, it is not possible to perform the limits in this order because one can only obtain a *finite* number of terms in any perturbation expansion for a nontrivial problem. Thus, the question is this: How can one make sense of (2.6) when there are only a limited number of terms known in the series? A number of solutions to this problem have been suggested and studied [**14, 15, 29**]; all of the approaches involve the use of Padé approximations.

The simplest approach is to treat the perturbation parameter ε as small ($\varepsilon \ll 1$) and to conduct a sequence of algebraic manipulations whose objective is to change the form of the right side of (2.6) to one that has a limit as $\varepsilon \to \infty$. With each manipulation we retain the terms of the series in (2.6) to order ε^N, where N is the number of terms that we have calculated in perturbation theory. First, we square the right side of (2.6) to eliminate the fractional power of ε. This gives a structure of the form $\varepsilon \sum_{k=0}^{N} A_k \varepsilon^k$, where the numbers A_k are obtained by squaring the series in (2.6) term-by-term. We emphasize that consistency demands that we truncate the squared series after the ε^N term. Next, we invert the A-series term-by-term and again truncate the resulting series after the ε^N term: $\varepsilon \,/\, \sum_{k=0}^{N} B_k \varepsilon^k$. We now

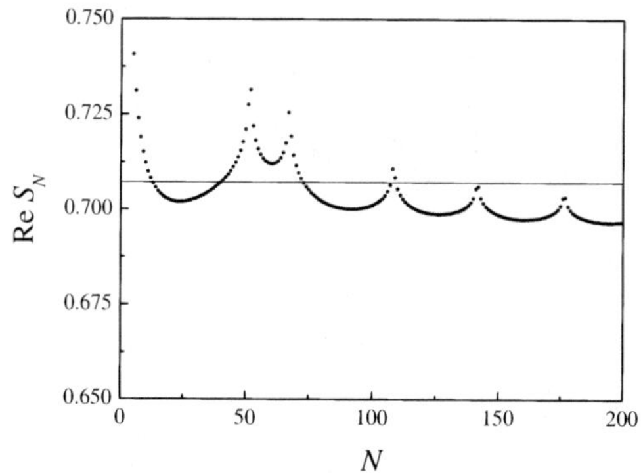

FIGURE 1. Padé approximants S_N up to 200th order. For large N the approximants do not converge to the exact solution, which is represented by the horizontal solid line. For a small range of N near each of the peaks the approximants become complex, so we plot the real part of S_N.

raise the B series to the power N:

$$\varepsilon^N \Big/ \sum_{k=0}^{N} C_k \varepsilon^k. \tag{2.7}$$

The continuum limit $\varepsilon \to \infty$ now exists! In this limit (2.7) becomes $1/C_N$. Finally, we compensate for having raised the right side of (2.6) to the powers 2 and N by taking the $2N$th root of this limit; we *define* the Nth *extrapolant* S_N by

$$S_N \equiv (C_N)^{1/(2N)}. \tag{2.8}$$

Clearly, the hope is that as $N \to \infty$, the Nth extrapolant S_N will tend to the correct limit $1/\sqrt{2} = 0.70711\ldots$. However, what we actually observe is that the extrapolants approach the correct limit for a while and then veer away from this limit. Specifically, as N increases from 1, the extrapolants S_N seem to be approaching the correct limit monotonically: $S_1 = 1.0$, $S_2 = 0.84090$, $S_3 = 0.78193$, $S_4 = 0.75724$, $S_5 = 0.74076$, $S_6 = 0.73121$, $S_7 = 0.72393$, $S_8 = 0.71905$, $S_9 = 0.71515$, $S_{10} = 0.71231$. The extrapolants continue to decrease until they *undershoot* the exact value. Eventually, the extrapolants reach a broad, flat minimum in 24th order: $S_{24} = 0.70198$. The relative error between this value and the exact answer is less than 1%. Then the extrapolants gradually rise; they recross the value 0.70711 at 41st order and continue rising until they reach a peak. They then come down from the peak, reach a new broad minimum, and rise again. This behavior apparently continues, with successive minima decreasing [**20**]. The behavior of S_N up to $N = 200$ is shown on Fig. 1. Note that for a small range of N near each of the peaks the extrapolants become complex so only the real part of S_N is plotted. Evidently, there is no advantage to going to higher order in powers of ε. We believe that underlying this complicated behavior is the fact that we are solving a boundary-layer problem, which is a *singular* perturbation problem.

In this talk we show how to remedy this serious divergence problem by using a superior lattice approximation. We begin by using the well-known fact that a lattice approximation to a differential equation may be regarded as a *higher-order derivative perturbation* of that differential equation. An example of a higher-order derivative perturbation of the differential equation $y''(x) = [y(x)]^3 - y(x)$ is

$$\varepsilon^2 y''''(x) + y''(x) = [y(x)]^3 - y(x), \tag{2.9}$$

where ε is a small parameter.

Introducing a lattice is merely another way to perturb an equation. On a lattice of lattice spacing a we make the replacement

$$\begin{aligned} y''(x) \to D^2 y(x) &\equiv a^{-2}[y(x+a) - 2y(x) + y(x-a)] \\ &= y''(x) + \tfrac{1}{12}a^2 y''''(x) + \tfrac{1}{360}a^4 y''''''(x) + \cdots . \end{aligned} \tag{2.10}$$

Either of the above equations is a *singular perturbation* of $y''(x) = [y(x)]^3 - y(x)$ because higher derivatives are multiplied by powers of the small parameter. However, we can use the Lie symmetry of the underlying unperturbed nonlinear equation $y''(x) = [y(x)]^3$ to eliminate small parameter factors multiplying the higher derivatives. This equation is invariant under

$$x \to \alpha x, \quad y(x) \to \alpha^{-1} y(x). \tag{2.11}$$

Thus, choosing $\alpha = \varepsilon$ in (2.9) or $\alpha = a$ in $D^2 y(x) = [y(x)]^3 - y(x)$, we get

$$y''(x) + y''''(x) = [y(x)]^3 - \varepsilon^2 y(x) \tag{2.12}$$

and

$$y''(x) + \tfrac{1}{12}y''''(x) + \tfrac{1}{360}y''''''(x) + \cdots = [y(x)]^3 - a^2 y(x). \tag{2.13}$$

These equations appear to be *regular* perturbation problems because the higher derivatives are no longer multiplied by factors of the perturbation parameter. However, this is not really so because the singular behavior has been shifted to the point $z = \infty$, where $x = \varepsilon z$.

It can be proved[1] that *all* solutions except the trivial solution $y(x) = 0$ to these equations are *not* regular at $z = \infty$. Indeed, in the (y, y')-phase plane [the two-dimensional projection of the full $(y, y', y'', y''', \dots)$-phase space] the phase portrait of (2.13) exhibits chaotic behavior [**1, 28, 31, 32**].

Similarly, if the cubic nonlinear term in (2.9) or in the corresponding discretized equation is replaced by the quadratic nonlinear term $[y(x)]^2$. For this case the nonlinear equation $y''(x) = [y(x)]^2$ is invariant under the Lie symmetry

$$x \to \alpha x, \quad y(x) \to t\alpha^{-2} y(x). \tag{2.14}$$

Again there is no nontrivial regular solution at $z = \infty$. [A central-difference discretization of (1.2) is equivalent to the Hénon map that is known to exhibit chaotic behavior.]

A necessary condition for there to be a smooth continuum limit is that at $z = \infty$ there exist a nontrivial regular solution (in the form of a Taylor series in powers of $1/z$); such a solution would not exhibit chaotic behavior. It is the rigorously demonstrated nonexistence of such a regular solution at $z = \infty$ that probably prevents us from extrapolating smoothly to the continuum limit of the

---

[1] A rigorous of splitting separatrices in general can be found; for two-dimensional systems ooo [**28**]. Chaotic behavior of systems with with a quadratic nonlinearity is demonstrated in [**1, 30–32**]. The cubic case is considered in [**27**].

lattice approximation in (2.10). However, if we could find a discretization scheme that respects the discrete Lie symmetry, then the Taylor series at $z = \infty$ would truncate and we would have a regular solution. (Of course, the mere existence of a solution that is regular at $z = \infty$ does not in itself guarantee that the solution we seek will be regular. However, we immediately reject other discretization schemes because they are associated with chaotic behavior.) We summarize these remarks as a general criterion [**22**]:

CRITERION 1. One must discretize the nonlinear term in the differential equation so that the resulting discrete difference equation has the same (singular) scaling solution (that is, the same Lie symmetry) as the original differential equation.

We apply this criterion to (2.1) as follows: Neglect the linear term, which in the scaled equation (2.13) is multiplied by the small parameter a^2. Observe that there is a singular scaling solution to $\delta^2 y''(x) = [y(x)]^3$ of the form $y(x) = c/x$, where c is a constant. We seek to discretize the nonlinear term $[y(x)]^3$ so that the difference equation has a solution of the form c/n. The choice of difference equation is unambiguous; there is a unique cubic term that respects the Lie symmetry:

$$\varepsilon(y_{n+1} - 2y_n + y_{n-1}) = (y_{n+1} + y_{n-1})y_n^2/2. \tag{2.15}$$

[As we show in Section 3 for the case of a quadratic nonlinearity, we seek an exact solution of the form $y_n = c/[n(n+1)]$.]

The above criterion does not provide any guidance about how to discretize the linear term in the differential equation. Lacking guidance, we might choose

$$\varepsilon(y_{n+1} - 2y_n + y_{n-1}) = (y_{n+1} + y_{n-1})y_n^2/2 - y_n, \tag{2.16}$$

which is the simplest scheme. However, this choice is unacceptable as we now show. Consider the ultralocal ($\varepsilon = 0$) equation:

$$(y_{n+1} + y_{n-1})y_n^2/2 - y_n = 0. \tag{2.17}$$

The natural solution to (2.17) at $n = 0$ is $y_0 = 0$ because it incorporates the boundary condition $y(0) = 0$. Next, we examine (2.17) at $n = 1$; this equation reads $y_2 y_1^2 = 2y_1$. Let $y_1 = \alpha$, where α is arbitrary. Then, $y_2 = 2/\alpha$. Next, we examine (2.17) at $n = 2$; this equation reads $(y_3 + y_1)y_2^2 = 2y_2$, whence $y_3 = 0$. If we continue this process, we obtain $y_4 = \beta$, where β is arbitrary, $y_5 = 2/\beta$, $y_6 = 0$, $y_7 = \gamma$, where γ is arbitrary, $y_8 = 2/\gamma$, $y_9 = 0$, and so on. We reject the discretization in (2.16) because it gives a sequence of y_n at $\varepsilon = 0$ that does not have a continuum limit; rather, it has a choppy, fluctuating structure.

To determine an acceptable discretization of (2.1), we formulate a second general criterion [**22**]:

CRITERION 2. The unperturbed ultralocal solution to the difference equation in the outer region must be smooth so that it will have a continuum limit.

Imposing this criterion *uniquely* determines a discretization for (2.1):

$$\varepsilon(y_{n+1} - 2y_n + y_{n-1}) = (y_{n+1} + y_{n-1})y_n^2/2 - (y_{n+1} + y_{n-1})/2. \tag{2.18}$$

The ultralocal solution to this equation satisfies Criterion 2:

$$y_0 = 0, \quad y_n = 1 \ (n > 0). \tag{2.19}$$

Although this ultralocal solution is not smooth at $n = 0$, this jump discontinuity does not violate Criterion 2 because this jump is in the boundary-layer (inner) region and not in the outer region.

Next, we expand y_n for each n as a series in powers of ε and obtain the analog of (2.5):

$$\begin{aligned} y_0 &= 0, \\ y_1 &= 1 - \varepsilon + \tfrac{3}{2}\varepsilon^2 - \tfrac{5}{2}\varepsilon^3 + \tfrac{35}{8}\varepsilon^4 + \cdots, \\ y_2 &= 1 - \tfrac{1}{2}\varepsilon^2 + \varepsilon^3 - \tfrac{13}{8}\varepsilon^4 + \cdots, \\ y_3 &= 1 - \tfrac{1}{4}\varepsilon^3 + \tfrac{3}{4}\varepsilon^4 + \cdots, \\ y_4 &= 1 - \tfrac{1}{8}\varepsilon^4 + \cdots, \end{aligned} \tag{2.20}$$

and so on. If we now determine the extrapolants S_N from these new series, we obtain a dramatic improvement; not only does S_N converge to the exact answer as $N \to \infty$, but S_N *equals the exact answer* for all N:

$$S_N = 1/\sqrt{2} \quad (\text{all } N). \tag{2.21}$$

We are able to verify this result because while it is extremely rare to find an exact solution to a nonlinear equation, we have succeeded in solving the difference equation (2.18) *exactly and in closed form* for all values of ε:

$$y_0 = 0, \quad y_1 = \frac{1}{\sqrt{2\varepsilon+1}}, \quad y_2 = \frac{\sqrt{2\varepsilon+1}}{\varepsilon+1}, \quad y_3 = \frac{3\varepsilon+2}{(\varepsilon+2)\sqrt{2\varepsilon+1}},$$

$$y_4 = \frac{(2\varepsilon+2)\sqrt{2\varepsilon+1}}{\varepsilon^2+4\varepsilon+2}, \quad y_5 = \frac{5\varepsilon^2+10\varepsilon+4}{(\varepsilon^2+6\varepsilon+4)\sqrt{2\varepsilon+1}}, \quad y_6 = \frac{(3\varepsilon^2+8\varepsilon+4)\sqrt{2\varepsilon+1}}{\varepsilon^3+9\varepsilon^2+12\varepsilon+4},$$

and so on. In general, for all n we have

$$y_n = \frac{(\sqrt{2\varepsilon+1}+1)^n - (\sqrt{2\varepsilon+1}-1)^n}{(\sqrt{2\varepsilon+1}+1)^n + (\sqrt{2\varepsilon+1}-1)^n}, \tag{2.22}$$

which is the lattice version of the hyperbolic tangent function in (2.2) [**27**].

3. Oscillator with a quadratic nonlinearity

In this section we show how to use the criteria formulated in Section 2 to find the correct discretization of the nonlinear equation

$$\delta^2 y''(x) = [y(x)]^2 - 2y(x) \tag{3.1}$$

subject to the boundary conditions $y(\pm\infty) = 2$. Because (3.1) has a translation symmetry, there is an infinite number of soliton solutions to this problem, all parameterized by the location of the center (minimum value) of the wave. To eliminate this translation ambiguity, we impose the further condition that $y(x)$ be an *even* function of x. There are exactly two solutions that satisfy this additional requirement, a trivial constant solution, $y(x) \equiv 2$, and an exact nontrivial solution in terms of the hyperbolic tangent function

$$y(x) = -1 + 3\tanh^2[x/(\delta\sqrt{2})]. \tag{3.2}$$

Our objective will be to find the numerical value of $y(0)$; from the above two equations we know that the exact answers are $y(0) = 2$ and $y(0) = -1$.

Like the boundary-value problem considered in Section 2, the differential equation in (3.1) is a boundary-layer problem. From the exact solution in (3.2) we know that the boundary layer occurs at the origin and has thickness δ.

We introduce a lattice according to the principles formulated in Section 2. First, we observe that the nonlinear differential equation $\delta^2 y''(x) = [y(x)]^2$ has a scale invariance (a Lie symmetry) and admits singular, double-pole solutions of the form

$$y(x) = cx^{-2}, \tag{3.3}$$

where c is a constant. The lattice equivalent of a double pole is $[n(n+1)]^{-1}$ (or, more generally, $[(n+\alpha)(n+1+\alpha)]^{-1}$) and *not* n^{-2}, as one might naively think. In general, the lattice equivalent of the continuum function x^{-k} is

$$f_k = [(n+\alpha)(n+1+\alpha)(n+2+\alpha)(n+3+\alpha)\cdots(n+k-1+\alpha)]^{-1}. \tag{3.4}$$

To understand this equivalence note that the discrete analog of $d/dx x^{-k} = -kx^{-k-1}$ is $Df_k = -kf_{k+1}$, where D is the discrete derivative (first difference) operator.

Now we apply Criterion 1 of Section 2. Since we are looking for a symmetric solution to (3.1), we represent $y''(x)$ by a *symmetric* double difference: $a^{-2}(y_{n+1} - 2y_n + y_{n-1})$. Taking the second difference of the solution $y_n = 1/[(n+\alpha)(n+1+\alpha)]$, we get $1/[(n-1+\alpha)(n+\alpha)(n+1+\alpha)(n+2+\alpha)]$, apart from a multiplicative constant. There are exactly *two* symmetric quadratic lattice structures that we could use on the right side of the equation to produce this result from $y_n = 1/[(n+\alpha)(n+1+\alpha)]$; the structure $y_{n+1}y_{n-1}$ gives $1/[(n-1+\alpha)(n+\alpha)(n+1+\alpha)(n+2+\alpha)]$, and $y_n(y_{n+1}+y_n+y_{n-1})$ gives $3/[(n-1+\alpha)(n+\alpha)(n+1+\alpha)(n+2+\alpha)]$. (Criterion 1 does not place any requirements on the linear term.) We conclude that the most general symmetric difference equation satisfying Criterion 1 is

$$\varepsilon(y_{n+1} - 2y_n + y_{n-1}) = +Qy_{n+1}y_{n-1} - Ry_{n+1} + y_{n-1} - Sy_n, \tag{3.5}$$

where $\varepsilon = \delta^2 a^{-2}$. The arbitrary constants P, Q, R, and S in (3.5) obey two constraints: Since the coefficient of the quadratic term in (3.1) is 1, we have

$$3P + Q = 1, \tag{3.6}$$

and since the coefficient of the linear term in (3.1) is -2, we have

$$2R + S = 2. \tag{3.7}$$

Next we impose Criterion 2 (smoothness of the ultralocal solution in the outer region). Criterion 2 states that outside the boundary layer at $x = 0$ ($n = 0$ on the lattice) the unperturbed solution must have a slowly varying continuum limit. We therefore seek an ultralocal ($\varepsilon = 0$) solution of the form

$$y_0 = \xi, \quad y_n = 2 \ (n \neq 0). \tag{3.8}$$

For $n = 0$ this gives the constraint

$$0 = \xi(4+\xi)P + 4Q - 4R - \xi S, \tag{3.9}$$

and for $n = 1$ this gives

$$0 = 2(4+\xi)P + 2\xi Q - (2+\xi)R - 2S. \tag{3.10}$$

When $n > 1$, we obtain the condition $0 = 6P + 2Q - 2R - S$, which is already true by virtue of Eqns. (3.6) and (3.7).

The equations (3.6), (3.7), (3.9), and (3.10) determine uniquely the arbitrary constants in the lattice equation (3.5):

$$P=\frac{2}{2-\xi},\quad Q=\frac{\xi+4}{\xi-2},\quad R=\frac{2\xi+4}{\xi-2},\quad S=\frac{2\xi+12}{2-\xi}. \tag{3.11}$$

It appears as if all that remains is for us to substitute (3.11) into (3.5) and to expand y_n for each n as series in powers of ε. However, as we now show, this procedure is not quite as straightforward as in Section 2 because here we encounter a subtlety with regard to the form of the regular perturbation expansion. In particular, if we attempt to expand y_n as a series in powers of ε, we reach an immediate contradiction! Let

$$y_0=\xi+a_0\varepsilon+\mathrm{O}(\varepsilon^2),\quad y_n=2+a_n\varepsilon+\mathrm{O}(\varepsilon^2)\ (n\neq 0). \tag{3.12}$$

Now, for $n=0$ we obtain

$$2-\xi+a_0+a_1+a_{-1}=0, \tag{3.13}$$

and for $n=1$ we have

$$\xi-2+(\xi+4)a_2=0. \tag{3.14}$$

For $n>1$ we get simply

$$a_2=0,\quad a_3=0,\quad a_4=0, \tag{3.15}$$

and so on. Hence, from Eqs. (3.13)–(3.15) we have $\xi=2$ and we are forced to conclude that there is *no boundary layer* at $x=0$.

There are two ways to avoid this problem. The first is to recognize that because the boundary layer in the continuum differential equation is thick [of size $\delta=\mathrm{O}(\sqrt{\varepsilon})$ and not $\mathrm{O}(\varepsilon)$], the boundary layer on the lattice must be made thicker. Thus, we replace (3.8) by

$$y_0=\xi_1,\quad y_{\pm1}=\xi_2,\quad y_n=2\ (n\neq 0,\pm1). \tag{3.16}$$

We have examined this approach in detail and have verified that it works successfully; in fact, with the choice in (3.16) we are able to find all the terms in the perturbation series and to sum the perturbation series to all orders in closed form. However, we do not pursue this approach further here because there is a second procedure that is simpler and more natural. Since the thickness of the boundary layer in the continuum is of order δ, we seek an expansion of y_n for each n as a series in powers of $z=\sqrt{\varepsilon}$. Thus, we replace (3.12) by

$$\begin{aligned} y_0&=\xi+a_0z+b_0z^2+c_0z^3+\mathrm{O}(z^4),\\ y_n&=2+a_nz+b_nz^2+c_nz^3+\mathrm{O}(z^4)\ (n\neq 0). \end{aligned} \tag{3.17}$$

With this *ansatz* we obtain the following series representations in powers of $z=\sqrt{\varepsilon}$ on the lattice:

$$\begin{aligned} y_0&=\xi+a_0z+\frac{(\xi-2)(\xi+12)}{4}z^2+\frac{(\xi-2)^2(\xi+8)}{2a_0}z^3\\ &\qquad+\frac{(\xi-2)(\xi+4)(\xi+6)}{16}z^4+\cdots,\\ y_{\pm1}&=2-\frac{a_0}{2}z+\frac{(2-\xi)}{2}z^2-\frac{(\xi-2)^2(\xi+6)}{4a_0}z^3-\frac{(\xi-2)(\xi+6)}{8}z^4+\cdots, \end{aligned} \tag{3.18}$$

$y_{\pm 2} = 2 - a_0 z^3/4 + \mathrm{O}(z^4)$, $y_{\pm 3} = 2 + \mathrm{O}(z^5)$, $y_{\pm 4} = 2 + \mathrm{O}(z^7)$, $y_{\pm 5} = 2 + \mathrm{O}(z^9)$, $y_{\pm 6} = 2 + \mathrm{O}(z^{11})$, and so on, where $(a_0)^2 = 2(\xi - 2)^2$.

We have been able to carry out this analysis to *all* orders in powers of z and to sum the series in closed form. We are especially interested in y_0. For this case the odd terms and the even terms in the series have completely different structures. For even powers in z we have

$$\begin{aligned}
\{y_0\}_{\text{even}} &= \xi + \frac{(\xi-2)(\xi+12)}{4} z^2 + \frac{(\xi-2)(\xi+4)(\xi+6)}{16} z^4 \\
&\quad + \frac{(\xi-2)^2(\xi+4)(\xi+6)}{64} z^6 + \frac{(\xi-2)^3(\xi+4)(\xi+6)}{256} z^8 \\
&\quad + \frac{(\xi-2)^4(\xi+4)(\xi+6)}{1024} z^{10} + \cdots, \\
&= \xi + \frac{(\xi-2)(\xi+12)}{4} z^2 + \frac{(\xi-2)(\xi+4)(\xi+6)}{16} z^4 \sum_{k=0}^{\infty} \left[\frac{(\xi-2)z^2}{4}\right]^k \\
&= \xi + \frac{(\xi-2)(\xi+12)}{4} z^2 + \frac{(\xi-2)(\xi+4)(\xi+6) z^4}{16 - 4(\xi-2)z^2}.
\end{aligned}$$

The odd powers of z in the series for y_0 are somewhat more complicated:

$$\{y_0\}_{\text{odd}} = z a_0 \sum_{k=0}^{\infty} \mathcal{P}_k(\xi) \left(\frac{z^2}{4}\right)^k, \tag{3.19}$$

where $\mathcal{P}_k$ is the set of polynomials in the variable ξ:

$$\begin{aligned}
\mathcal{P}_0 &= 1, \\
\mathcal{P}_1 &= \xi + 8, \\
\mathcal{P}_2 &= \xi^2 + 6\xi, \\
\mathcal{P}_3 &= \xi^3 + 4\xi^2 - 12\xi - 16, \\
\mathcal{P}_4 &= \xi^4 + 2\xi^3 - 20\xi^2 + 8\xi + 64, \\
\mathcal{P}_5 &= \xi^5 - 24\xi^3 + 48\xi^2 + 48\xi - 192,
\end{aligned}$$

and so on. These polynomials satisfy the simple difference equation

$$\mathcal{P}_{k+1} = (\xi - 2)\mathcal{P}_k + T_k, \tag{3.20}$$

where T_k $(k = 0, 1, 2, 3, 4, \dots)$ is the sequence of numbers 10, 16, -16, 32, -64, 0, 1536, -16896, 146432, $-1171456, \dots$. An exact, closed-form expression for T_k is

$$T_k = \frac{(k-5)(-8)^k}{\sqrt{\pi}(k+1)!} \Gamma\left(k - \frac{1}{2}\right). \tag{3.21}$$

Hence, an explicit formula for $\mathcal{P}_k$ is

$$\mathcal{P}_k = (\xi - 2)^k + \sum_{j=0}^{k-1} T_j (\xi - 2)^{k-j-1}. \tag{3.22}$$

If we substitute (3.22) into (3.19) and perform the double summation, we obtain

$$\{y_0\}_{\text{odd}} = \frac{2 a_0 z (2 + 3z^2)\sqrt{1 + 2z^2}}{4 - (\xi - 2) z^2}. \tag{3.23}$$

Finally, combining the even and odd parts of y_0 and eliminating z and a_0 in favor of ε, we have

$$y_0 = \xi + \frac{(\xi - 2)\varepsilon}{4 - \varepsilon(\xi - 2)}\left[12\varepsilon + \xi + 12 - (8 + 12\varepsilon)\sqrt{1 + \frac{1}{2\varepsilon}}\right]. \tag{3.24}$$

Observe that when $\xi = 2$, we have $y_0 = 2$ for any ε; thus, we have found the trivial solution to the problem. On the other hand, when $\xi \neq 2$, in the limit of zero lattice spacing we obtain

$$y_0 = -1 - \frac{19\xi + 58}{(8\xi - 16)\varepsilon} + \mathrm{O}(\varepsilon^{-2}) \quad (\varepsilon \to \infty). \tag{3.25}$$

Thus, in the limit as $\varepsilon \to \infty$ we recover the nontrivial boundary-layer solution to the problem.

It is startling indeed that we have managed once again to find the exact, closed-form solution to a nonlinear second-order difference equation. It is quite remarkable that imposing Criteria 1 and 2 on the nonlinear differential equations in Sections 2 and 3 has led us to formulate discrete nonlinear difference equations that can be solved exactly and in closed form; any other discretizations yield difference equations for which there is virtually no hope of finding exact solutions.

4. Blasius equation

The Blasius equation is a nonlinear third-order differential equation that arises in the description of a fluid flowing along a flat plate:

$$\delta y'''(x) + y(x)y''(x) = 0, \tag{4.1}$$

where $y(x)$ satisfies the boundary conditions

$$y(0) = 0, \quad y'(0) = 0, \quad y'(\infty) = 1. \tag{4.2}$$

Our goal is to show that

$$y''(0) = 0.46960\ldots/\sqrt{\delta}. \tag{4.3}$$

This is an extremely difficult boundary-value problem that not only has no known analytical solution but also is extremely difficult to solve numerically.

In Refs. [**2, 9**] this problem was considered but the discretization scheme used there gave only moderately good and not excellent results. The discretization was exactly what one might choose in the absence of the criteria developed in Section 2:

$$\begin{aligned} y(x) &\to y_n, \\ y''(x) &\to (y_{n+1} - 2y_n + y_{n-1})/a^2, \\ y'''(x) &\to (y_{n+1} - 3y_n + 3y_{n-1} - y_{n-2})/a^3. \end{aligned}$$

Using as boundary conditions $y_{-1} = y_0 = 0$, a lattice series for y_1 in powers of the scaled small parameter δ was obtained to 38th order. From this series a sequence of lattice extrapolants was obtained. The lattice extrapolants E_N form a monotonically decreasing sequence: $E_1 = 0.7071/\sqrt{\delta}$, $E_2 = 0.5945/\sqrt{\delta}$, $E_3 = 0.5583/\sqrt{\delta}$, $E_4 = 0.5401/\sqrt{\delta}$, $E_5 = 0.5292/\sqrt{\delta}$, and so on. Although this sequence continues to decrease, it is not obvious whether it approaches the correct answer in (4.3); indeed it appears to level off at a value that is about 5% too high: $E_{25} = 0.4953/\sqrt{\delta}$, $E_{26} = 0.4950/\sqrt{\delta}$, $E_{37} = 0.4928/\sqrt{\delta}$, $E_{38} = 0.4927/\sqrt{\delta}$.

Let us discretize the Blasius equation following the procedures described in Sections 2 and 3. First, we make the replacement

$$y'''(x) \to (f_{n+3} - 3f_{n+2} + 3f_{n+1} - f_n)/a^2. \tag{4.4}$$

Here, a is the lattice spacing and for convenience we perform a scaling $y_n = af_n$. Next, we consider the nonlinear term and make the replacement:

$$\begin{aligned} y(x)y''(x) \to\ & Pf_{n+2}f_{n+3} + (1-2Q-P)f_nf_{n+1} + (1-3Q)f_nf_{n+3} \\ & - 3Qf_{n+1}f_{n+2} + (6Q+2P-2)f_nf_{n+2} + (2Q-2P)f_{n+1}f_{n+3}, \end{aligned} \tag{4.5}$$

where P and Q are arbitrary parameters. This expression is the *most general* discretization of $y(x)y''(x)$ using quadratic terms at the lattice points n, $n+1$, $n+2$, $n+3$ such that the following two requirements are satisfied:

(1) Criterion 1: The Blasius equation has the Lie symmetry scaling solution $y = c/x$. If we substitute the lattice equivalent $f_n = 1/n$ into (4.5), we get $2/[n(n+1)(n+2)(n+3)]$ for all P and Q. This is the same structure that one gets when one substitutes $f_n = 1/n$ into the right side of (4.4).
(2) Criterion 2: For $f_n = An + B$ (any linear function), the result vanishes; hence, we obtain a smooth ultralocal solution in the outer region (away from the boundary layer at $n = 0$).

We now take $\varepsilon = \delta/a^2$. Thus, we must solve

$$\begin{aligned} 0 = \varepsilon(f_{n+3} - 3f_{n+2} + 3f_{n+1} - f_n) + Pf_{n+2}f_{n+3} + (1-2Q-P)f_nf_{n+1} \\ + (1-3Q)f_nf_{n+3} - 3Qf_{n+1}f_{n+2} + (6Q+2P-2)f_nf_{n+2} \\ + (2Q-2P)f_{n+1}f_{n+3}, \end{aligned} \tag{4.6}$$

subject to the three boundary conditions $f_0 = 0$, $f_1 = 0$, $\lim_{n\to\infty} f_{n+1} - f_n = 1$, and our objective is to show that

$$\lim_{\varepsilon\to\infty} \sqrt{\varepsilon} f_2 = 0.46960. \tag{4.7}$$

We begin our analysis by looking at the case $n = 0$. This gives

$$\varepsilon(f_3 - 3f_2) + Pf_2f_3 = 0. \tag{4.8}$$

Assuming that f_2 and f_3 are nonzero when $\varepsilon = 0$, we are compelled to choose $P = 0$. This fixes one of the two arbitrary parameters. With $P = 0$ we have

$$\begin{aligned} 0 = \varepsilon(f_{n+3} - 3f_{n+2} + 3f_{n+1} - f_n) + (1-2Q)f_nf_{n+1} + (1-3Q)f_nf_{n+3} \\ - 3Qf_{n+1}f_{n+2} + (6Q-2)f_nf_{n+2} + (2Q)f_{n+1}f_{n+3}. \end{aligned} \tag{4.9}$$

To determine the parameter Q we note that there is one more symmetry of the Blasius equation that we have not yet used, namely, reflection symmetry. Equation (4.1) is symmetric under the discrete symmetry $x \to -x$, $y \to -y$. While reflection symmetry is not a Lie symmetry, it does seem natural to impose this symmetry here because it uniquely fixes the value of the remaining arbitrary parameter Q. On the lattice this symmetry takes the form

$$f_n \to -f_{n+3}, \quad f_{n+1} \to -f_{n+2}, \quad f_{n+2} \to -f_{n+1}, \quad f_{n+3} \to -f_n.$$

Requiring that (4.9) be invariant under this symmetry gives the value $Q = \frac{1}{2}$.

For this case, let

$$f_n = a_n + \varepsilon b_n + \varepsilon^2 c_n + \varepsilon^3 d_n + \cdots. \tag{4.10}$$

In the ultralocal limit ($\varepsilon = 0$), a_n satisfies

$$a_{n+3} = a_{n+2}\frac{3a_{n+1} - 2a_n}{2a_{n+1} - a_n}. \tag{4.11}$$

This equation gives the sequence $a_0 = 0$, $a_1 = 0$, $a_2 = x$, $a_3 = 3x$, $a_4 = 9x/2$, $a_5 = 63x/10$, $a_6 = 63x/8$, $a_7 = 77x/8$, where x is an arbitrary parameter. The exact formula for a_n is

$$\begin{aligned} a_{2n} &= \frac{2x\Gamma(1/4)\Gamma(n-1/4)\Gamma(n+1/2)}{\Gamma(3/4)\Gamma(1/2)\Gamma(n-3/4)\Gamma(n)}, \\ a_{2n+1} &= \frac{2x\Gamma(1/4)\Gamma(n+3/4)\Gamma(n+1/2)}{\Gamma(3/4)\Gamma(1/2)\Gamma(n+1/4)\Gamma(n)}. \end{aligned} \tag{4.12}$$

Imposing the requirement that $\lim_{n\to\infty} a_{n+1} - a_n = 1$ fixes the value of x: $x = \Gamma(3/4)\Gamma(1/2)/\Gamma(1/4) = 0.5990701173677961$.

Next, we examine the solution to first order in ε. Note that b_n satisfies

$$\begin{aligned} 0 = {} & a_{n+3} - 3a_{n+2} + 3a_{n+1} - a_n + a_{n+1}b_{n+3} + a_{n+3}b_{n+1} + a_n b_{n+2} \\ & + a_{n+2}b_n - \tfrac{1}{2}a_n b_{n+3} - \tfrac{1}{2}a_{n+3}b_n - \tfrac{3}{2}a_{n+1}b_{n+2} - \tfrac{3}{2}a_{n+2}b_{n+1}. \end{aligned} \tag{4.13}$$

The first few values of b_n are $b_0 = 0$, $b_1 = 0$, $b_2 = y$, $b_3 = 3y$, $b_4 = 9y/2 + 3/2$, $b_5 = 63y/10 + 89/50$, $b_6 = 63y/8 + 51/16$, $b_7 = 77y/8 + 4697/1296$, where y is an arbitrary parameter.

The sequence b_n consists of two parts. The part multiplying y is just a_n (the homogeneous solution). To isolate the second part, we write $b_n = y/xa_n + z_n$. Now, as $n \to \infty$, we can show that

$$b_n \sim \frac{y}{x}n + nL, \tag{4.14}$$

where $L = 0.9272586576$. Hence, $y = -x(0.9272586576)$. Thus, to first order in powers of ε we have

$$f_2 = x(1 - 0.9272586576\varepsilon + \cdots). \tag{4.15}$$

We use this result to obtain the *first* extrapolant E_1 to the value of $y''(0)$:

$$\begin{aligned} E_1 &= \frac{1}{\sqrt{\delta}} \lim_{\varepsilon\to\infty} x\sqrt{\varepsilon}(1 - 0.9272586576\varepsilon) \\ &= \frac{x}{\sqrt{\delta}} \lim_{\varepsilon\to\infty} \frac{\sqrt{\varepsilon}}{1 + 2(0.9272586576)\varepsilon} \\ &= \frac{x}{\sqrt{\delta}}(0.7343187) = \frac{1}{\sqrt{\delta}}(0.43990837). \end{aligned} \tag{4.16}$$

When we compare this number with 0.46960 (exact) we see that our result is 6.3% low. This first-order result is already comparable in accuracy with that obtained in Refs. [**2**, **9**] in 38th order!

References

1. C. J. Amick and J. B. McLeod, *A singular perturbation problem in water waves*, Stability Appl. Anal. Contin. Media **1** (1991), no. 2, 127–148.
2. C. M. Bender, *A novel approach to boundary layer problems*, Los Alamos Science **2** (1981), 76–91.
3. C. M. Bender and S. Boettcher, *Dimensional expansion for the ising limit of quantum field theory*, Phys. Rev. D **48** (1993), 4919–4923.

4. ———, *Eleventh-order calculation of Green's functions in the ising limit for arbitrary space-time dimension D*, Phys. Rev. D **51** (1994), 1875–1879.
5. C. M. Bender, S. Boettcher, and L. Lipatov, *Dimensional expansions*, Phys. Rev. Lett. **68** (1992), no25, 3674–3677.
6. ———, *Almost zero-dimensional quantum field theories*, Phys. Rev. D (3) **46** (1992), no. 12, 5557–5573.
7. C. M. Bender, P. H. Burchard, A. Das, H.-A. Lim, and J. A. Shapiro, *High-order strong-coupling calculation of the ground-state energy density in supersymmetric field theory*, Phys. Rev. Lett. **54** (1985), 2481–2484.
8. C. M. Bender, F. Cooper, and A. Das, *Continuum limit of supersymmetric field theories on a lattice*, Phys. Rev. Lett. **50** (1983), 397–399.
9. C. M. Bender, F. Cooper, G. S. Guralnik, E. Mjolsness, H. Rose, and D. H. Sharp, *A novel approach to the solution of boundary-layer problems*, Adv. in Appl. Math. **1** (1980), 22–36.
10. C. M. Bender, F. Cooper, G. S. Guralnik, H. Moreno, R. Roskies, and D. H. Sharp, *Effective potential for a renormalized d-dimensional $g\varphi^4$ field theory in the limit $g \to \infty$*, Phys. Rev. Lett. **45** (1980), 501–504.
11. C. M. Bender, F. Cooper, G. S. Guralnik, R. Roskies, and D. H. Sharp, *Strong-coupling expansion for the effective potential on a lattice*, Phys. Rev. D **23** (1981), 2976–2998.
12. ———, *Numerical computation of the renormalized effective potential in the strong-coupling limit*, Phys. Rev. D **23** (1981), 2999–3009.
13. ———, *Higher-order terms in the strong-coupling expansion for the renormalized effective potential*, Phys. Rev. D **24** (1981), 2683–2692.
14. ———, *Improvement of an extrapolation scheme for strong-coupling expansions in quantum field theory*, Phys. Rev. Lett. **43** (1979), 537–540.
15. C. M. Bender, F. Cooper, G. S. Guralnik, and D. H. Sharp, *Strong-coupling expansion in quantum field theory*, Phys. Rev. D **19** (1979), 1865–1881.
16. C. M. Bender, A. Das, and H.-A. Lim, *On higher-order computations in strong coupling*, Pramana **23** (1984), 695–701.
17. C. M. Bender, A. Das, H.-A. Lim, and L. M. Simmons, *Strong-coupling approximation in supersymmetric field theory*, Phys. Lett. B **134** (1984), 225–229.
18. ———, *Strong-coupling calculation of mass ratios in supersymmetric field theory*, Phys. Lett. B **154** (1985), 381–385.
19. C. M. Bender and S. A. Orszag, *Advanced mathematical methods for scientists and engineers*, Chapter 9, McGraw-Hill, New-York, 1978.
20. C. M. Bender, A. Pelster, and F. Weissbach, *Boundary-layer theory, strong-coupling series, and large- order behavior*, J. Math. Phys. **43** (2002), 4202–4220.
21. C. M. Bender and D. H. Sharp, *Determination of $f(\infty)$ from the asymptotic series for $f(x)$ about $x = 0$*, Phys. Rev. D **24** (1981), 1691–1705.
22. C. M. Bender and A. Tovbis, *Continuum limit of lattice approximation schemes*, J. Math. Phys. **38** (1997), 3700–3717.
23. E. R. Caianiello, *On quantum field theory.* I. *Explicit solution of Dyson's equation in electrodynamics without use of Feynman graphs*, Nuovo Cimento (9) **10** (1953), 1634–1652.
24. ———, *On quantum field theory.* II. *Non-perturbative equations and methods*, Nuovo Cimento (9) **11**(1954), 492–529.
25. E. R. Caianiello and G. Scarpetta, *The $g\Phi^4$ quantum field theory in the static ultralocal approximation*, Nuovo Cimento A **22** (1974), 448–460.
26. ———, *Exact asymptotic behaviour of $g\Phi^4$ theory from static ultralocal approximation*, Lett. Nuovo Cimento **11** (1974), 283–286.
27. B. M. Herbst and M. J. Ablowitz, *Numerical chaos, symplectic integrators, and exponentially small splitting distances*, J. Comput. Phys. **105** (1993), 122–132.
28. V. F. Lazutkin, *Splitting of complex separatrices*, Funktsional. Anal. i Prilozhen. **22** (1988), no. 2, 83–84 (Russian); English transl. Funct. Anal. Appl. **22** (1988), no. 2, 154–156.
29. R. J. Rivers, *Comments on an approximation scheme for strong-coupling expansions*, Phys. Rev. D **20** (1979), 3425–3430.
30. A. Tovbis, *Asymptotics beyond all orders and analytic properties of inverse Laplace transforms of solutions*, Comm. Math. Phys. **163** (1994), no. 2, 245–255.
31. A. Tovbis, M. Tsuchiya, and C. Jaffé, *Chaos-integrability transition in nonlinear dynamical systems: exponential asymptotics approach*, Differential Equations and Applications to

Biology and to Industry (Claremont, 1994), World Sci. Publishing, River Edge, NJ, 1996, pp. 495–506.

32. ______, *Exponential asymptotics and approximation of stable and unstable manifolds in singularly perturbed nonlinear systems*, Hamiltonian Systems with Three or More Degrees of Freedom (S'Agaró, 1995), NATO Adv. Sci. Inst. Ser. C Math. Phys. Sci., vol. 533, Kluwer Acad. Publ., Dordrecht, 1999, pp. 610–613.

DEPARTMENT OF PHYSICS, CAMPUS BOX 1105, WASHINGTON UNIVERSITY ST. LOUIS, MO 63130, USA.

E-mail address: cmb@wuphys.wustl.edu

Centre de Recherches Mathématiques
CRM Proceedings and Lecture Notes
Volume **39**, 2005

Algebraic Structures on Ordered Rooted Trees and Their Significance to Lie Group Integrators

H. Berland and B. Owren

ABSTRACT. Most Lie group integrators can be expanded in series indexed by the set of ordered rooted trees. One can associate to each tree two distinct higher order derivation operators, which we call frozen and unfrozen operators. Composition of frozen operators induces a concatenation product on the trees, whereas composition of unfrozen operators induces a somewhat more complicated product known as the Grossman–Larson product. Both of these algebraic structures can be supplemented by the same coalgebraic structure and an antipode, the result being two distinct cocommutative graded Hopf algebras. We discuss the use of these structures and characterize subsets of the Hopf algebras corresponding to vector fields and mappings on manifolds. This is relevant for deriving order conditions for a general class of Lie group integrators and for deriving the modified vector field in backward error analysis for these integrators.

1. Introduction

The derivation of high order Runge–Kutta methods was revolutionized by Butcher's discovery of the beautiful connection between their series expansion in terms of the stepsize and the set T of rooted trees [**2**]. Virtually overnight, long and tedious calculations were replaced by elegant recursion formulas expressed in terms of trees. Later on, in 1972 Butcher published a paper [**3**] where he showed that Runge–Kutta methods form a group under composition, and derived explicit expressions for the group operations as induced on the trees. Hairer and Wanner named it the Butcher group and contributed substantially to the theory in [**10**]. The group is defined on the dual of the tree space by using a bialgebra structure on the space of rooted trees.

More recently, Kreimer [**13**] used a Hopf algebra of rooted trees to deal with the combinatorics of renormalization in quantum field theory, the connection to the work of Butcher was nicely presented in [**1**].

In the last few years, new classes of integrators have been subjected to order analysis by means of trees. In [**19**] order conditions for composition methods are studied by means of so called ∞-trees, see also [**8**]. Another class of novel schemes

2000 *Mathematics Subject Classification.* Primary: 65L05.
This is the final form of the paper.

is the one based on Lie group actions on manifolds. Such integrators, which generalize classical Runge–Kutta methods, are now commonly referred to as Lie group integrators. Early contributors to this class of schemes include Crouch and Grossman [**6**] as well as Lewis and Simo [**14, 15**]. The Lie group schemes were later subjected to a more systematic treatment by many authors, see the survey [**12**]. In [**16**] Munthe–Kaas showed how a certain subclass of the Lie group integrators could be expanded in a series for the purpose of order analysis, but his approach was not pursued any further at that time since he discovered in [**17**] that a suitable change of variable would allow him to use the classical Butcher theory for deriving the order conditions. But the schemes of Crouch and Grossman [**6**] did not fit into this framework, and so Owren and Marthinsen [**21**] developed a slightly different framework based on *ordered rooted trees* for deriving the general order conditions for such schemes.

In this note, we will see how the algebraic structure introduced by Butcher can be extended to the set of ordered rooted trees. In particular we will present two different Hopf algebras, the first one was introduced by Grossman and Larson in [**7**] and the second is related to the one presented by Reutenauer in [**22**]. We will discuss their relevance to order conditions and backward error analysis for a general class of Lie group integrators.

2. Lie group integrators

An ordinary differential equation on a manifold has various different formulations,

$$y' = F(y) = f(y) \cdot y = \sum_i f^i(y) E_i(y). \tag{2.1}$$

F is here a smooth vector field on $\mathcal{M}$. The second equality tacitly refers to a transitive action on $\mathcal{M}$ by a Lie group G, and $f\colon \mathcal{M} \to \mathcal{V} \subset \mathfrak{g}$ where $\mathfrak{g}$ is the Lie algebra of G, and $\mathcal{V}$ is a subspace of $\mathfrak{g}$. The notation $v \cdot p$, $v \in \mathfrak{g}$, $p \in \mathcal{M}$ signifies the derivative of the group action in the sense that

$$v \cdot p = \left.\frac{\mathrm{d}}{\mathrm{d}t}\right|_{t=0} \exp(tv) \cdot p.$$

In the last equality of (2.1), we have used a set of frame vector fields $E_1, \ldots, E_d$ that may be defined as $E_i(p) = e_i \cdot p$ for some basis $e_1, \ldots, e_d$ of $\mathfrak{g}$ (or $\mathcal{V}$). The functions $f^i\colon \mathcal{M} \to k$ are then given such that $f(y) = \sum f^i(y) e_i \cdot y$. To the end of this note, we shall always assume that the field k is either $\mathbf{R}$ or $\mathbf{C}$. Note that the action is usually not assumed to be free. For our purposes, the language of actions and frames can be used interchangeably, but for this present exposition we find it slightly advantageous to use frames. By a minor abuse of notation, we shall therefore denote by $\mathcal{V}$ also the linear span of the frame vector fields

$$\mathcal{V} = \operatorname{span}\{E_1(y), \ldots, E_d(y)\}.$$

We will as usual interpret vector fields as derivations acting on functions $\psi\colon \mathcal{M} \to k$, and denote this action by $F[\psi]$ for any $F \in \mathfrak{X}(\mathcal{M})$.

In order to develop numerical integrators, we approximate the vector field by the *freeze* map $\mathrm{Fr}\colon \mathcal{M} \times \mathfrak{X}(\mathcal{M}) \to \mathcal{V}$, in which the vector fields E_i are assumed to be simpler to integrate than F,

$$\left(p, \sum_i f^i(y)E_i(y)\right) \mapsto \sum_i f^i(p)E_i(y). \tag{2.2}$$

We propose a scheme for Lie group integrators which generalizes both the Crouch–Grossman (CG) type of schemes [**6, 21**], and the Runge–Kutta–Munthe–Kaas (RKMK) methods [**16, 17**]. The scheme uses the point $p \in \mathcal{M}$ as input, together with a stepsize $h \in \mathbf{R}$ and produces as output a point $y_1 \in \mathcal{M}$ such that $y_1 \approx \exp(hF)p$.

$$g_i = \exp\Big(\sum_r \alpha^r_{i,J} K_r\Big) \cdots \exp\Big(\sum_r \alpha^r_{i,1} K_r\Big) p, \tag{2.3}$$

$$\overline{K}_i = h\mathrm{Fr}(g_i, F) = h \sum_\ell f^\ell(g_i) E_\ell, \tag{2.4}$$

$$K_i = P_i(\overline{K}_1, \ldots, \overline{K}_s) \quad \text{(Lie polynomial)}, \tag{2.5}$$

$$y_1 = \exp\Big(\sum_r \beta^r_J K_r\Big) \cdots \exp\Big(\sum_r \beta^r_1 K_r\Big) p. \tag{2.6}$$

In (2.3) to (2.5) the index i runs from 1 to s, where s is the number of stages in the step.

This scheme belongs to the RKMK class if $J = 1$ in (2.3) and (2.6) (only one exponential used for g_i and y_1). If $\alpha^r_{i,j} = 0$ when $r \neq j$ in (2.3), $\beta^r_j = 0$ when $r \neq j$ in (2.6), and $K_i = \overline{K}_i$ for all i, then this is a Crouch–Grossman scheme. Requiring only $\overline{K}_i = K_i$ for all i, one obtains a commutator-free Lie group scheme [**4**].

The scheme is explicit if P_i in (2.5) depends only on $\overline{K}_1, \ldots, \overline{K}_i$ and $\alpha^r_{i,j} = 0$ when $r \geq i$, $1 \leq j \leq J$.

In the sequel we shall introduce an algebraic formalism which will provide a unified setting for analysis of Lie group integrators of the form (2.3)–(2.6).

3. Algebraic structures on trees

In this section we shall impose algebraic structures on the space of ordered rooted trees. We will define two distinct associative products and a unit element and thereby introduce two different algebra structures. We next define a coassociative coproduct with counit to obtain one coalgebra which can be used with each of the two algebras to form two distinct bialgebras. Each of these can be equipped with an antipode, the result being two distinct cocommutative graded Hopf algebras. A good reference for the theory of Hopf algebras is Sweedler's book [**23**], to which we will refer frequently.

The set of all ordered rooted trees is denoted by T_O, see e.g. [**5**] for a rigorous treatment of such trees and their combinatorial properties. We shall work recursively with trees, and our notation is based on the fact that a tree $t \in T_O$ is either the one-node tree $\bullet \in T_O$ (the identity tree), or obtained by connecting the roots of an ordered set of trees $t_1, \ldots, t_\mu$ to a new common root, where each $t_i \in T_O$. We use the notation $t = B_+(t_1, \ldots, t_\mu)$ for this operation. Conversely, $B_-(t)$ will denote the (ordered) set of trees obtained by deleting the root of t, and we call this set the subtrees of t. Let $\sigma(t)$ be the underlying set of nodes of t. The number of nodes in a tree t is denoted $|t|$, and we let the grading be $\nu(t) = |t| - 1$. So $\nu(\bullet) = 0$, and for any other tree $t = B_+(t_1, \ldots, t_\mu)$ one has $\nu(t) = \mu - 1 + \sum_{i=1}^{\mu} \nu(t_i)$. It is well known, see e.g. [**5**], that the number ν_r of trees such that $\nu(t) = r$ is given by the Catalan number

$$\nu_r = \frac{1}{r+1}\binom{2r}{r}. \tag{3.1}$$

The linear space kT_O is obtained by forming finite linear combinations of trees over the field k. The homogeneous components of grade m is the subspace of kT_O spanned by the trees $\{t \in T_O : \nu(t) = m\}$.

We shall later make use of the subset of T_O consisting of trees with only one subtree at the first level, we denote this subset by T_O^1, and let kT_O^1 be the corresponding subspace of kT_O.

$$T_O^1 = \{t \in T_O : t = B_+(t'), t' \in T_O\}.$$

3.1. Grossmann–Larson product on trees. An *attachment map* is a map which associates to any element of an ordered finite subset S of T_O a unique element of another set M, which will typically be the nodes of a tree, $\sigma(w)$. We write $d\colon S \to \sigma(w)$ for this map. By the notation

$$w \leftharpoondown_d S \tag{3.2}$$

we mean an augmented tree t where each tree $s \in S$ has been attached to the node $d(s)$ of w by adding an edge from the root of s to $d(s)$. Thus, s becomes a subtree of the tree rooted at $d(s)$. This subtree will be ordered *before* any of the existing subtrees at $d(s)$. Moreover, if $d(s) = d(s')$ where $s < s'$ then s will be ordered before s' as subtrees at $d(s)$ in t. By convention, we depict this ordering by grafting elements of S to the left of the already existing subtrees at nodes in t. An example attachment is

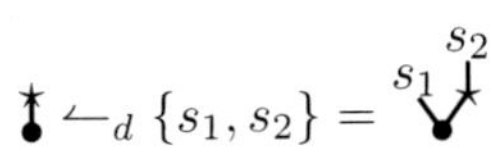

where we have $d(s_1) = \bullet$ and $d(s_2) = \star$. s_1 and s_2 are arbitrary trees in T_O.

The sum over all possible attachment maps d from S to $\sigma(w)$ is written as

$$w \leftharpoondown S := \sum_d w \leftharpoondown_d S.$$

Definition 3.1 (Grossman–Larson algebra). The identity element is the one-node tree $\bullet$, and the product of two trees is

$$\mu_{\mathrm{GL}}(v \otimes w) = w \leftharpoondown B_-(v), \quad v, w \in T_O.$$

The unit in the algebra is $u\colon k \to kT_O$ given by $u(\alpha) = \alpha\bullet$ for $\alpha \in k$.

This product is non-commutative. An example is

(3.3) 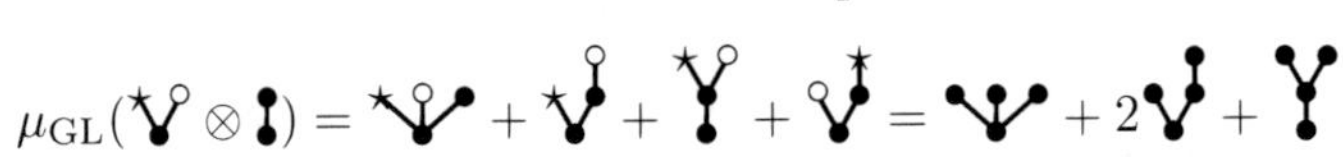

where we have temporarily inserted a star and a circle for the grafted nodes to indicate the various attachment maps d involved in the sum.

Grossmann and Larson prove in [**7**, Section 3] that this product is indeed associative.

3.2. Concatenation algebra on trees. We define a simpler product on the trees in T_O.

Definition 3.2 (Concatenation algebra). The concatenation product of two trees v and w results from joining all the subtrees of both v and w to a new common root,

$$\mu_{\mathrm{M}}(v \otimes w) = B_+\big(B_-(v) \cup B_-(w)\big), \quad v, w \in T_O$$

such that the order is preserved. The identity is the one-node tree $\bullet$.

Alternatively, if $v = B_+(v_1, \dots, v_\mu)$ and $w = B_+(w_1, \dots, w_\nu)$, then $\mu_{\mathrm{M}}(v \otimes w) = B_+(v_1, \dots v_\mu, w_1, \dots w_\nu)$.

Note that this product has a subset of the terms arising from the product in the Grossman–Larson algebra. We may write $\mu_M(v \otimes w) = w \leftharpoonup_{d_0} B_-(v)$ where d_0 is the map that sends all elements of the set $B_-(v)$ to the root $r \in \sigma(w)$.

3.3. Coalgebra on trees. The two Hopf algebras on trees we will present, share the same coproduct $\Delta\colon kT_O \to kT_O \otimes kT_O$. We here follow the presentation in [**7**], but we refer to [**23**] for the basic theory on algebras and coalgebras.

Definition 3.3 (Coalgebra). The coproduct on the trees in T_O is given by

$$\Delta(t) = \sum_{\mathcal{X} \subseteq B_-(t)} B_+(\mathcal{X}) \otimes B_+(\mathcal{X}^c)$$

which extends linearly to kT_O. The subsets $\mathcal{X}$ inherit the ordering from t as do the complements $\mathcal{X}^c$. We include the empty set $\emptyset$ as well as $B_-(t)$ in the sum, using the convention $B_+(\emptyset) = \bullet$. The counit is the linear map $\varepsilon\colon kT_O \to k$ such that

$$\varepsilon(\bullet) = 1 \quad \text{and} \quad \varepsilon(t) = 0, \qquad t \in T_O,\ t \neq \bullet.$$

We refer to [**7**] for a proof that the coproduct defined above is coassociative, meaning that $(I \otimes \Delta) \circ \Delta = (\Delta \otimes I) \circ \Delta$.

A bialgebra has both an algebra structure and a coalgebra structure which are compatible in the sense that the coproduct must be an algebra homomorphism (or equivalently, the product is a coalgebra homomorphism). That is

$$\Delta\big(\mu_{\mathrm{GL}}(v \otimes w)\big) = \mu_{\mathrm{GL}\otimes\mathrm{GL}}\big(\Delta(v) \otimes \Delta(w)\big). \tag{3.4}$$

We refer to Grossman and Larson [**7**, Section 3] for a proof of Eq. (3.4). Note that the product structure $\mu_{\mathrm{GL}\otimes\mathrm{GL}}$ on $kT_O \otimes kT_O$ is naturally constructed using $\mu_{\mathrm{GL}\otimes\mathrm{GL}} = (\mu_{\mathrm{GL}} \otimes \mu_{\mathrm{GL}}) \circ (I \otimes T \otimes I)$ where $T\colon a \otimes b \mapsto b \otimes a$ is the twist map.

It is easier to prove that Δ is an algebra homomorphism with respect to the concatenation algebra.

Proposition 3.4. *The coproduct Δ given in Definition* 3.3 *is a concatenation algebra homomorphism, that is*

$$\Delta\big(\mu_{\mathrm{M}}(t_1 \otimes t_2)\big) = \mu_{\mathrm{M}\otimes\mathrm{M}}\big(\Delta(t_1) \otimes \Delta(t_2)\big)$$

Proof. From the left we have

$$\begin{aligned}\Delta\big(\mu_{\mathrm{M}}(t_1 \otimes t_2)\big) &= \Delta\Big(B_+\big(B_-(t_1) \cup B_-(t_2)\big)\Big) \\ &= \sum_{\mathcal{X} \subseteq B_-(t_1) \cup B_-(t_2)} B_+(\mathcal{X}) \otimes B_+(\mathcal{X}^c)\end{aligned}$$

and from the right, using $\mu_{\mathrm{M}\otimes\mathrm{M}} = (\mu_\mathrm{M} \otimes \mu_\mathrm{M}) \circ (I \circ T \circ I)$.

$$\begin{aligned}\mu_{\mathrm{M}\otimes\mathrm{M}}\big(\Delta(t_1)\otimes\Delta(t_2)\big) &= \mu_{\mathrm{M}\otimes\mathrm{M}}\Big(\sum_{\mathcal{X}_1,\mathcal{X}_2} B_+(\mathcal{X}_1)\otimes B_+(\mathcal{X}_1^c)\otimes B_+(\mathcal{X}_2)\otimes B_+(\mathcal{X}_2^c)\Big)\\ &= (\mu_\mathrm{M}\otimes\mu_\mathrm{M})\Big(\sum_{\mathcal{X}_1,\mathcal{X}_2} B_+(\mathcal{X}_1)\otimes B_+(\mathcal{X}_2)\otimes B_+(\mathcal{X}_1^c)\otimes B_+(\mathcal{X}_2^c)\Big)\\ &= \sum_{\mathcal{X}_1,\mathcal{X}_2} B_+(\mathcal{X}_1\cup\mathcal{X}_2)\otimes B_+(\mathcal{X}_1^c\cup\mathcal{X}_2^c)\end{aligned}$$

which is a sum equivalent to the one above. □

It is also apparent from the definition of the coproduct that the coalgebra is cocommutative, that is $\Delta = \Delta \circ T$.

3.4. Grossmann–Larson Hopf algebra. A mapping $\mathcal{S}\colon kT_O \to kT_O$ in a bialgebra, is an antipode [**23**] if it satisfies

$$\mu \circ (\mathcal{S} \otimes \mathrm{Id}) \circ \Delta = u \circ \varepsilon = \mu \circ (\mathrm{Id} \otimes \mathcal{S}) \circ \Delta. \tag{3.5}$$

Note that $u \circ \varepsilon$ is zero on all trees in T_O except from $\bullet$, where one has $u \circ \varepsilon(\bullet) = \bullet$. One may apply (3.5) with μ replaced by μ_{GL} recursively to obtain $\mathcal{S}_{\mathrm{GL}}(t)$ for any $t \in T_O$, noting that one gets $\mathcal{S}_{\mathrm{GL}}(\bullet) = \bullet$ as well as identities of the form

$$\mathcal{S}_{\mathrm{GL}}(t) = -t - \sum \mu_{\mathrm{GL}}\big(\mathcal{S}_{\mathrm{GL}}(t_{i_1}) \otimes t_{i_2}\big), \quad |t_{i_1}| < |t|.$$

An explicit formula for $\mathcal{S}_{\mathrm{GL}}(t)$ seems hard to derive or use in general. But there are some simple cases. For instance,

$$\mathcal{S}_{\mathrm{GL}}(t) = -t, \quad \text{for} \quad t \in T_O^1. \tag{3.6}$$

For trees with exactly two branches emanating from the root, one has

$$\mathcal{S}_{\mathrm{GL}}\big(B_+(t_1, t_2)\big) = B_+(t_2, t_1) + B_+(t_1 \leftharpoondown t_2 + t_2 \leftharpoondown t_1)$$

where B_+ has been extended to a linear map.

Example 3.5 (Grossmann–Larson antipode).

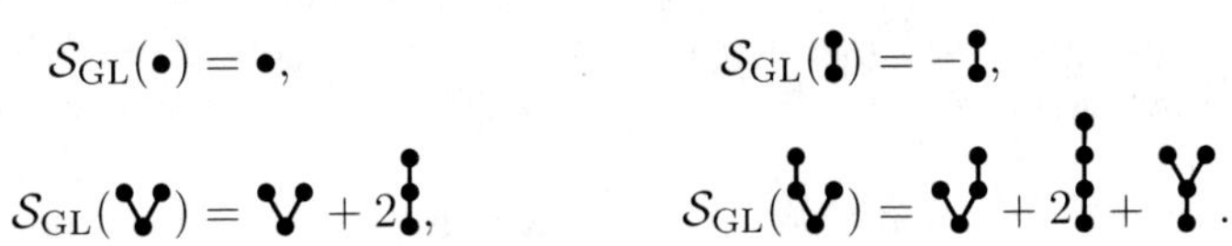

3.5. Concatenation Hopf algebra. We define the antipode for the concatenation algebra in the same way, but the trivial product provides an easy factorization which facilitates a formula for the antipode.

The antipode of a tree $t = B_+(t_1, \ldots, t_\mu)$ in the concatenation algebra takes the form

$$\mathcal{S}(t) = (-1)^\mu B_+(t_\mu, \ldots, t_1)$$

which follows from the fact that the antipode of a Hopf algebra is an antiautomorphism [**23**, Proposition 4.0.1, p. 74] and from Eq. (3.6).

Because the coproduct for both Hopf algebras is cocommutative, the antipodes have the property that $\mathcal{S}^2$ is the identity map on kT_O [**23**, Proposition 4.0.1, p. 74].

3.6. Infinite series and their subsets. We may define formal series on T_O as infinite sums

$$S = \sum_{t \in T_O} S_t t$$

where $S_t \in k$ is the coefficient of the tree t in the series S. The set of all such series is denoted $k_\infty T_O$ of which kT_O is clearly a subspace. We can extend the bialgebras to this space by setting the coefficients of the product of two series S and T to

$$\big(\mu(S \otimes T)\big)_t = \sum_{\mu(v \otimes w) = t} S_v T_w \tag{3.7}$$

the sum being finite since the homogeneous components of kT_O are finite dimensional. The coproduct Δ is extended in a similar way. We may now define the commutator between two series in either algebra as

$$[S, T] = \mu(S \otimes T - T \otimes S).$$

The primitive elements of the extended Hopf algebras are those which satisfy

$$\Delta(S) = S \otimes \bullet + \bullet \otimes S. \tag{3.8}$$

The linear space of primitive elements is from now on denoted by $\mathfrak{g}$, it is closed under the commutator and thus forms a Lie algebra. Later, we shall see that $\mathfrak{g}$ plays an important role as its members represent formal expansions of the vector fields used in the integration schemes. In particular, we observe from the definition of the coproduct that $k_\infty T_O^1 \subset \mathfrak{g}$. Moreover, it follows immediately from (3.8) and the general definition of the antipode that

$$\mathcal{S}(S) = -S, \quad S \in \mathfrak{g}.$$

The Milnor–Moore theorem ensures that the universal enveloping algebra of $\mathfrak{g}$ is isomorphic to $k_\infty T_O$.

The series $T \in k_\infty T_O$ with coefficient $T_\bullet = \langle T, \bullet \rangle = 1$ form a group under the product in (3.7). Taking the formal exponential of all Lie series $\mathfrak{g}$, we obtain a subgroup $G = \exp(\mathfrak{g})$ with the property

$$\Delta(T) = T \otimes T. \tag{3.9}$$

if $T = \exp(S)$ for $S \in \mathfrak{g}$. The proof is simple, and can be found in [**22**, Thm. 3.2].

The group G is also invariant under the antipode, the antipode represents the group inverse, as we get

$$\mu\big(\mathcal{S}(T) \otimes T\big) = \bullet, \quad \text{for all} \quad T \in G = \exp(\mathfrak{g}) \tag{3.10}$$

from the defining equation (3.5) of the antipode. Later, we shall see that the elements of G represent expansions of maps defined as part of the integration schemes.

4. Elementary high order derivations

In this section we return to the Lie group integration schemes. Suppose that a frame set (action) is given together with the vector field F in (2.1). Letting vector fields be derivation operators acting on the functions on the manifold, we shall recursively define high order derivation operators which will feature in the expansions of the various quantities appearing in the Lie group integrators. The algebra obtained by composing these operators will be seen to coincide with the tree algebras of the previous section.

4.1. High order derivations.

Definition 4.1 (Elementary High Order Derivation, EHOD). Let $F \in \mathfrak{X}(\mathcal{M})$ and $E_1, \ldots, E_d \in \mathfrak{X}(\mathcal{M})$ be given. We define $\mathbf{F}$ to be the k-linear map from $k_\infty T_O$ to the vector space of high order derivations on the manifold such that for $t \in T_O$

$$\mathbf{F}(t) \colon \psi \mapsto \mathbf{F}(t)[\psi]$$

where $\mathbf{F}(\bullet) = \mathrm{Id}$ and if $t = B_+(t_1, \ldots, t_\mu)$,

$$\mathbf{F}(t) = \sum_{i_1, \ldots, i_\mu} \mathbf{F}(t_1)[f^{i_1}] \cdots \mathbf{F}(t_\mu)[f^{i_\mu}] E_{i_1} \cdots E_{i_\mu}.$$

We now give some motivation for the above definition. Formally, one has the expansion of the flow of $F \in \mathfrak{X}(\mathcal{M})$ relative to the point $p \in \mathcal{M}$,

$$\psi(\exp(hF)p) = \psi(p) + F[\psi](p) + \tfrac{1}{2}F^2[\psi](p) + \cdots = \exp(hF)[\psi](p). \tag{4.1}$$

The powers of F are obtained by repeated application of F as a derivation operator. The expansion (4.1) is sometimes called a *pullback series.*

We may substitute Eq. (2.1) ($F = f^i E_i$) into (4.1) and use Leibniz' rule repeatedly to generate terms that are EHODs. We use the convention that indices appearing more than once are summed over, and from [**21**] we find

(4.2) $F^1 = F = f^i E_i = \mathbf{F}($$)$,

$F^2 = F[F] = f^i E_i[f^j E_j] = f^i E_i[f^j]E_j + f^i f^j E_i[E_j]$

$= \mathbf{F}($$) + \mathbf{F}($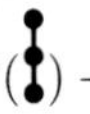$)$

$F^3 = f^i f^j f^k E_i E_j E_k + f^i f^k E_i[f^j]E_j E_k + 2f^i f^j E_i[f^k]E_j E_k$
$\qquad + f^i f^j E_i E_j[f^k]E_k + f^i E_i[f^j]E_j[f^k]E_k$

$= \mathbf{F}($$) + \mathbf{F}($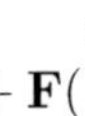$) + 2\mathbf{F}($$) + \mathbf{F}($$) + \mathbf{F}($$)$.

Proposition 4.2. *The map* $\mathbf{F}$ *in Definition* 4.1 *is an algebra homomorphism from the Grossman–Larson algebra to the algebra of* EHOD*s under composition,*

$$\mathbf{F}\big(\mu_{\mathrm{GL}}(v \otimes w)\big) = \mathbf{F}(v) \circ \mathbf{F}(w).$$

We first prove the following lemma.

Lemma 4.3. *Let* $v = B_+(t) \in T_O^1$, $t \in T_O$ *and* $w \in T_O$. *Then*

$$\mathbf{F}(v) \circ \mathbf{F}(w) = \mathbf{F}(w \leftharpoonup t).$$

Proof. The proof is by induction on $|w|$. Suppose first that $w = \bullet$, then $\mathbf{F}(v) \circ \mathbf{F}(\bullet) = \mathbf{F}(\bullet \leftharpoonup_{d_\bullet} \{t\}) = \mathbf{F}(v)$. Next suppose that the lemma holds for each w such that $|w| \leq k$ where $k \geq 1$. Any tree with $k+1$ nodes is of the form $w = B_+(w_1, \ldots, w_\omega)$ where each $|w_i| \leq k$. We calculate

$$\mathbf{F}(v) \circ \mathbf{F}(w) = \sum_i \mathbf{F}(t)[f^i]E_i \Bigg[\sum_{\mathbf{j}} \mathbf{F}(w_1)[f^{j_1}] \cdots \mathbf{F}(w_\omega)[f^{j_\omega}]E_{j_1} \cdots E_{j_\omega}\Bigg].$$

We use the Leibniz rule and split the result in two parts, $\mathbf{F}(v) \circ \mathbf{F}(w) = T_1 + T_2$ where

$$T_1 = \mathbf{F}\big(B_+(t, w_1, \ldots, w_\omega)\big) = \mathbf{F}(w \leftharpoonup_{d_r} \{t\})$$

where r is the root of w. T_1 is the part where E_i above acts as a derivation on the part $E_{j_1}\cdots E_{j_\omega}$. The second part is where E_i acts on the coefficient functions,

$$\begin{aligned} T_2 &= \sum_{q=1}^{\omega}\sum_{i,\mathbf{j}} \mathbf{F}(t)[f^i]E_i\left[\mathbf{F}(w_q)[f^{j_q}]\right]\prod_{k\neq q}\mathbf{F}(w_k)[f^{j_k}]E_{j_1}\cdots E_{j_\omega} \\ &= \sum_{q=1}^{\omega}\sum_{\mathbf{j}} \mathbf{F}(v)\circ\mathbf{F}(w_q)[f^{j_q}]\prod_{k\neq q}\mathbf{F}(w_k)[f^{j_k}]E_{j_1}\cdots E_{j_\omega} \\ &= \sum_{\mathbf{j}}\sum_{q=1}^{\omega}\sum_{x_q\in\sigma(w_q)} \mathbf{F}(w_q \leftharpoondown_{d_{x_q}} \{t\})[f^{j_q}]\prod_{k\neq q}\mathbf{F}(w_k)[f^{j_k}]E_{j_1}\cdots E_{j_\omega} \\ &= \sum_{x\in\sigma(w)\setminus\{r\}} \mathbf{F}(w\leftharpoondown_{d_x}\{t\}). \end{aligned}$$

□

PROOF OF PROPOSITION 4.2. If v is the unit tree $\bullet$, the result is obvious. Suppose that $v = B_+(v_1,\ldots,v_\nu)$, each $v_i \in T_O$, and let y be the independent variable for the EHODs. Then

$$\begin{aligned} \mathbf{F}(v)\circ\mathbf{F}(w) &= \sum_{\mathbf{i}} \mathbf{F}_s(v_1)[f^{i_1}]\cdots\mathbf{F}_s(v_1)[f^{i_\nu}]\Big|_{s=y} E_{i_1}\cdots E_{i_\nu}[\mathbf{F}(w)] \\ &= \sum_{\mathbf{i}} (\mathbf{F}_s(B_+(v_1))[f^{i_1}]E_{i_1})\cdots(\mathbf{F}_s(B_+(v_\nu))[f^{i_\nu}]E_{i_\nu})[\mathbf{F}(w)]\Big|_{s=y}. \end{aligned}$$

Each of the ν trees $B_+(v_i)$ are of the form $v = B_+(t)$ required by Lemma 4.3. The trees are all attached only to the nodes of w, there is no accumulation since the coefficient functions of the attached subtrees do not depend on y. We get according to the lemma that

$$\mathbf{F}(v)\circ\mathbf{F}(w) = \sum_{k=1}^{\nu}\sum_{x\in\sigma(w)} \mathbf{F}(w\leftharpoondown_{d_x}\{v_k\}) = \sum_{d}\mathbf{F}\big(w\leftharpoondown_d B_-(v)\big).$$

□

Example 4.4. We use the same trees as in the example in Eq. (3.3)

$\mathbf{F}($  $)\circ\mathbf{F}($  $) = \mathbf{F}($ 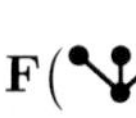 $+ 2$  $+$  $).$

The corresponding EHODs composed with each other results in

$$\begin{aligned} f^if^jE_iE_j\circ f^kE_k &= f^if^jE_iE_j[f^kE_k] \\ &= f^if^jE_iE_j[f^k]E_k + f^if^jE_j[f^k]E_iE_k \\ &\qquad + f^if^jE_i[f^k]E_jE_k + f^if^jf^kE_iE_jE_k \end{aligned}$$

which we see correspond to the correct trees as in Eq. (4.2).

4.2. Frozen elementary high order derivations. In (2.2) we introduced the freeze map which assigns to a pair (F,p) the vector field $F_p \in \operatorname{span}\{E_1,\ldots,E_d\}$ which coincides with F at p, $F|_p = F_p|_p$. This amounts to freezing the coefficient functions f^i at the point p in the representation $F = f^iE_i$ in terms of the frame.

Definition 4.5 (Frozen elementary high order derivation, FEHOD). Let $F \in \mathfrak{X}(\mathcal{M})$, $p \in \mathcal{M}$ and $E_1,\ldots,E_d \in \mathfrak{X}(\mathcal{M})$ be given. We define $\mathbf{F}_p$ to be the k-linear map from $k_\infty T_O$ to the vector space of high order derivations on the manifold such that for $t \in T_O$, one has: $\mathbf{F}_p(\bullet) = \mathrm{Id}$, and if $t = B_+(t_1, \ldots, t_\mu)$, then

$$\mathbf{F}_p(t) = \sum_{i_1,\ldots,i_\mu} \mathbf{F}(t_1)[f^{i_1}](p) \cdots \mathbf{F}(t_\mu)[f^{i_\mu}](p)\, E_{i_1} \cdots E_{i_\mu}.$$

Proposition 4.6. *The map* $\mathbf{F}_p$ *is an algebra homomorphism from the concatenation algebra to the algebra of* FEHODs *under composition,*

$$\mathbf{F}_p\big(\mu_{\mathrm{M}}(v \otimes w)\big) = \mathbf{F}_p(v) \circ \mathbf{F}_p(w).$$

PROOF. Let $v = B_+(t_1, \ldots, t_\mu)$ and $w = B_+(t_{\mu+1}, \ldots, t_\nu)$, then

$$\begin{aligned}
\mathbf{F}_p(v) \circ \mathbf{F}_p(w) &= \sum_{i_1,\ldots,i_\mu} \mathbf{F}(t_1)[f^{i_1}](p) \cdots \mathbf{F}(t_\mu)[f^{i_\mu}](p)\, E_{i_1} \cdots E_{i_\mu} \\
&\qquad \circ \Big(\sum_{i_{\mu+1},\ldots,i_\nu} \mathbf{F}(t_{\mu+1})[f^{i_{\mu+1}}](p) \cdots \mathbf{F}(t_\nu)[f^{i_\nu}](p)\, E_{i_{\mu+1}} \cdots E_{i_\nu} \Big) \\
&= \sum_{i_1,\ldots,i_\nu} \mathbf{F}(t_1)[f^{i_1}](p) \cdots \mathbf{F}(t_\nu)[f^{i_\nu}](p)\, E_{i_1} \cdots E_{i_\nu} \\
&= \mathbf{F}_p\big(\mu_{\mathrm{M}}(v \otimes w)\big)
\end{aligned}$$

as each of the $\mathbf{F}(t_j)[f^{i_j}](p)$ is just a constant and thereby unaffected by the E_i's. □

5. B-series

Let kT_O^* be the algebraic dual of the space kT_O. For any $\mathbf{a} \in kT_O^*$, $p \in \mathcal{M}$ we associate a formal series of operators

$$B_p(\mathbf{a}) = \sum_{t \in T_O} h^{|t|-1} \langle \mathbf{a}, t \rangle \mathbf{F}_p(t) \tag{5.1}$$

where $\langle \cdot, \cdot \rangle$ is the duality pairing. We can think of (5.1) as a generalization of the B-series discussed in [**9**]. Some authors include symmetry coefficients $\sigma(t)$, and then take $\langle \mathbf{a}', t \rangle = \langle \mathbf{a}, t \rangle / \sigma(t)$ instead of $\langle \mathbf{a}, t \rangle$ in the definition.

Keeping the algebra homomorphism $\mathbf{F}_p$ from Definition 4.5 in mind, we employ the shorthand notation

$$\sum_{t \in T_O} \langle \mathbf{a}, t \rangle t \tag{5.2}$$

for the B-series (5.1) and this is now a series in $k_\infty T_O$. This formula also shows the natural identification of kT_O^* with $k_\infty T_O$.

Suppose that a map $\phi_{\mathbf{a}}$ has a B-series expansion with coefficients from $\mathbf{a} \in kT_O^*$ and relative to the point $p \in \mathcal{M}$. This means that formally

$$\psi\big(\phi_{\mathbf{a}}(y)\big) = B_p(\mathbf{a})[\psi](y) \tag{5.3}$$

for any function $\psi \in C^\infty(\mathcal{M}, k)$ and $y \in \mathcal{M}$. Suppose that we freeze the vector field F at the point $\phi_{\mathbf{a}}(p) \in \mathcal{M}$, we then compute the series of $hF_{\phi_{\mathbf{a}}(p)}$ using (5.3)

$$\begin{aligned}
h \sum_i f^i\big(\phi_{\mathbf{a}}(p)\big) E_i &= h \sum_i \sum_{t \in T_O} h^{|t|-1} \langle \mathbf{a}, t \rangle \mathbf{F}_p(t)[f^i](p) E_i \\
&= h \sum_{t \in T_O} h^{|t|-1} \langle \mathbf{a}, t \rangle \mathbf{F}_p\big(B_+(t)\big) = \sum_{t \in T_O^1} h^{|t|-1} \langle \mathbf{a}, B_-(t) \rangle \mathbf{F}_p(t).
\end{aligned}$$

In other words, the B-series of a frozen vector field is associated to the space $k_\infty T_O^1$. In Section 3.6 we defined the Lie algebra $\mathfrak{g} \subset k_\infty T_O$, that is, those $S \in \mathfrak{g}$ which satisfy $\Delta(S) = S \otimes \bullet + \bullet \otimes S$. This Lie algebra contains $k_\infty T_O^1$. We may therefore conclude that the commuted vector fields $K_i = P_i(\overline{K_1}, \ldots, \overline{K_s})$ for each of the Lie polynomials P_i in (2.5) belongs to $\mathfrak{g}$. Now, the quantities g_i (2.3) and y_1 (2.6) are, owing to the Baker–Campbell–Hausdorff (BCH) formula, exponentials of elements in $\mathfrak{g}$ and therefore belong to the group G. Note that the BCH formula can also be used in a formal way here so that no convergence criterion is needed.

The fact that series in $\mathfrak{g}$ satisfy (3.8) imposes restrictions on the dual elements $\mathbf{a}$ which represent such series. From [**22**, Theorem 3.1] we find that $\langle \mathbf{a}, v ⧢ w\rangle = 0$ for all $v, w \in T_O \backslash \{\bullet\}$ where $v ⧢ w$ is the shuffle product denoting the sum of all possible ordered insertions of w into v on the first subtree level [**22**, pp. 23–24]. For instance, the shuffle product of the two trees $B_+(t_1, t_2)$ and $B_+(t_1, t_3)$ is

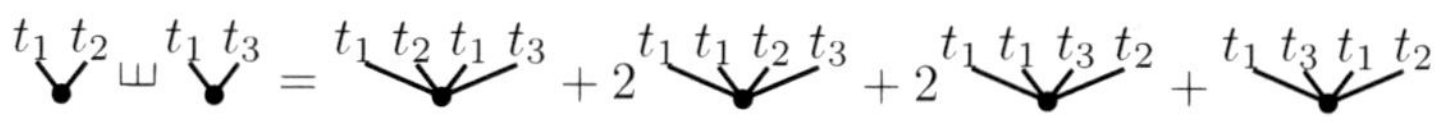

where $t_1, t_2, t_3 \in T_O$.

Similarly, (3.9) can now be used to characterize the coefficient forms in kT_O^* of a series which belongs to G. Suppose that such a $\mathbf{b} \in kT_O^*$ is representing a series in G. We find that $\langle \mathbf{b}, \bullet\rangle = 1$ and that $\langle \mathbf{b}, v ⧢ w\rangle = \langle \mathbf{b}, v\rangle\langle \mathbf{b}, w\rangle$, which is called a shuffle relation. These relations were derived in a different way in [**21**], see also [**24**].

From [**21**] we find that the exact flow $\psi(\exp(hF)y)$ of the differential equation $\dot{y} = F(y)$ can be expressed in a B-series $B_p(\mathbf{a})$ where $\langle \mathbf{a}, t\rangle = \alpha(t)/(|t| - 1)!$ and where $\alpha(t)$ is defined recursively as

$$(5.4) \qquad \alpha(\bullet) = 1 \quad \text{and} \quad \alpha\big(B_+(t_1, \ldots, t_\mu)\big) = \prod_{\ell=1}^{\mu} \binom{\sum_{i=1}^{\ell} |t_i| - 1}{|t_\ell| - 1} \alpha(t_\ell).$$

The linear form $\mathbf{a} \in kT_O^*$ obeys the shuffle relation $(\mathbf{a}, v ⧢ w) = (\mathbf{a}, v)(\mathbf{a}, w)$.

As argued above, the numerical integration schemes presented in Section 2 admit a B-series expansion as well. Thus, for each such scheme, there exists a $\mathbf{b} \in kT_O^*$ with which (5.1) holds formally. In [**20**] one can find the details of how the B-series is obtained for commutator-free schemes.

Given two mappings on a manifold, say $\phi_\mathbf{a}, \phi_\mathbf{b} \colon \mathcal{M} \to \mathcal{M}$ with corresponding B-series $B(\mathbf{a})$ and $B(\mathbf{b})$ in $k_\infty T_O$. The composition $\phi_\mathbf{c} = \phi_\mathbf{b} \circ \phi_\mathbf{a}$ also has a B-series, with coefficients from $\mathbf{c}$. By applying (5.3) twice we get that $B_p(\mathbf{c}) = B_p(\mathbf{a}) \circ B_p(\mathbf{b})$ (note the usual reversal of order, passing from composition of mappings to composition of operators). The concatenation product on $k_\infty T_O$ now yields

$$\begin{aligned}(5.5) \qquad \mu_{\mathrm{M}}\Big(\sum_{v \in T_O} \langle \mathbf{a}, v\rangle v \otimes \sum_{w \in T_O} \langle \mathbf{b}, w\rangle w\Big) &= \sum_{v, w \in T_O} \langle \mathbf{a}, v\rangle \langle \mathbf{b}, w\rangle\, \mu_{\mathrm{M}}(v \otimes w) \\ &= \sum_{t \in T_O} \sum_{\mu_{\mathrm{M}}(v \otimes w) = t} \langle \mathbf{a}, v\rangle\, \langle \mathbf{b}, w\rangle t = \sum_{t \in T_O} \langle \mathbf{c}, t\rangle t.\end{aligned}$$

The resulting B-series has coefficients $\langle \mathbf{c}, \bullet \rangle = 1$ and for $t = B_+(t_1, \ldots, t_\mu)$,

$$\langle \mathbf{c}, t \rangle = \sum_{k=0}^{\mu} \langle \mathbf{a}, B_+(t_1, \ldots, t_k) \rangle \, \langle \mathbf{b}, B_+(t_{k+1}, \ldots, t_\mu) \rangle. \tag{5.6}$$

In view of the identification of $k_\infty T_O$ with kT_O^*, we can think of μ_{M} as a product on kT_O^* and simply write

$$\mathbf{c} = \mu_{\mathrm{M}}(\mathbf{a} \otimes \mathbf{b}).$$

Taking the adjoint operator $\Delta_M := \mu_{\mathrm{M}}^* \colon kT_O \to kT_O \otimes kT_O$ (using the usual identification of $(kT_O^* \otimes kT_O^*)^*$ with $kT_O \otimes kT_O$) we obtain

$$\Delta_M(B_+(t_1, \ldots, t_\mu)) = \sum_{k=0}^{\mu} B_+(t_1, \ldots, t_k) \otimes B_+(t_{k+1}, \ldots, t_\mu),$$

and $\langle \mathbf{c}, t \rangle = \langle \mu_{\mathrm{M}}(\mathbf{a} \otimes \mathbf{b}), t \rangle = \langle \mathbf{a} \otimes \mathbf{b}, \Delta_M(t) \rangle$ which is precisely what (5.6) says.

The antipode in the concatenation algebra $\mathcal{S}_{\mathrm{M}}$ has an immediate application to the B-series $B_p(\mathbf{a})$ of maps $\phi_{\mathbf{a}}$ on the manifold. The antipode of a series $T \in G = \exp(\mathfrak{g})$ is the corresponding series of the inverse map $\phi_{\mathbf{a}}^{-1}$. This is evident from Eq. (3.10) and from the fact that the B-series for the identity map on $\mathcal{M}$ has its coefficients from the counit $\varepsilon \in kT_O^*$ where $\langle \varepsilon, \bullet \rangle = 1$, $\langle \varepsilon, t \rangle = 0$ for all other trees $t \neq \bullet$.

6. Applications

6.1. Order conditions for integration schemes. The paper [**20**] presents the order conditions for a subclass of the schemes introduced in Section 2. We will show here how one can use the algebra on trees to count the order conditions for any order of the numerical method.

The algebra kT_O is graded by $\nu(t) = |t| - 1$ as in the introduction to Section 3. We can decompose the algebra kT_O and the Lie algebra $\mathfrak{g}$ of primitive elements in kT_O into their respective graded components

$$kT_O = \sum_{k=0}^{\infty} \mathcal{B}_k \quad \text{and} \quad \mathfrak{g} = \sum_{k=0}^{\infty} \mathfrak{g}_k.$$

The dimensions of each graded component in $\mathfrak{g}$ up to grade q are added to yield the number of order conditions for a qth-order integration scheme. From (3.1) we have $\dim \mathcal{B}_k = \nu_k = 1/k + 1\binom{2k}{k}$. It is well known [**25**, Theorem 3.2.8] that

$$kT_O = \mathrm{UEA}(\mathfrak{g})$$

where $\mathrm{UEA}(\mathfrak{g})$ is the universal enveloping algebra of $\mathfrak{g}$. Comparing generating functions for the dimensions of kT_O and $\mathrm{UEA}(\mathfrak{g})$ we get

$$\prod_{n=1}^{\infty} (1 - T^n)^{-\dim \mathfrak{g}_n} = \sum_{k=0}^{\infty} \nu_k T^k.$$

When solved for $\dim \mathfrak{g}_k$ one obtains

$$\dim \mathfrak{g}_k = \frac{1}{2k} \sum_{d|k} \mu(k) \binom{2k/d}{k/d} \tag{6.1}$$

where $\mu(k)$ is the Möbius function defined for any positive integer as $\mu(1) = 1$, $\mu(k) = (-1)^p$ when k is the product of p distinct primes, and $\mu(k) = 0$ otherwise. The sum is over all positive integers d which divide k, including 1 and k.

The number of order conditions of an integration scheme of order of consistency q is then

$$N_q = \sum_{k=1}^{q} \dim \mathfrak{g}_k.$$

and the first ten numbers are

$$1, 1, 3, 8, 25, 75, 245, 800, 2700, 9225.$$

The formula (6.1) is well-known in the literature. For example, it counts the number of balanced Lyndon words [**18**] and also has an application to double bracket flows [**11**].

6.2. Backward error analysis. Suppose a numerical method applied to the differential equation

$$y' = F(y) \tag{6.2}$$

is the map $\phi_{h,F} \colon \mathcal{M} \to \mathcal{M}$. If there exists an h-dependent vector field $\widetilde{F}$ such that

$$\phi_{h,F} = \exp(h\widetilde{F}),$$

then the numerical method for Eq. (6.2) solves the differential equation $y' = \widetilde{F}(y)$ exactly. We call $\widetilde{F}$ the *modified vector field* for $\phi_{h,F}$.

One may follow for example Hairer, Lubich and Wanner [**8**] and formally expand the modified vector field in powers of h

$$\widetilde{F} = F + hF_2 + h^2F_3 + \cdots . \tag{6.3}$$

The numerical method $\phi_{h,F}$ as a mapping on the manifold has an expansion in a B-series, say $B(\mathbf{a})$. Then we require that our modified vector field $\widetilde{F}$ must obey

$$\psi(\phi_{h,F}(p)) = \psi(\exp(h\widetilde{F})p) = B(\mathbf{a})[\psi](p).$$

where $B(\mathbf{a})$ is the series defined in (5.1).

We calculate $\exp(h\widetilde{F})$ according to the expansion in (6.3) and get

$$\exp(h\widetilde{F}) = I + hF + h^2(F_2 + \tfrac{1}{2}F^2) + h^3(F_3 + \tfrac{1}{2}(FF_2 + F_2F) + \tfrac{1}{6}F^3) + \cdots .$$

To calculate F_2 we compare coefficients of h^2 in the equation above and in $B(\mathbf{a})$. We get

$$\begin{aligned} \mathbf{a}(\treeII)\mathbf{F}(\treeII) + \mathbf{a}(\treeV)\mathbf{F}(\treeV) &= F_2 + \tfrac{1}{2}F^2 = F_2 + \tfrac{1}{2}\mathbf{F}(\treeI)^2 \\ &= F_2 + \frac{1}{2}\Big(\mathbf{F}(\treeII) + \mathbf{F}(\treeV)\Big) \end{aligned}$$

where we have used consistency of the method, $F = \mathbf{F}(\treeI)$ and the Grossman–Larson product of $\treeI$ and $\treeI$. Consistency of the numerical method requires $\mathbf{a}(\treeI) = 1$, and by a shuffle relation we also have that $\mathbf{a}(\treeV) = \frac{1}{2}\mathbf{a}(\treeI)^2 = \frac{1}{2}$. In the end we solve for F_2 and get

$$F_2 = \left(\mathbf{a}(\treeII) - \frac{1}{2}\right)\mathbf{F}(\treeII) \tag{6.4}$$

The same approach will for the h^3 coefficients lead to

(6.5)

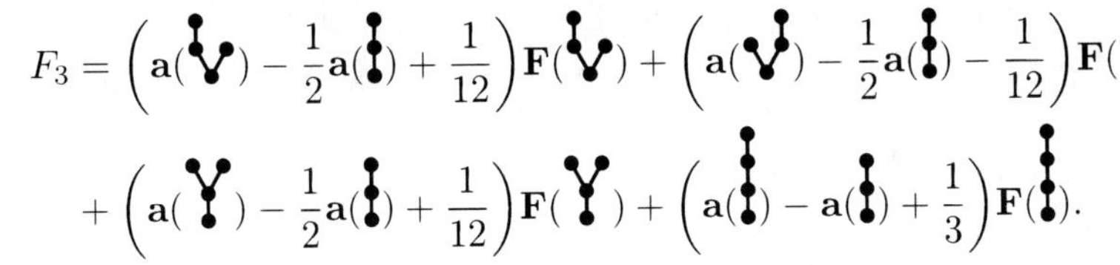

Again, putting this in a more general perspective, let $\varepsilon + \mathbf{a} \in kT_O^*$ represent the B-series of the numerical method, where ε is the counit in the Grossman–Larson Hopf algebra, and $\langle \mathbf{a}, \bullet \rangle = 0$. We now find that the modified vector field $\widetilde{F}$ has a B-series in $\mathfrak{g}$ associated to $\mathbf{b} \in kT_O^*$ such that

$$\mathbf{b} = \log_{\mathrm{GL}}(\varepsilon + \mathbf{a}) = \mathbf{a} - \tfrac{1}{2}\mu_{\mathrm{GL}}(\mathbf{a} \otimes \mathbf{a}) + \cdots .$$

Here, μ_{GL} is induced on kT_O^* from its identification with $k_\infty T_O$.

The expression for the vector fields F_k are obtained by projecting $\mathbf{b}$ onto the kth graded component of $k_\infty T_O$. One obtains that the series associated to $\mathbf{b}$ belongs to $\mathfrak{g}$.

7. Conclusion

We have discussed Hopf algebra structures on the space of ordered rooted trees, and shown how these structures are related to the expansions of Lie group integration schemes for ordinary differential equations on manifolds. The theory presented here is fairly general, in the sense that it accounts for most of the known Lie group integrators which are based on exponentials. It also allows for the analysis of schemes that are hybrids between for instance the RKMK methods [**17**] and the commutator-free schemes [**4**]. Although we have emphasized the algebraic aspects of the theory, we believe that there is a potential for using these aspects in developing integration schemes with high accuracy and good long-time behaviour.

References

1. Ch. Brouder, *Runge–Kutta methods and renormalization*, Eur. Phys. J. C **12** (2000), 521–534.
2. J. C. Butcher, *Coefficients for the study of Runge–Kutta integration processes*, J. Austral. Math. Soc. **3** (1963), 185–201.
3. ———, *An algebraic theory of integration methods*, Math. Comp. **26** (1972), 79–106.
4. E. Celledoni, A. Marthinsen, and B. Owren, *Commutator-free Lie group methods*, Future Generation Computer Systems **19** (2003), 341–352.
5. D. I. A. Cohen, *Basic Techniques of Combinatorial Theory*, John Wiley & Sons, 1978.
6. P. E. Crouch and R. Grossman, *Numerical integration of ordinary differential equations on manifolds*, J. Nonlinear Sci. **3** (1993), 1–33.
7. R. Grossman and R. G. Larson, *Hopf-algebraic structure of families of trees*, J. Algebra **126** (1989), no. 1, 184–210.
8. E. Hairer, Ch. Lubich, and G. Wanner, *Geometric numerical integration*, Structure-Preserving Algorithms for Ordinary Differential Equations, Springer Ser. Comput. Math., vol. 31. Springer-Verlag, Berlin, 2002.
9. E. Hairer, S. P. Nørsett, and G. Wanner, *Solving ordinary differential equations.* I, Nonstiff Problems, Springer, Berlin, 1987.
10. E. Hairer and G. Wanner, *On the Butcher group and general multivalue methods*, Computing **13** (1974), 1–15.
11. A. Iserles, *On the discretization of double-bracket flows*, Found. Comput. Math. **2** (2002), 305–329.
12. A. Iserles, H. Z. Munthe–Kaas, S. P. Nørsett, and A. Zanna, *Lie-group methods*, Acta Numerica **9** (2000), 215–365.

13. D. Kreimer, *On the Hopf algebra structure of perturbative quantum field theories*, Adv. Theor. Math. Phys. **2** (1998), 303–334.
14. D. Lewis and J. C. Simo, *Conserving algorithms for the dynamics of Hamiltonian systems of Lie groups*, J. Nonlinear Sci. **4** (1994), 253–299.
15. ———, *Conserving algorithms for the n dimensional rigid body*, Fields Inst. Com. **10** (1995).
16. H. Munthe-Kaas, *Runge–Kutta methods on Lie groups*, BIT **38** (1998), no. 1, 92–111.
17. ———, *High order Runge–Kutta methods on manifolds*, Appl. Numer. Math. **29** (1999), 115–127.
18. H. Munthe-Kaas and S. Krogstad, *On enumeration problems in Lie–Butcher theory*, Future Generation Computer Systems **19** (2003), no. 7, 1197–1205.
19. A. Murua and J. M. Sanz-Serna, *Order conditions for numerical integrators obtained by composing simpler integrators*, Phil. Trans. Royal Soc. A **357** (1999), 1079–1100.
20. B. Owren, *Order conditions for commutator-free Lie group methods* (in preparation).
21. B. Owren and A. Marthinsen, *Runge–Kutta methods adapted to manifolds and based on rigid frames*, BIT **39** (1999), no. 1, 116–142.
22. Ch. Reutenauer, *Free Lie algebras*, Oxford Univ. Press, UK, 1993.
23. M. Sweedler, *Hopf algebras*, W. A. Benjamin, Inc., New York, 1969.
24. S. Tracogna and B. Welfert, *Notes on riffle shuffles*, Available upon request, 1998.
25. V. S. Varadarajan, *Lie groups, Lie algebras, and their representations*, Grad. Texts in Math., vol. 102, Springer-Verlag, New York, 1984.

Centre for Advanced Study and Department of Mathematical Sciences, Alfred Getz vei 1, N-7034 Trondheim, Norway.

E-mail address: `Havard.Berland@math.ntnu.no`

URL: `http://www.math.ntnu.no/~berland`

Centre for Advanced Study and Department of Mathematical Sciences, Alfred Getz vei 1, N-7034 Trondheim, Norway.

E-mail address: `Brynjulf.Owren@math.ntnu.no`

URL: `http://www.math.ntnu.no/~bryn`

Centre de Recherches Mathématiques
CRM Proceedings and Lecture Notes
Volume **39**, 2005

Aspects of Generalized Double-Bracket Flows

A. M. Bloch and A. Iserles

ABSTRACT. In this paper we analyse generalizations of double-bracket flows to more complex isospectral equations of the form $\boldsymbol{Y}' = [\boldsymbol{Y}, [\boldsymbol{G}(\boldsymbol{Y}-\boldsymbol{N}), \boldsymbol{Y}]]$, $t \geq 0$, over real symmetric matrices. The new flows possess a nontrivial equilibrium structure which we explore. We discuss the stability of these equilibria. We also discuss the relationship of the new flows to the structure of (convex) Horn–Schur polytopes.

1. Introduction

A *double-bracket flow* (DBF) is defined as a matrix differential equation of the form

$$\boldsymbol{Y}' = [[\boldsymbol{N}, \boldsymbol{Y}], \boldsymbol{Y}], \quad t \geq 0, \qquad \boldsymbol{Y}(0) = \boldsymbol{Y}_0 \in \mathrm{Sym}(n), \tag{1.1}$$

where $\boldsymbol{N} \in \mathrm{Sym}(n)$, $\mathrm{Sym}(n)$ is the symmetric space of $n \times n$ real symmetric matrices, while $[\,\cdot\,,\,\cdot\,]$ is the standard matrix commutator. Without loss of generality we may assume that $\boldsymbol{N}$ is a diagonal matrix. Such flows were introduced by Brockett [**5**] and by Chu and Driessel [**8**] in connection with control theory and with inverse eigenvalue problems in numerical linear algebra, respectively, and their Lie-algebraic generalization was analysed in [**3, 4**]. Related equations are discussed in [**6, 9**].

Double-bracket flows possess a number of interesting features. They are a special case of an *isospectral flow*

$$\boldsymbol{Y}' = [\boldsymbol{B}(t, \boldsymbol{Y}), \boldsymbol{Y}], \quad t \geq 0, \qquad \boldsymbol{Y}(0) = \boldsymbol{Y}_0 \in \mathrm{Sym}(n), \tag{1.2}$$

where $\boldsymbol{B}\colon \mathbb{R}_+ \times \mathrm{Sym}(n) \to \mathfrak{so}(n)$, the set $\mathfrak{so}(n)$ being the Lie algebra of $n \times n$ real, skew-symmetric matrices. Since $\boldsymbol{Y}(t) = \boldsymbol{Q}(t)Y_0\boldsymbol{Q}^\top(t)$, $t \geq 0$, where $\boldsymbol{Q}' = \boldsymbol{B}(t, \boldsymbol{Q}\boldsymbol{Y}_0\boldsymbol{Q}^\top)$, $t \geq 0$, $\boldsymbol{Q}(0) = \boldsymbol{I}$, it follows easily that $\boldsymbol{Y}(t)$ for all $t \geq 0$ is similar to $\boldsymbol{Y}_0$. In other words, the eigenvalues of $\boldsymbol{Y}_0$ are integrals of the (1.2). (Hence the name "isospectral flows".) This feature makes isospectral flows of interest in many areas of mathematics including integrable systems and is also crucial in their numerical solution [**7, 10**]. This isospectral property of the double-bracket flow is

---

2000 *Mathematics Subject Classification.* Primary: 34C40; Secondary: 51M20.
The first author acknowledges partial support of the National Science Foundation.
This is the final form of the paper.

highly relevant in the special case when the system reduces to the Toda lattice flow (see [**4**] and references therein).

Another important feature of the DBF (1.1) is that it is a *gradient flow* with respect to the so-called *normal metric* on an appropriate adjoint orbit of a compact Lie group and possesses a global Lyapunov function [**4, 5**]. Note that this feature is in general false for general isospectral flows (1.2). Construction of alternative Lyapunov functions for (1.1) was addressed in [**1**].

A key feature of DBF, originally observed by Brockett [**5**] and analysed extensively in [**1**], is its optimality. Roughly speaking, $\widehat{\boldsymbol{Y}} = \lim_{t\to\infty} \boldsymbol{Y}(t)$ (which exists because (1.1) is a gradient flow) minimizes the distance from $\boldsymbol{N}$ along the isospectral orbit with regard to various unitarily-invariant norms, in particular Schatten p-norms for $p \geq 1$.

The theme of the present paper is a generalization of the double-bracket flow to instances when the flow is the gradient of a higher-power trace. This leads a richer equilibrium and stability structure which we proceed to examine.

We also discuss the role of the Schur–Horn polytope in the dynamics of these systems, generalizing the work in [**4**].

The result of Schur is as follows: Let A be a Hermitian matrix with eigenvalues λ_j, $j = 1, \ldots, n$. Denote by $\lambda = (\lambda_1, \ldots, \lambda_n)$ the vector of eigenvalues of A and by $A^0 = (A_{11}, \ldots, A_{nn})$ the diagonal of A. Then $A^0 \in \widehat{(\Sigma_n \lambda)}$. Here the "wide hat" notation denotes the convex hull under the action of the permutation group.

Conversely the result of Horn is: Let $\lambda \in \mathbb{R}^n$, with components arranged in non-increasing order. If $A^0 \in \widehat{(\Sigma_n \lambda)}$, there is a Hermitian matrix A with eigenvalues λ whose diagonal is A^0.

As we show in the sequel, the structure of the Schur–Horn polytope plays an important role in understanding the equilibria and stability of the generalized double-bracket flows.

This work will be further elaborated and extended in [**2**].

2. Generalized double-bracket equations

2.1. Main equations. Let $\boldsymbol{Y}_0, \boldsymbol{N} \in \mathrm{Sym}(n)$ and suppose that $g\colon \mathrm{Sym}(n) \to \mathbb{R}$ is a smooth function. In particular, we are interested in the functions

$$g_m(\boldsymbol{X}) = \mathrm{tr}\, \boldsymbol{X}^m, \qquad m \in \mathbb{Z}_+.$$

We denote by $\boldsymbol{G} : \mathrm{Sym}(n) \to \mathrm{Sym}(n)$ the Jacobian of g. In particular,

$$[\boldsymbol{G}_m(\boldsymbol{X})]_{k,l} = \frac{\partial g_m(X)}{\partial \boldsymbol{X}_{k,l}}, \quad 1 \leq k,\ l \leq n.$$

It is an easy matter to prove that

$$\boldsymbol{G}_m(X) = m\boldsymbol{X}^{m-1}, \quad m \in \mathbb{Z}_+. \tag{2.1}$$

The *generalized double-bracket equation* (GDBE) is

$$\boldsymbol{Y}' = \big[\boldsymbol{Y}, [\boldsymbol{G}(\boldsymbol{Y} - \boldsymbol{N}), \boldsymbol{Y}]\big], \quad t \geq 0, \qquad \boldsymbol{Y}(0) = \boldsymbol{Y}_0. \tag{2.2}$$

More specifically, in the present paper we focus on $\boldsymbol{G} = \boldsymbol{G}_m$, in which case we drop the multiplicative factor m (which merely corresponds to reparametrisation) and consider

$$\boldsymbol{Y}' = \big[\boldsymbol{Y}, [(\boldsymbol{Y} - \boldsymbol{N})^{m-1}, \boldsymbol{Y}]\big], \quad t \geq 0, \qquad \boldsymbol{Y}(0) = \boldsymbol{Y}_0. \tag{2.3}$$

This, clearly, makes sense and is nontrivial only for $m \geq 2$.

We are interested in the stability and optimality of the equilibria of this generalized system. For $m \geq 3$, as we shall discuss below, not all equilibria are diagonal. The diagonal equilibria are of great interest however and we begin by considering these.

Given

$$\boldsymbol{Y}' = \big[\boldsymbol{Y}, [\boldsymbol{Y}, (\boldsymbol{Y} - \boldsymbol{N})^m]\big]$$

the linearization at $\boldsymbol{Y} = \boldsymbol{D} = \operatorname{diag} \boldsymbol{d}$ is

$$(\delta \boldsymbol{Y})' = \big[\boldsymbol{D}, [\delta \boldsymbol{Y}, (\boldsymbol{D} - \boldsymbol{N})^m]\big] \\ + \big[\boldsymbol{D}, \big[\boldsymbol{D}, \delta \boldsymbol{Y}(\boldsymbol{D} - \boldsymbol{N})^{m-1} + (\boldsymbol{D} - \boldsymbol{N})\delta \boldsymbol{Y}(\boldsymbol{D} - \boldsymbol{N})^{m-2} + \cdots\big]\big].$$

Writing this in components with $\eta_j = d_j - \lambda_j$ we have

$$(\delta y_{ij})' = \delta y_{ij}(\eta_j^m - \eta_i^m)(d_i - d_j) + (\delta y_{ij}\eta_j^{m-1} + \eta_i \delta y_{ij} \eta_j^{m-2} + \cdots)(d_i - d_j)^2$$

or

$$(\delta y_{ij})' = \delta y_{ij} \left[(\eta_j^m - \eta_i^m)(d_i - d_j) + \frac{\eta_j^m - \eta_i^m}{\eta_j - \eta_i}(d_i - d_j)^2 \right].$$

Hence for stability we require

$$(d_i - d_j)(\eta_j^m - \eta_i^m)\left(1 + \frac{d_i - d_j}{\eta_j - \eta_i}\right) < 0$$

or

$$(d_i - d_j)(\lambda_i - \lambda_j)\frac{\eta_j^m - \eta_i^m}{\eta_j - \eta_i} < 0, \qquad 1 \leq i,\ j \leq n, \quad i \neq j.$$

3. The dynamics of GDBE

3.1. The case $m = 3$. The stability conditions are

$$(d_k - d_l)(\lambda_k - \lambda_l)(\lambda_k + \lambda_l - d_l - d_k) > 0, \quad 1 \leq k,\ l \leq n. \tag{3.1}$$

In other words, given distinct real $\lambda_1, \ldots, \lambda_n, d_1, \ldots, d_n$, we seek a permutation of $d_1, d_2, \ldots, d_n$ so that (3.1) holds.

The combinatorial problem (3.1) has a solution for $n = 3$, as can be demonstrated in a direct manner, by considering a welter of special cases. Such direct demonstration is, unfortunately, fairly complicated and it is highly unlikely that it can be extended even to $n = 4$. It is more natural and promising to define the function $h(t) = \operatorname{tr}(\boldsymbol{Y} - \boldsymbol{N})^3$ and hope that it is strictly monotone for all non-diagonal $\boldsymbol{Y}$. Numerical computations confirm this but, so far, analysis is *very* messy. The beauty of this approach is that, once we prove it, we can claim that the above combinatorial problem has solution for every n, a pleasing, roundabout, homotopy argument for the solution of a discrete problem.

A weaker, yet interesting, alternative course of action is to show that $h'(t) < 0$ for an open, nonempty set of initial values: at the very least, this will imply the existence of an attractive fixed point, i.e. of a solution to the set of inequalities (3.1).

Let $\boldsymbol{\lambda} \in \mathbb{R}^n$ be given and assume that its components are distinct. We say that a vector $\boldsymbol{v} \in \mathbb{R}^n$ is *consistent with* $\boldsymbol{\lambda}$ if

$$(\lambda_k - \lambda_l)(v_k - v_l)(v_k + v_l - \lambda_k - \lambda_l) > 0, \quad 1 \leq k,\ l \leq n,\ k \neq l. \tag{3.2}$$

We denote the set of all such vectors by $\mathscr{C}_{\boldsymbol{\lambda}}$.

Lemma 1. *The set $\mathscr{C}_{\boldsymbol{\lambda}}$ is nonempty.*

PROOF. We commence by assuming that $\lambda_k > 0$, $1 \le k \le n$. Let $\rho > 0$ and set $v_k = \rho\lambda_k$, $1 \le k \le n$. Then

$$(\lambda_k - \lambda_l)(v_k - v_l)(v_k + v_l - \lambda_k - \lambda_l) = \rho(\rho - 1)(\lambda_k - \lambda_l)^2(\lambda_k + \lambda_l) > 0$$

and (3.2) is satisfied.

We next observe that, for any $\alpha \in \mathbb{R}$,

$$\boldsymbol{v} \in \mathscr{C}_{\boldsymbol{\lambda}} \iff \boldsymbol{v} + \alpha\mathbf{1} \in \mathscr{C}_{\boldsymbol{\lambda}+\alpha\mathbf{1}}.$$

We can thus always choose α so that $\lambda_k + \alpha > 0$, $1 \le k \le n$ and reduce everything to the framework of the previous paragraph. □

The last lemma is less useful than it sounds, because we are not interested in arbitrary members of $\mathscr{C}_{\boldsymbol{\lambda}}$. Given $\boldsymbol{Y}_0 \in \mathrm{Sym}(n)$ and $\boldsymbol{N} = \mathrm{diag}\,\boldsymbol{\lambda}$, we interpret $\boldsymbol{v}$ as $\mathrm{diag}\,\boldsymbol{Y}_0$, therefore we wish $\boldsymbol{v}$ to live in the Horn–Schur polytope of $\boldsymbol{d} = \sigma(\boldsymbol{Y}_0)$: according to the Horn lemma, in that case, for every $\boldsymbol{v}$ in the polytope we can always find $\boldsymbol{Y}_0 \in \mathrm{Sym}(n)$ with $\sigma(\boldsymbol{Y}_0) = \boldsymbol{\lambda}$ and $\mathrm{diag}\,\boldsymbol{Y}_0 = \boldsymbol{v}$.

3.2. Stability sectors in the Horn–Schur polytope. What is going on in the Horn–Schur polytope? We start from the case $m = 3$, $n = 3$ and note that, since every point in the polytope is normal to $\mathbf{1}$, it lives in $\mathbb{R}^2$. Thus, plotting it, we can project it on a plane: Given $\boldsymbol{x} \in \mathbb{R}^3$ such that (as is the case in the HSP) $\boldsymbol{x}^\top\mathbf{1} = c$, where $c = \boldsymbol{d}^\top\mathbf{1}$, we plot the coordinates

$$\mathbb{R}^2 \ni \boldsymbol{z} = \begin{bmatrix} \frac{1}{2}(1+\sqrt{3}) & \frac{1}{2}(-1+\sqrt{3}) & 0 \\ \frac{1}{2}(-1+\sqrt{3}) & \frac{1}{2}(1+\sqrt{3}) & 0 \end{bmatrix} (\boldsymbol{x} - \tfrac{1}{3}c\mathbf{1}).$$

Our first collection of plots takes two vectors, $\boldsymbol{d}, \boldsymbol{\lambda} \in \mathbb{R}^3$ and colours differently the set of all points in the HS polytope according to the following scheme: Take a point $\boldsymbol{y}$ therein and check for which permutation $\boldsymbol{\eta} = \boldsymbol{\pi}(\boldsymbol{\lambda})$ of $\boldsymbol{\lambda}$ it is true that

$$(\eta_k - \eta_l)(y_k - y_l)(y_k + y_l - \eta_k - \eta_k) \ge 0, \quad k, l = 1, 2, 3.$$

We use the following colour code, consistent with RGB colouring:

$$\begin{aligned} \boldsymbol{\pi} &= [1,2,3]: && \text{red}, \\ \boldsymbol{\pi} &= [1,3,2]: && \text{yellow}, \\ \boldsymbol{\pi} &= [2,3,1]: && \text{green}, \\ \boldsymbol{\pi} &= [3,2,1]: && \text{cyan}, \\ \boldsymbol{\pi} &= [3,1,2]: && \text{blue}, \\ \boldsymbol{\pi} &= [2,1,3]: && \text{magenta}. \end{aligned}$$

In monochrome, "colours" correspond to different shades of gray, ordered from the lightest to the darker as

$$\text{yellow} \prec \text{cyan} \prec \text{green} \prec \text{magenta} \prec \text{red} \prec \text{blue}.$$

This scheme of permutations is consistent with 'travelling' along the faces of the HS polytope.

The meaning of the experiment is as follows: the red points are arranged consistently with the stable attractor for $N = \mathrm{diag}[\lambda_1, \lambda_2, \lambda_3]$, the magenta points consistently with $N = \mathrm{diag}[\lambda_2, \lambda_1, \lambda_3]$ and so on. Of course, it does not mean (or, at least, we did not prove it, unlike the case $m = 2$, and it might well be false)

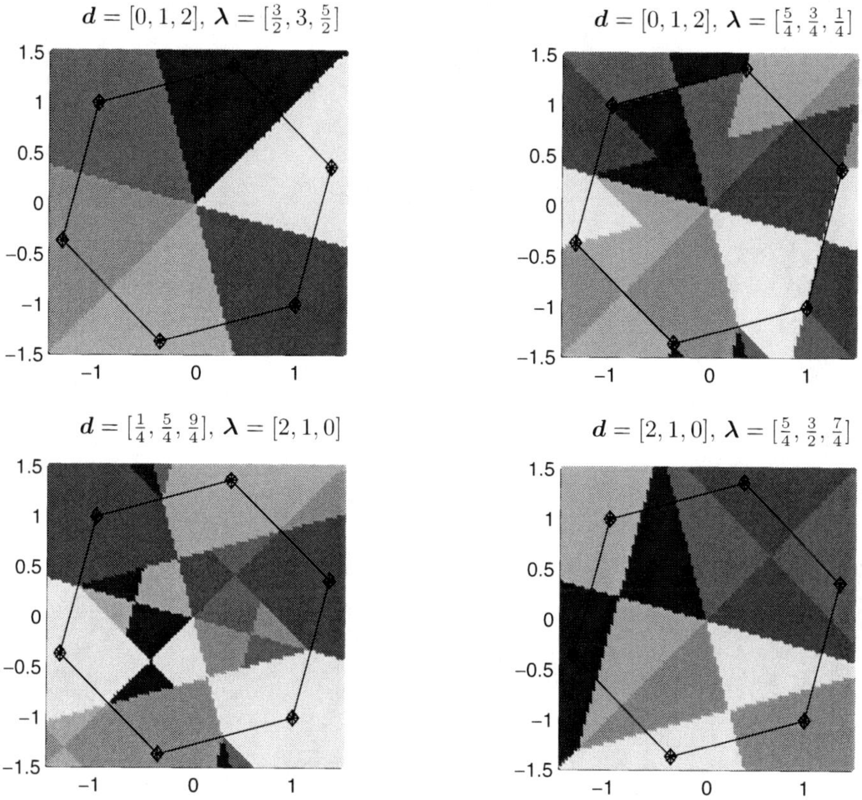

FIGURE 1. Colouring of the HSP.

that, once you solve the ODE with an initial condition in the red region (and with $N = \mathrm{diag}[\lambda_1, \lambda_2, \lambda_3]$) then you *must* tend to the corresponding vertex. All it means is that if a vertex is bordered completely by a red region and we start near enough to it, attractivity of the fixed point will bring us to the vertex in question.

In Fig. 1 we display four such plots, for different choices of $\boldsymbol{d}$ and $\boldsymbol{\lambda}$.

Since the λ_k's are distinct, the colour changes once one of six equalities holds for the vector $\boldsymbol{\eta}$,

$$\eta_k = \eta_l, \quad \eta_k + \eta_l = \lambda_k + \lambda_l, \qquad 1 \le k < l \le 3.$$

Recall that everything has been rotated in Fig. 1 to a plane, hence we need to express the above equalities as linear equations in the (x, y) plane. Moreover, the one consequence of being in the HS polytope is that $\mathbf{1}^\top \boldsymbol{\eta} = c$ for $c = \mathbf{1}^\top \boldsymbol{d}$.

For example, the condition $\eta_2 = \eta_3$, in tandem with $\eta_1 + \eta_2 + \eta_3 = c$, after rotation, yields

$$x = \tfrac{1}{4}[(3 + \sqrt{3})\eta_1 - (1 + 3\sqrt{3})c], \quad y = \tfrac{1}{4}[(-3 + \sqrt{3})\eta_1 + (1 - 3\sqrt{3})c].$$

Eliminating η_1 from the above equations results in the line

$$y(x) = (\sqrt{3} - 2)[x + 8(1 + \sqrt{3})c].$$

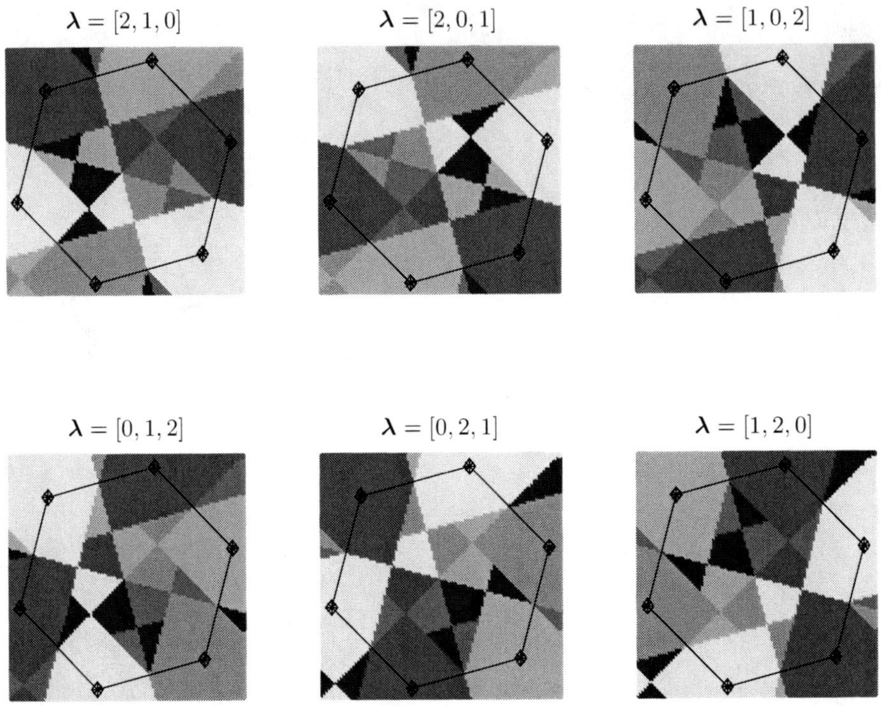

FIGURE 2. $\boldsymbol{d} = [\frac{1}{4}, \frac{5}{4}, \frac{9}{4}]$ and different permutations of $\lambda = [2,1,0]$.

In general, we obtain the following lines along which colour might change:

$$\begin{aligned}
\eta_1 = \eta_2 :&\quad y(x) = x,\\
\eta_1 = \eta_3 :&\quad y(x) = -(2+\sqrt{3})[x + 4(\sqrt{3}-1)c],\\
\eta_2 = \eta_3 :&\quad y(x) = (\sqrt{3}-2)[x + 8(1+\sqrt{3})c],\\
\eta_1 + \eta_2 = \lambda_1 + \lambda_2 :&\quad y(x) = -x - 2\sqrt{3}c + \sqrt{3}(\lambda_1 + \lambda_2),\\
\eta_1 + \eta_3 = \lambda_1 + \lambda_3 :&\quad y(x) = (2-\sqrt{3})[x - (6+2\sqrt{3})(\lambda_1 + \lambda_3)],\\
\eta_2 + \eta_3 = \lambda_2 + \lambda_3 :&\quad y(x) = (2+\sqrt{3})[x + (6-2\sqrt{3})(\lambda_2 + \lambda_3)].
\end{aligned}$$

Note that the first three lines are always parallel to a face of the (rotated) HS polytope. Moreover, the *slopes* of all six lines are independent of the choice of $\boldsymbol{d}$ and $\boldsymbol{\lambda}$, all that changes is the affine shift.

In Fig. 2 we display the coloured HS polytope for fixed $\boldsymbol{d}$ and all permutations of $\boldsymbol{\lambda}$. Several symmetries stare us in the face. For example, any permutation of the form $(a,b,c) \mapsto (a,c,b)$ corresponds to red $\mapsto$ magenta, blue $\mapsto$ green and cyan $\mapsto$ yellow. There are other symmetries, which might include both colour swaps and reflections (or rotations?) of the picture. Are they universal or special for this particular choice of $\boldsymbol{d}$ and $\boldsymbol{\lambda}$? We intend to discuss this further in [**2**].

4. General equilibria

As mentioned earlier, it is entirely possible for the matrices $\boldsymbol{Y} \in \mathrm{Sym}(n)$ and $\boldsymbol{N} = \mathrm{diag}\,\boldsymbol{\lambda}$ to be such that $[(\boldsymbol{Y}-\boldsymbol{N})^2, \boldsymbol{Y}] = O$, yet $(\boldsymbol{Y}-\boldsymbol{N})^2$ is *not* diagonal. The very important consequence is that *we might have fixed points—even attractive fixed points*—inside *the HS polytope*

4.1. Nontrivial solutions: the solution for $n = 3$. We consider now solutions to the equilibrium condition

$$[\boldsymbol{N}^2, \boldsymbol{Y}] = [\boldsymbol{N}, \boldsymbol{Y}^2] \tag{4.1}$$

The idea is to commence by specifying the *diagonal* of $\boldsymbol{Y} \in \mathrm{Sym}(3)$. Thus, suppose that

$$\boldsymbol{Y} = \begin{pmatrix} \alpha & a & b \\ a & \beta & c \\ b & c & \gamma \end{pmatrix},$$

where α, β, γ are fixed and we seek to determine a, b, c so that (4.1) is satisfied.

The matrix equation (4.1) is equivalent to

$$(\lambda_k + \lambda_l)Y_{k,l} = (\boldsymbol{Y}^2)_{k,l}, \quad 1 \le k < l \le n \tag{4.2}$$

(in our case $n = 3$) and the latter can be used to find the conditions on a, b, c: they are

$$(\lambda_1+\lambda_2)a = (\alpha+\beta)a+cb, \quad (\lambda_1+\lambda_3)b = (\alpha+\gamma)b+ac, \quad (\lambda_2+\lambda_3)c = (\beta+\gamma)c+ab.$$

An obvious solution of the above equations is $a = b = c = 0$, which corresponds to a diagonal matrix $\boldsymbol{Y}$ (which, needless to say, obeys (4.1)). Being interested in nontrivial solutions, we exclude this case, whereby

$$a^2(\lambda_1+\lambda_2-\alpha-\beta) = b^2(\lambda_1+\lambda_3-\alpha-\gamma) = c^2(\lambda_2+\lambda_3-\beta-\gamma) = abc.$$

Note that in the present case $a, b, c \neq 0$. We obtain

$$a = (-1)^{k_a} b\sqrt{\frac{\lambda_1+\lambda_3-\alpha-\gamma}{\lambda_1+\lambda_2-\alpha-\beta}}, \quad c = (-1)^{k_c} b\sqrt{\frac{\lambda_1+\lambda_3-\alpha-\gamma}{\lambda_2+\lambda_3-\beta-\gamma}},$$

where $k_a, k_b \in \{0,1\}$. Moreover,

$$abc = (\lambda_1+\lambda_3-\alpha-\gamma)b^2 = (-1)^{k_a+k_c}\frac{\lambda_1+\lambda_3-\alpha-\gamma}{\sqrt{(\lambda_1+\lambda_2-\alpha-\beta)(\lambda_2+\lambda_3-\beta-\gamma)}}b^3.$$

Therefore, we deduce that

$$\begin{aligned} a &= (-1)^{k_a}\sqrt{(\lambda_1+\lambda_3-\alpha-\gamma)(\lambda_2+\lambda_3-\beta-\gamma)}, \\ b &= (-1)^{k_a+k_c}\sqrt{(\lambda_1+\lambda_2-\alpha-\beta)(\lambda_2+\lambda_3-\beta-\gamma)}, \\ c &= (-1)^{k_c}\sqrt{(\lambda_1+\lambda_2-\alpha-\beta)(\lambda_1+\lambda_3-\alpha-\gamma)}. \end{aligned} \tag{4.3}$$

This exhausts all possible solutions, except that we must ensure that they are real. This will happen if and only if the three numbers

$$\lambda_1+\lambda_2-\alpha-\beta, \lambda_1+\lambda_3-\alpha-\gamma, \lambda_2+\lambda_3-\beta-\gamma$$

are all of the same sign. Therefore, (α,β,γ) must lie in one of two cones: either

$$\alpha+\beta \ge \lambda_1+\lambda_2, \quad \alpha+\gamma \ge \lambda_1+\lambda_3, \quad \beta+\gamma \ge \lambda_2+\lambda_3$$

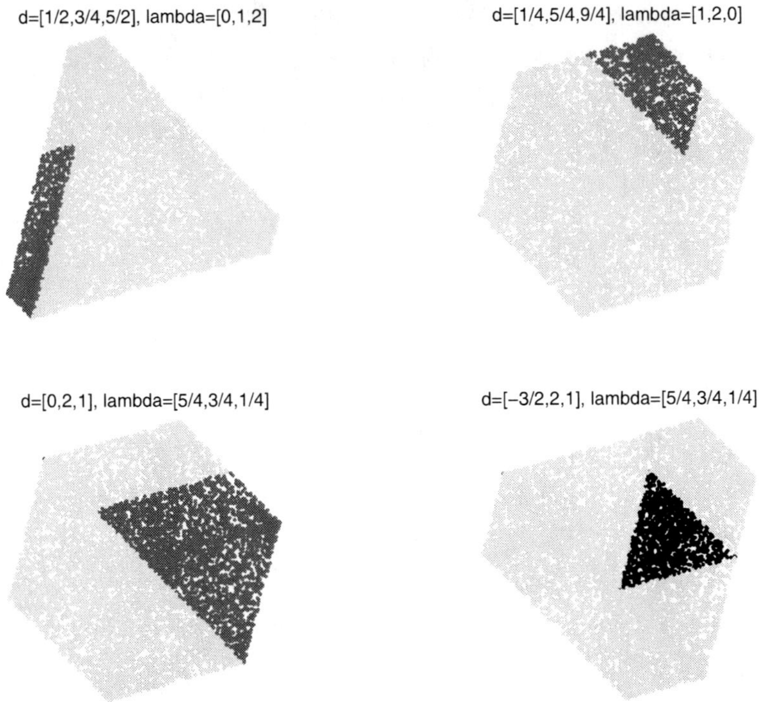

FIGURE 3. Different nontrivial solution regimes. Magenta denotes points in $\boldsymbol{\lambda}+\mathcal{C}$, black points are in $\boldsymbol{\lambda}-\mathcal{C}$ and yellow in neither cone. (In monochrome, "yellow" is the brightest and "black" the darkest.)

or

$$\alpha+\beta \le \lambda_1+\lambda_2, \quad \alpha+\gamma \le \lambda_1+\lambda_3, \quad \beta+\gamma \le \lambda_2+\lambda_3.$$

Letting $\bar{\alpha}=\alpha-\lambda_1$, $\bar{\beta}=\beta-\lambda_2$, $\bar{\gamma}=\gamma-\lambda_3$, the first cone, $\mathcal{C}$, say, becomes

$$\bar{\alpha}+\bar{\beta}\ge 0, \quad \bar{\alpha}+\bar{\gamma}\ge 0, \quad \bar{\beta}+\bar{\gamma}\ge 0,$$

while the second cone is $-\mathcal{C}$. Therefore the condition for the existence of a nontrivial solution is that

$$\operatorname{diag} Y \in \boldsymbol{\lambda}+\mathcal{C} \quad \text{or} \quad \operatorname{diag} Y \in \boldsymbol{\lambda}-\mathcal{C}. \tag{4.4}$$

Of course, the whole point is that $\operatorname{diag} Y \in \mathcal{HS}$, the Horn–Schur polytope. Thus, nontrivial solutions exist if and only if

$$\text{either} \quad \mathcal{HS}\cap\{\boldsymbol{\lambda}+\mathcal{C}\}\ne\emptyset \quad \text{or} \quad \mathcal{HS}\cap\{\boldsymbol{\lambda}-\mathcal{C}\}\ne\emptyset.$$

Recall that all $\boldsymbol{x}\in\mathcal{HS}$ lie on the plane $\mathbf{1}^\top\boldsymbol{x}=\mathbf{1}^\top\boldsymbol{d}$. The intersection of this plane with the union of the two cones is always nonempty, but this is not to say that the intersection of the HS polytope itself with the union of the cones is nonempty. The easiest special case is when $\mathbf{1}^\top\boldsymbol{d}=\mathbf{1}^\top\boldsymbol{\lambda}$, whereby the intersection of the plane with the cones is just the point set $\{\boldsymbol{\lambda}\}$. Therefore, in that case, either $\boldsymbol{\lambda}\in\mathcal{HS}$ and we have a nontrivial solution,[1] or $\boldsymbol{\lambda} \notin \mathcal{HS}$ and all solutions are trivial. For all other values of $\mathbf{1}^\top \boldsymbol{d}$ the set of nontrivial solutions in the HS polytope is itself a polytope. Moreover, every point in this polytope corresponds to four different solutions, with different sign configurations.

4.2. General $n \geq 2$. Let us commence by setting the stage for general n. Denoting $\operatorname{diag} \boldsymbol{Y} = \boldsymbol{x}$, it follows from (4.2) that

$$(\lambda_k + \lambda_l) Y_{k,l} = \sum_{i=1}^{k-1} Y_{i,k} Y_{i,l} + x_k Y_{k,l} + \sum_{i=k+1}^{l-1} Y_{k,i} Y_{i,l} + x_l Y_{k,l} + \sum_{i=l+1}^{n} Y_{k,i} Y_{l,i}, \quad 1 \leq k < l \leq n.$$

Therefore, letting

$$\mu_{k,l} = \lambda_k + \lambda_l - x_k - x_l, \quad 1 \leq k < l \leq n,$$

we have

$$\mu_{k,l} Y_{k,l} = \sum_{i=1}^{k-1} Y_{i,k} Y_{i,l} + \sum_{i=k+1}^{l-1} Y_{k,i} Y_{i,l} + \sum_{i=l+1}^{n} Y_{k,i} Y_{l,i}, \quad 1 \leq k < l \leq n. \tag{4.5}$$

Consider now the general case. There is just a single solution for $n = 2$, namely a diagonal matrix ($Y_{1,2} = 0$). We denote the number of solutions in the $n \times n$ case by ψ_n, therefore

$$\psi_2 = 1, \quad \psi_3 = 5, \quad \psi_4 = 25.$$

(It might be tempting by this stage to conjecture that $\psi_n = 5^{n-2}$, but, as will be clear later, is probably false.)

Clearly, for $n \geq 3$ there are lower-dimensional problems hidden in the matrix and they can be obtained simply by taking rows and columns of off-diagonal zeros. Our real interest is in the solutions which cannot be obtained from lower-dimensional cases. We denote the number of such solutions by $\psi_n^\star$: $\psi_2^\star = 1$, $\psi_3^\star = 4$, $\psi_4^\star = 16$. It follows trivially that

$$\psi_n = \sum_{k=0}^{n-3} \binom{n}{k} \psi_{n-k}^\star + 1 = \sum_{k=3}^{n} \binom{n}{k} \psi_k^\star + 1 \tag{4.6}$$

(the "+1" corresponds to the diagonal solution). From now on, we concentrate only on the "truly n-dimensional" solutions.

We say that the symmetric matrix $E = (e_{k,l})_{k,l=1}^n$ is *tri-balanced* if

$$e_{k,l} \in \{0, 1\}, \quad 1 \leq k, l \leq n \quad \text{and} \quad e_{i,k} + e_{i,l} = e_{k,l} \bmod 2, \qquad 1 \leq i, k, l \leq n$$

(note that automatically $e_{k,k} = 0$, but we will not use diagonal elements). Assuming that E is tri-balanced, we set

$$Y_{k,l} = (-1)^{e_{k,l}} \sqrt{(\sigma - \lambda_k + x_k)(\sigma - \lambda_l + x_l)}, \qquad 1 \leq k, l \leq n, \quad k \neq l, \tag{4.7}$$

where

$$\sigma = \frac{1}{n-2} \sum_{i=1}^{n} (\lambda_i - x_i) = \frac{1}{n-2} \operatorname{tr}(\boldsymbol{N} - \boldsymbol{Y}). \tag{4.8}$$

[1]Recall our assumption that the spectra are distinct, hence $\boldsymbol{\lambda}$ cannot be a vertex of the HS polytope.

We claim that (4.7) represents a solution of (4.5), thereby of (4.1):

$$\sum_{\substack{i=1\\ i\neq k,l}}^{n} Y_{k,i}Y_{l,i} = \sum_{\substack{i=1\\ i\neq k,l}}^{n} (-1)^{e_{i,k}+e_{i,l}}(\sigma-\lambda_i+x_i)\sqrt{(\sigma-\lambda_k+x_k)(\sigma-\lambda_l+x_l)}$$
$$= \sum_{\substack{=1\\ i\neq k,l}}^{n} (-1)^{e_{i,k}+e_{i,l}-e_{k,l}}(\sigma-\lambda_i+x_i)Y_{k,l}$$
$$= \left[\sum_{i=1}^{n}(\sigma-\lambda_i+x_i) - 2\sigma + (\lambda_k+\lambda_l) - (x_k+x_l)\right] Y_{k,l} = \mu_{k,l}Y_{k,l}$$

for all $k \neq l$. Therefore, for every tri-balanced matrix there corresponds a solution (4.7).

But how many tri-balanced matrices are there? Suppose that $e_{k,1} = e_{1,k}$ is given for all $1 \le k \le n$ (of course, $e_{1,1} = 0$) and set

$$e_{k,l} = e_{1,k} + e_{1,l}, \quad 1 \le k, l \le n.$$

It is easy to see that, with the above definition,

$$e_{k,i} + e_{i,l} = (e_{1,k} + e_{1,i}) + (e_{1,l} + e_{1,i}) = e_{1,k} + e_{1,l} = e_{k,l} \bmod 2,$$

hence we have a tri-balanced matrix. Moreover, no other choice of $e_{k,l}$ is tri-balanced. This is true because suppose that there exist k, l such that $e_{i,k} \neq e_{1,i} + e_{1,k} \bmod 2$. Then necessarily $e_{i,k} \neq e_{1,i} + e_{1,k} + 1 \bmod 2$, hence for every i

$$\begin{aligned} & e_{i,k} + e_{i,l} = e_{1,k} + e_{1,l} \bmod 2 \\ \Longrightarrow\quad & e_{i,l} = e_{1,l} - (e_{1,i} + e_{1,k} + 1) = e_{1,i} + e_{1,l} + 1 \bmod 2 \\ \stackrel{i=l}{\Longrightarrow}\quad & 2e_{1,l} + 1 = 0 \bmod 2, \end{aligned}$$

2 which is a contradiction. Therefore, we have precisely 2^{n-1} tri-balanced matrices and $\psi_n^\star \ge 2^{n-1}$.

We deduce that

$$\psi_n \ge 1 + \sum_{k=3}^{n} \binom{n}{k} 2^{k-1} = \tfrac{1}{2}(3^n + 1 - 2n^2).$$

Stranger number than 5^{n-2}, yet this upper bound coincides with ψ_n for $n = 1, 2, 3$. However, it is presently unclear whether it represents *all* solutions of (4.1) or is just a lower bound.

Again, solutions are valid only if either

$$x_k \ge \lambda_k - \sigma, \quad 1 \le k \le n,$$

or

$$x_k \le \lambda_k - \sigma, \quad 1 \le k \le n.$$

A cone and its negative, again.... Thus, we have a solution (given eigenvalues $\boldsymbol{d}$, rather than the diagonal $\boldsymbol{x}$) if and only if one of these cones intersects the HS polytope.

Clearly, the above analysis falls short of a comprehensive description of the dynamics of generalized double-bracket flows and its connection with the Horn–Schur polytope. Yet, even these partial and often tentative results demonstrate that the underlying problem is fascinating, nontrivial and leads in some highly interesting directions.

References

1. Anthony M. Bloch and Arieh Iserles, *On the optimality of double-bracket flows*, DAMTP, Tech. Report, no. NA 2003/04, University of Cambridge, 2003.
2. ———, *Generalized double bracket equations*, DAMTP, University of Cambridge, 2003 (to appear).
3. Anthony M. Bloch, Roger W. Brockett, and T. S. Ratiu, *A new formulation of the generalized Toda lattice equations and their fixed-point analysis via the moment map*, Bull. Amer. Math. Soc. (N.S.) **23** (1990), 447–456.
4. ———, *Completely integrable gradient flows*, Comm. Math. Phys. **147** (1992), 57–74.
5. Roger W. Brockett, *Dynamical systems that sort lists, diagonalize matrices, and solve linear programming problems*, Linear Algebra Appl. **146** (1991), 79–91.
6. ———, *Differential geometry and the design of gradient algorithms*, Proc. Sympos. Pure Math. **54** (1993), 69–92.
7. M. P. Calvo, Arieh Iserles, and A. Zanna, *Numerical solution of isospectral flows*, Math. Comp. **66** (1997), 1461–1486.
8. Moody T. Chu and Kenneth R. Driessel, *Constructing symmetric nonnegative matrices with prescribed eigenvalues by differential equations*, SIAM J. Math. Anal. **22** (1991), 1372–1387.
9. U. Helmke and J. Moore, *Optimization and dynamical systems*, Springer Verlag, New York, 1994.
10. Arieh Iserles, *On the discretization of double-bracket flows*, Found. Comput. Maths. **2** (2002), 305–329.

Department of Mathematics, University of Michigan, Ann Arbor, MI 48109, USA.
E-mail address: abloch@umich.edu

Department of Applied Mathematics and Theoretical Physics, Centre for Mathematical Sciences, University of Cambridge, Wilberforce Road, Cambridge CB3 0WA, England.
E-mail address: A.Iserles@damtp.cam.ac.uk.

Centre de Recherches Mathématiques
CRM Proceedings and Lecture Notes
Volume **39**, 2005

Eulerian and Semi-Lagrangian Schemes Based on Commutator-Free Exponential Integrators

E. Celledoni

Abstract. We consider commutator-free exponential integrators combined with splitting and applied to convection diffusion problems semidiscretized in space. Numerical results show improved accuracy of the new approach compared to other methods. Numerical dispersion is however experienced due to the classical Eulerian formulation of the problem. A semi-Lagrangian approach is proposed to take care of this drawback.

1. Introduction

Commutator-free exponential integrators were proposed in [**3**]. These are a new class of Lie group integrators which do not require the use of commutators for achieving numerical approximations of order higher then two. In [**3**] the good performance of these methods in dealing with stiff problems has been substantiated by numerical tests. Krogstad in [**11**] shows the relationship between the commutator-free exponential integrators and the methods developed by Cox and Matthews in [**5**], and by Kassam and Trefethen in [**7**], and the good properties of the commutator-free methods in dealing with stiffness is further investigated and explained. Another class of exponential integrators for large stiff systems of differential equations was introduced by Hochbruck and Lubich [**9**], and Hochbruck Lubich and Selhofer [**10**]. These methods are also 'commutator-free', but their relationship with the classes of methods mentioned above remains to be clarified.

In this paper we focus on the use of commutator-free exponential integrators on convection diffusion problems of the type

$$\frac{\partial}{\partial t}u(\mathbf{x},t) + \mathbf{V}\cdot\nabla u(\mathbf{x},t) = \Gamma\nabla^2 u + f(\mathbf{x}), \tag{1.1}$$

with $\mathbf{x}\in\Omega\subset\mathbb{R}^d$ and $\mathbf{V}\colon\mathbb{R}^d\times[0,T]\to\mathbb{R}$ is a given vector field, $u\colon\mathbb{R}^d\times[0,T]\to\mathbb{R}^d$, and $u(\mathbf{x},0)=u_0(\mathbf{x})$. We can distinguish between two important special cases of the above equation. The first case arises when u is a conserved passive scalar,

2000 *Mathematics Subject Classification.* Primary: 65L06, 76R99; Secondary: 65N35, 65N30.

The author is grateful to Einar Rønquist, Robert McLachlan and Brynjulf Owren for useful discussions.

This is the final form of the paper.

$f(x) = 0$, u can represent temperature, or salt concentration in the water, $\mathbf{V}$ is known a priori. The second case is given by the Navier–Stokes equations where u is a vector field, $\mathbf{V} = u$, and $f(x) = -\nabla p$, with the constraint $\nabla \cdot u = 0$.

The source of stiffness in these problems is due to the discretization of the diffusion. If we semidiscretize (1.1) in space with a finite element or finite difference method we obtain a system of ordinary differential equations of the type

$$By_t - Cy = Ay, \quad y(0) = y_0.$$

Here C is the discretized convection operator, which in some cases is a skew-symmetric matrix, A corresponds to the linear diffusion term, often symmetric and negative definite, and B is a mass matrix, which we assume symmetric and positive definite and we get

$$y_t - B^{-1}Cy = B^{-1}Ay, \quad y(0) = y_0. \tag{1.2}$$

Note that if C is skew symmetric then $B^{-1}C$ is skew-symmetric with respect to the inner product $\langle \cdot, B\cdot \rangle$, where $\langle \cdot, \cdot \rangle$ is the Euclidean inner product. In general the discretized convection operator can be nonlinear, $C = C(y)$. A typical approach for solving numerically the semidiscretized equations is to treat convection and diffusion separately, the diffusion with an implicit approach and the convection with an explicit integrator possibly of higher order, see for example [**1, 2**]. We here will follow the approach proposed in [**13**] by Maday, Patera and Rønquist. In this paper the authors propose a method called Operator Integrating Factor (OIF), based on a splitting technique, this method coincides with other more classical Integrating Factor approaches [**14**] in the linear case. Let us consider the following change of variables

$$y(t) = W(t) \cdot z(t),$$

with

$$W_t = B^{-1}CW, \quad W(0) = I. \tag{1.3}$$

After differentiation we obtain the following equation for z

$$z_t = W^{-1}B^{-1}AWz, \quad z(0) = y_0. \tag{1.4}$$

The right-hand side of this new equation is therefore obtained as a similarity transformation of a semidiscretized pure diffusion problem.

The numerical solution of the transformed equation, (1.4), is then carried out and the numerical approximation is transformed back by applying $W(t)$. The matrix solution W of (1.3) is never computed explicitly, but rather one computes directly the result of applying $W(t)$ to some vector g, approximating then a semidiscretized pure convection problem of the type

$$w' = B^{-1}Cw, \quad w(0) = g. \tag{1.5}$$

In the case C is not depending on y and $\mathbf{V}$ is known and constant in time then $W = \exp(tB^{-1}C)$. Even when $\mathbf{V}$ is time dependent the solution of (1.3) is decoupled from the solution of (1.4). In the nonlinear case the two equations must instead be solved coupled.

One advantage of using such a change of variables can be seen in the following example. Suppose we apply a discretization method, e.g. the implicit Euler scheme directly to (1.2), where we assume for simplicity that both diffusion and convection are linear, then we will need to solve at each time step a non-symmetric linear system of equations of the type

$$\big(I - hB^{-1}(A + C)\big)y_1^{IE} = y_0,$$

while the same method applied to the reformulated equation (1.4) gives the linear system

$$(I - he^{-hB^{-1}C}B^{-1}Ae^{hB^{-1}C})z_1^{IE} = z_0.$$

When B is the identity matrix, C is skew-symmetric, and A is symmetric, this system is symmetric and positive definite. It is well known that there is a computational advantage in solving symmetric positive definite systems compared to general ones in terms of iterative techniques available.

In this paper we will restrict our attention to the linear case although generalizations to the case of nonlinear convection are possible.

We will here combine the approach of [**13**] with the use of commutator-free exponential integrators for PDEs as proposed in [**3**]. In [**13**] the equation (1.4) for z is solved implicitly, while we propose to treat this equation explicitly with the commutator-free methods. Our approach is competitive with the methods described in [**13**] and based on a *backward difference formula* for the implicit discretization of (1.4), or their variant obtained by using implicit Runge–Kutta integration of (1.4) using a Radau I A method. The classical implicit methods require typically the solution of symmetric and positive definite algebraic systems of equations involving the matrix $(I - hB^{-1}A)$ while the commutator-free methods require the computation of exponentials of the type $\exp(hB^{-1}A)v$, where h is the time step used for the integration. It is well known that using Krylov subspace methods for solving each of the two linear algebra problems results in a faster (superlinear) convergence in the case of the exponential. This at least in the case the linear system is not successfully preconditioned. We refer to [**4, 9**] and references therein for precise convergence estimates for Krylov subspace methods in the two cases.

Finally we propose a semi-Lagrangian version of the same method which shows higher performance in dealing with numerical dispersion. These methods arise naturally from the considered splitting, and are obtained by substituting the operator $W(t)$ in (1.4), with its approximation $\widetilde{W} = W + E$ obtained by computing the numerical solution of the undiscretized pure convection problem corresponding to (1.5). This is done directly by computing the characteristics originating in the grid points of the discretization.

2. Commutator-free methods

Consider the following differential equation

$$y' = F(y), \quad y_0 \in \mathbb{R}^n,$$

where F is a vector field on $\mathbb{R}^n$.

Assume there exists a set of frame vector fields,

$$E_1, E_2, \ldots, E_d,$$

such that $\forall y \in \mathbb{R}^n$

$$F(y) = \sum_{i=1}^{d} f_i(y)E_i(y),$$

and $f_i\colon \mathbb{R}^n \to \mathbb{R}$. In the commutator-free exponential integrators we consider linear combinations of vector fields of the type

$$F_p(y) = \sum_{i=1}^{d} f_i(p)E_i(y),$$

and $F_p \in \operatorname{span}\{E_1, \ldots, E_d\}$, and we say that F_p is the vector field F *frozen* at the point p.

For a given vector field F we denote with $\exp(hF)p$ the solution of the differential equation

$$y' = F(y), \quad y(0) = p,$$

on the interval $[0, h]$.

A commutator-free exponential integrator for the considered equation has then the following format:

Commutator-free method.

for $r = 1 : s$ do

$$Y_r = \exp\big(\textstyle\sum_k \alpha_{rJ}^k F_k\big) \cdots \exp\big(\sum_k \alpha_{r1}^k F_k\big)(p)$$

$$F_r = hF_{Y_r} = h\textstyle\sum_i f_i(Y_r)E_i$$

end

$y_1 = \exp\big(\sum_k \beta_J^k F_k\big) \cdots \exp\big(\sum_k \beta_1^k F_k\big)p.$

The integrator has s stages and parameters α_{rJ}^k, β_J^k. Each new stage value is obtained as a composition of J exponentials of linear combinations of vector fields frozen at the previously computed stage values.

In the following table we report the coefficients of a method of order 3 and a method of order 4. The method of order 3 requires the computation of one exponential of each internal stage value and the composition of two exponentials for updating the solution. In the order 4 method the first three stage values require one exponential each, while the fourth stage and the solution update require two exponentials.

$$\begin{array}{c|ccc}
0 & & & \\
1/3 & 1/3 & & \\
2/3 & 0 & 2/3 & \\
\hline
 & 1/3 & 0 & 0 \\
 & -1/12 & 0 & 3/4
\end{array}
\qquad
\begin{array}{c|cccc}
0 & & & & \\
1/2 & 1/2 & & & \\
1/2 & 0 & 1/2 & & \\
1 & 1/2 & 0 & 0 & \\
 & -1/2 & 0 & 1 & \\
\hline
 & 1/4 & 1/6 & 1/6 & -1/12 \\
 & -1/12 & 1/6 & 16 & 1/4
\end{array}$$

We want to apply these methods to (1.4), and then obtain numerical approximations for (1.2) by using the transformation $y_n = W(t_n)z_n$. Assuming linear convection constant in time, this means $W(t) = \exp(tB^{-1}C)z_n$ and $W(t)^{-1} = \exp(-tB^{-1}C)z_n$. The considered vector field is then

$$F(z) = W(t)^{-1}B^{-1}AW(t)z,$$

which we rewrite as

$$F(z) = W(t)^{-1}\tilde{A}W(t)z,$$

where $\tilde{A} = B^{-1}A$.

Instead of applying the methods in a straightforward manner we consider here a new *ad hoc* and computationally efficient format for the methods. This format exploits the linearity of the right-hand side of equation (1.4), and reduces the number of times one has to apply $W(t)$.

The vector fields frozen at the stage values are

$$F_{Z_r}(t, z) = W(t_0 + c_r h)^{-1}\tilde{A}W(t_0 + c_r h).$$

The exponential of any such vector field is the composition of three operators and it is of the type,

$$\exp(\alpha h F_{Z_r}) = W(t_0 + c_r h)^{-1}\exp(\alpha h\tilde{A})W(t_0 + c_r h).$$

The stage values are not directly involved in the computation of the frozen vector fields and for the method of order three for example we have

$$\begin{aligned}
F_1 &= hW(t_0)^{-1}\tilde{A}W(t_0),\\
F_2 &= hW\left(\left(t_0 + \frac{h}{3}\right)\right)^{-1}\tilde{A}W\left(t_0 + \frac{h}{3}\right),\\
F_3 &= hW\left(t_0 + \frac{2}{3}h\right)^{-1}\tilde{A}W\left(t_0 + \frac{2}{3}h\right).
\end{aligned}$$

The advancement of the solution is obtained by the composition of two exponentials, the first exponential corresponds to the first of the two rows of the β values and applied to the initial value gives the second stage Z_2,

$$Z_2 = W(t_0)^{-1}\exp\left(\frac{h}{3}\tilde{A}\right)W(t_0)\cdot p,$$

while the second exponential, corresponding to the second row of β values, arises from the linear combination of two frozen vector fields,

$$\begin{aligned}
-\frac{h}{12}F_1 + \frac{3}{4}hF_3 &= -\frac{h}{12}W(t_0)^{-1}\tilde{A}W(t_0) + \frac{3}{4}hW\left(t_0 + \frac{2}{3}h\right)^{-1}\tilde{A}W\left(t_0 + \frac{2}{3}h\right)\\
&= W(t_0)^{-1}\left(-\frac{h}{12}\tilde{A} + \frac{3}{4}hW\left(\frac{2}{3}h\right)^{-1}\tilde{A}W\left(\frac{2}{3}h\right)\right)W(t_0).
\end{aligned}$$

The numerical solution at $t_1 = t_0 + h$ is then

$$z_1 = \exp\left(-\frac{h}{12}F_1 + \frac{3}{4}hF_3\right)\cdot Z_2$$

and the corresponding numerical solution y_1 for the equation (1.1) is $y_1 = W(t_1)z_1$, hence

$$y_1 = W(h)\exp\left(-\frac{h}{12}\tilde{A} + \frac{3}{4}hW\left(\frac{2}{3}h)^{-1}\tilde{A}W\left(\frac{2}{3}h\right)\right)\cdot Y_2,$$

where $Y_2 = \exp(h/3\tilde{A})W(t_0)y_0$.

In summary we get to the following format for the implementation of the commutator-free third order method for the numerical solution of (1.2).

Order 3 CF.

$p = y_0$
for $n = 0 : N-1$ do
  $Y = \exp(\frac{h}{3}\tilde{A})p$
  $w_{n+1} = \exp\Big(-\frac{h}{12}\tilde{A} + \frac{3}{4}hW\big(\frac{2}{3}h\big)^{-1}\tilde{A}W\big(\frac{2}{3}h\big)\Big)\cdot Y$
  $y_{n+1} = W(h)w_{n+1}$
  $p = y_{n+1}$
end.

By analogous considerations we arrive to a similar expression for the implementation of the fourth order method.

Order 4 CF.

$p = y_0$
for $n = 0 : N-1$ do
  $Z = W(h)p$
  $Y = \exp\Big(\frac{h}{4}W(h)\tilde{A}W(h)^{-1} + \frac{h}{3}W\big(\frac{h}{2}\big)\tilde{A}W\big(\frac{h}{2}\big)^{-1} - \frac{h}{12}\tilde{A}\Big)\cdot Z$
  $y_{n+1} = \exp\Big(-\frac{h}{12}W(h)\tilde{A}W(h)^{-1} + \frac{h}{3}W\big(\frac{h}{2}\big)\tilde{A}W\big(\frac{h}{2}\big)^{1-} + \frac{h}{4}\tilde{A}\Big)\cdot Y$
  $p = y_{n+1}$
end.

The main cost of these methods is due to the computation of the exponentials giving w_{n+1} in the order 3 method, and computing Y and y_{n+1} in the order 4 method. We assume these exponentials are computed via Krylov subspace techniques which require only the repeated application of the operator to be exponentiated to a suitable vector.

As comparison methods we consider two *backward difference formulas* (BDF) of order 3 and 4 respectively. These methods are implicit as opposed to the methods described above, but specialized to our linear problem they require the inversion of just one linear system of equations per step. The BDF methods are multistep methods and as such they require the preliminary computation of the starting values $y_0 \approx y(0), \dots, y_k \approx y(kh)$, with $k-1$ the order of the BDF method, this is usually done by using a one step method. In our case we used the values of the exact solution as starting values for the BDF methods.

The BDF3 as applied to (1.4) gives rise to the following difference equation

$$W(t_n)^{-1}\Big(I - \frac{6}{11}\tilde{A}\Big)W(t_n)z_n = \frac{18}{11}z_{n-1} - \frac{9}{11}z_{n-2} + \frac{2}{11}z_{n-3},$$

and then transforming the equation in the y variables we get

$$\Big(I - \frac{6}{11}\tilde{A}\Big)y_n = \frac{18}{11}W(h)y_{n-1} - \frac{9}{11}W(2h)y_{n-2} + \frac{2}{11}W(3h)y_{n-3}.$$

The BDF4 has instead the following format

$$\left(I - \frac{12}{25}h\tilde{A}\right)y_n = b,$$

$$b = \frac{48}{25}W(h)y_{n-1} - \frac{36}{25}W(2h)y_{n-2} + \frac{16}{25}W(3h)y_{n-3} - \frac{3}{25}W(4h)y_{n-4}.$$

Since the CF methods are one step methods we consider for comparison also a Runge–Kutta method of order 3, the Radau I A method which is A-stable and implicit, ([**8**, p. 73]). By exploiting the linearity of the vector field this method is implemented as follows.

Radau I A, order 3.

$p = y_0$
for $n = 0 : N-1$ do
  $A_n = W(t_n)^{-1}\tilde{A}W(t_n)$
  $A_{\frac{2}{3}} = W(\frac{2}{3}h)^{-1}\tilde{A}W(\frac{2}{3})$
  $\left(I - h\left(\frac{5}{12}I - \frac{h}{16}A_n\left(I - \frac{h}{4}A_n\right)^{-1}\right)A_{2/3}\right)Z = \left(I - \frac{h}{4}A_n\right)^{-1}p$
  $R = \left(I - \frac{h}{4}A_n\right)^{-1}\left(p - \frac{h}{4}A_{2/3}Z\right)$
  $z_{n+1} = p + h\left(\frac{1}{4}A_nR + \frac{3}{4}A_{2/3}Z\right)$
  $p = z_{n+1}$
  $y_{n+1} = W(t_n)z_{n+1}$
end.

The operators A_n and $A_{2/3}$ are never computed explicitly, but just applied to suitable vectors while solving the linear systems of algebraic equations in the algorithm via iterative methods. These operators depend on A and $W(t)$ like in the CF case.

The results of the numerical experiments reported in Figs. 1, 2, 3 show the norm of the global error in the semidiscretized problem (1.2), for the methods, plotted against the stepsize in a logarithmic scale. The experiments have been repeated with different magnitudes of the viscosity $\Gamma = 0.1, 0.01, 0.001$ in (1.1), allowing for different regimes of stiffness in the problem.

We have considered the equation (1.1) in one space dimension, $x \in [-1, 1]$, $u(x, 0) = \operatorname{sech}(20(x - x_0)\pi)$, $(x_0 = 0.5)$, $\mathbf{V} = 1$. The discretization is carried out with a spectral element method with 10 elements and polynomials of degree 4. The interpolation and quadrature nodes are the Gauss–Lobatto Legendre nodes, i.e. the zeros of the polynomial

$$(1 - x^2)\frac{dL_p}{dx}(x) = 0,$$

where $L_p(x)$ is the Legendre polynomial of degree p.

The operator $W(t) \cdot g = \exp(-tB^{-1}C) \cdot g$, and it is computed to machine precision by the built in function of `Matlab`, *expm*. The exponentials involving the matrix $\tilde{A}$ in the commutator-free methods are computed iteratively using Krylov subspace methods. Also the linear systems of equations for the BDF methods and the Radau method, are solved numerically via iterative Krylov subspace techniques.

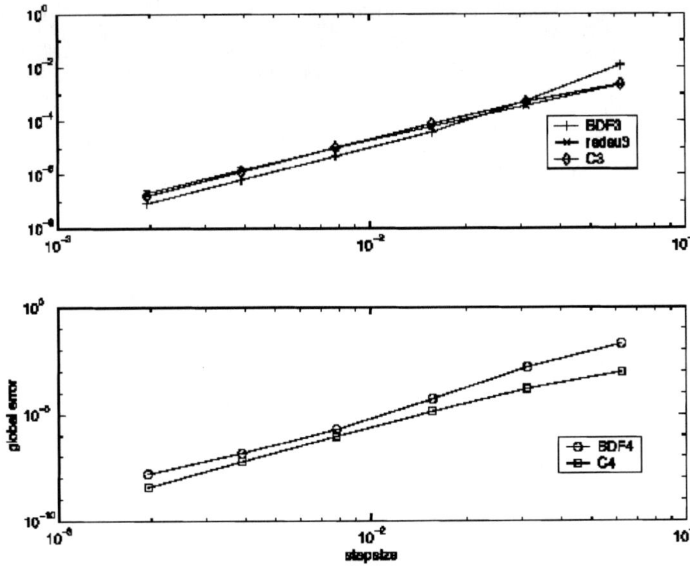

FIGURE 1. Global error versus stepsize for different methods applied to the semidiscretized convection diffusion linear equation: BDF, Commutator-free, Radau I A, $\Gamma = 0.1$.

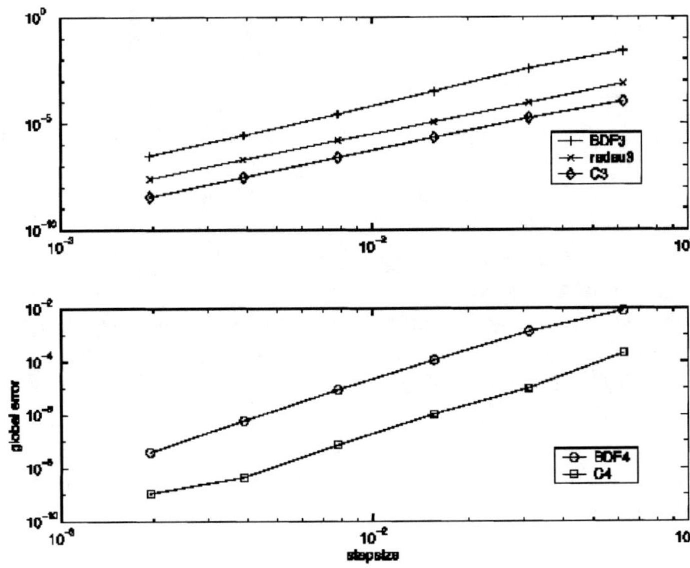

FIGURE 2. Global error versus stepsize for different methods applied to the semidiscretized convection diffusion linear equation: BDF, Commutator-free, Radau I A, $\Gamma = 0.01$.

The interval of integration is $[0, 1]$. The error is the Euclidean norm of the difference of the approximation and the exact solution of the semi-discretized problem.

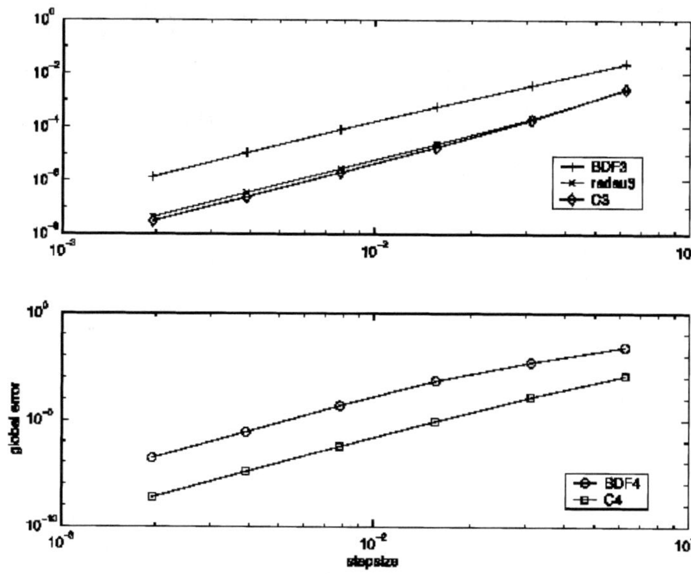

FIGURE 3. Global error versus stepsize for different methods applied to the semidiscretized convection diffusion linear equation: BDF, Commutator-free, Radau I A, $\Gamma = 0.001$.

In Fig. 1 we report the result of the experiments in the case of $\Gamma = 0.1$, all the methods perform similarly in this case, with a small advantage of the BDF methods. When $\Gamma = 0.01$, the CF methods achieve lower global error for the same size of the time step compared to BDF and Radau. Both Radau and CF, which are based on Runge–Kutta formulae, do better then the BDF methods for $\Gamma = 0.01$ and $\Gamma = 0.001$. As Γ decreased, the performance of the Radau method approaches that of the CF method.

In Fig. 4 we plot the values of the global error against the number of flops required for the different methods in our implementation. The commutator-free methods perform competitively with respect to the BDF methods, but we do not exclude improvements in the performance of the BDF methods if a good and cheap preconditioner is available for the solution of the linear systems.

3. Semi-Lagrangian version of the algorithm

Especially for small values of Γ the quality of the solution of (1.1) is polluted by the presence of oscillations, at least when the spatial discretization is not fine enough. The exact solution of the semidiscretized equation (1.2) does not propagate all the frequencies of the solution of (1.1) at the right speed. We can illustrate the phenomenon by comparing the solution to steady state of the unsteady convection-diffusion equation,

$$u_t + u_x = \nu u_{xx} + f, \quad u(x,0) = \cos(\pi/2x), \qquad f = 1,$$

$x \in [-1,1]$, $u(t,1) = u(t,-1) = 0$, with the exact solution of its discretized version

$$U_t = B^{-1}(A+C)U + f, \quad U_i(0) = u(x_i,0).$$

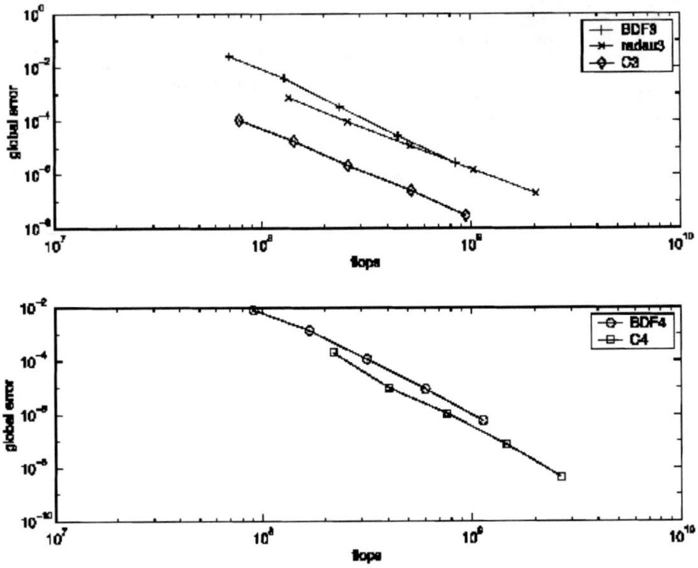

FIGURE 4. Global error versus number of flops for different methods applied to the semidiscretized convection diffusion linear equation: BDF, Commutator-free, Radau I A, $\Gamma = 0.01$

Here C is the discrete convection operator, and A is the discrete diffusion, obtained by considering a discretization in space by spectral methods on the Gauss–Lobatto Legendre nodes, and x_i is a node in the resulting grid, $i = 0, \ldots, p+1$. In the numerical experiments we considered $p = 16$ and $p = 32$.

In this case the forcing term f gives rise to a modification of the format of equation (1.4), which is now

$$z_t = W^{-1}(t)\tilde{A}W(t) + W^{-1}(t)f, \quad z(0) = y_0,$$

the corresponding frozen vector fields are then identified by the couples

$$F_{Z_r}(t, z) = \left(W(t_0 + c_r h)^{-1}\tilde{A}W(t_0 + c_r h), W(t_0 + c_r h)^{-1}f\right),$$

and their exponentials are the solutions of the differential equations

$$v_t = W(t_0 + c_r h)^{-1}\tilde{A}W(t_0 + c_r h) + W(t_0 + c_r h)^{-1}f,$$

with suitable initial values.

As we can note in Fig. 5 and Fig. 6, the oscillating exact solution of the discretized problem (dashed line) does not resemble qualitatively the features of the exact solution of the PDE especially when ν is small. Similar results for this example have been reported in [**6**].

We here consider semi-Lagrangian schemes based on the direct approximation of the pure convection problems (1.5) previously approximated by the exponentials $W(t)g = \exp(-tB^{-1}C)g$. We substitute the operator W with a perturbed operator $\widetilde{W} = W + E$ whose construction is discussed here below. This introduces an error in the solution of the discretized equation (1.2), but gives improved performance in the approximation of the PDE problem.

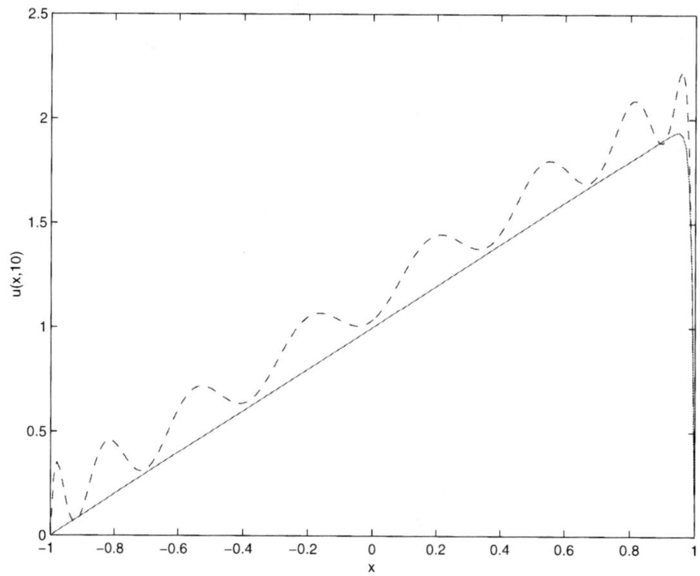

FIGURE 5. Exact solution of the semi-discretized problem (dashed line) and exact solution (solid line) at $t = 10$, $\nu = 0.01$, $p = 16$.

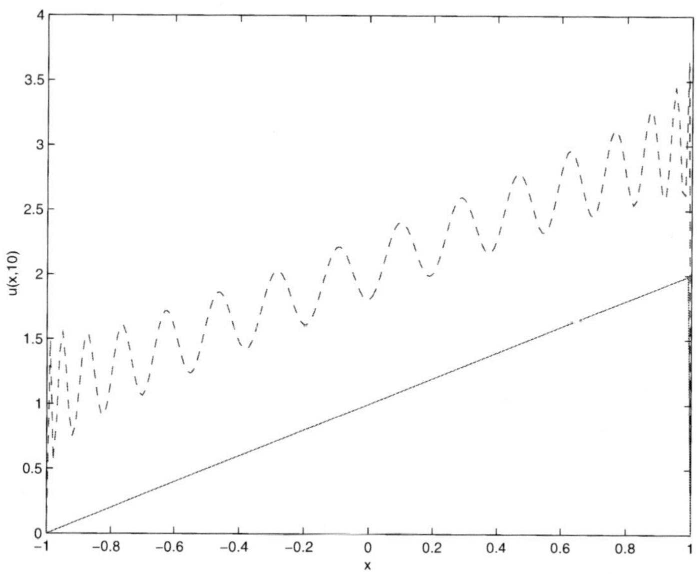

FIGURE 6. Exact solution of the semi-discretized problem (dashed line) and exact solution (solid line) at $t = 10$, $\nu = 0.001$, $p = 32$.

We consider functions $w(x,t)$ such that $w_t + w_x = 0$, $w(x,0) = w_0(x)$ and we assume $w_0(x_i) = g_i$, the boundary conditions are taken according to the boundary conditions of the original problem, we here consider $w(-1,t) = w(1,t) = 0$. We then have $w(x,t) = w(x-t,0)$ and in particular on the nodes of the discretization $w(x_i,t) = w(x_i-t,0)$. We assume $w \approx w_p(x,t) = \sum_{j=0}^{p} w_{p,j}(t)l_j(x)$, where $l_j(x)$ are the Lagrange basis functions defined on the nodes of the discretization, and $w_{p,j}(t)$ are the nodal values of the approximation at time t, i.e. $w_{p,j}(t) = w_p(x_j,t)$. Since $w(x,t) = w(x-t,0)$ we compute

$$w_{p,i}(t) = w_p(x_i - t, 0) = \sum_{j=0}^{p} w_{p,j}(0) l_j(x_i - t),$$

in other words we obtain the values of the approximation w_p at time t on the nodes of the grid, by interpolating the known values of the initial function on the nodes. If $\mathbf{g}(t)$ is a vector whose $p+1$ components are the values of $w_p(x,t)$ on the nodes then

$$\mathbf{g}(t) = \widetilde{W}(t)\mathbf{g}(0), \quad \widetilde{W}_{i,j}(t) = l_j(x_i - t),$$

with $\widetilde{W}_{i,j}(t)$ the (i,j) entry of the operator $\widetilde{W}$ written in a matrix form. The cost of evaluating each entry of the operator $\widetilde{W}$ is $\mathcal{O}(p)$, giving an overall cost of $\mathcal{O}(p^3)$. This cost should be compared with the cost of computing $\exp(tB^{-1}C)\mathbf{g}(0)$ which is $\mathcal{O}(p^2)$ if we use iterative methods for the approximation of the matrix exponential and we have fast convergence. The qualitative performance of the semi-Lagrangian methods is however clearly superior as can be seen in Figs. 7 and 8.

For more general convection diffusion problems in several space dimensions and with time dependent convecting vector fields $\mathbf{V}$, the considered pure convection problems are of the type $u_t + \mathbf{V}\cdot\nabla u = 0$ and we have that

$$\widetilde{W}_{i,j}(t) = l_j\big(X_i(t)\big),$$

where $X_i(t)$ is the characteristic originating in the grid point x_i, and satisfies the ordinary differential equation

$$\dot{X}_i = V\big(X_i(t)\big), \quad X_i(0) = x_i.$$

We report in Figs. 7 and 8 the results of applying the semi-Lagrangian version of the algorithm CF3 to the convection diffusion equation given at the beginning of this section. We also computed the error as the max norm of the difference between numerical approximation and the exact solution of the original undiscretized problem on the nodes of the discretization. This error is big and dominated by the error of the spatial discretization for the Eulerian version of the methods, while it is much smaller in the semi-Lagrangian case.

In this paper we have considered commutator-free exponential integrators for convection diffusion problems. We have applied the methods to a reformulated semidiscretized system of ordinary differential equations defined by a symmetric negative semi-definite operator. This problem is obtained after a change of variables which separates in a splitting fashion the semidiscrete convection operator from the semidiscrete diffusion operator. Exponential integrators are well suited for this kind of problems, because the computation of the matrix exponential using Krylov subspace methods is working well in the case of symmetric negative semi-definite problems. The comparison of the CF methods with BDF of order 3 and 4 and Radau I A of order 3, shows the competitiveness of the proposed approach. Improved performance has been obtained by considering a semi-Lagrangian version

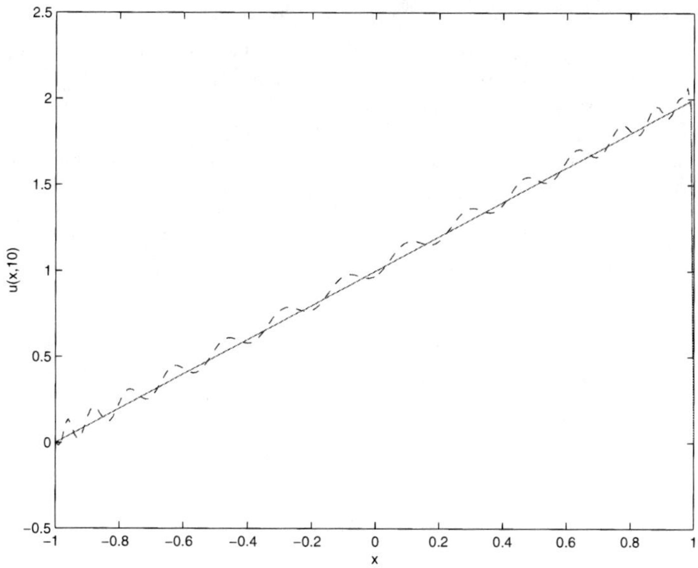

FIGURE 7. Numerical (dashed line) and exact (solid line) solution at $t = 10$, $\nu = 0.01$, $p = 16$.

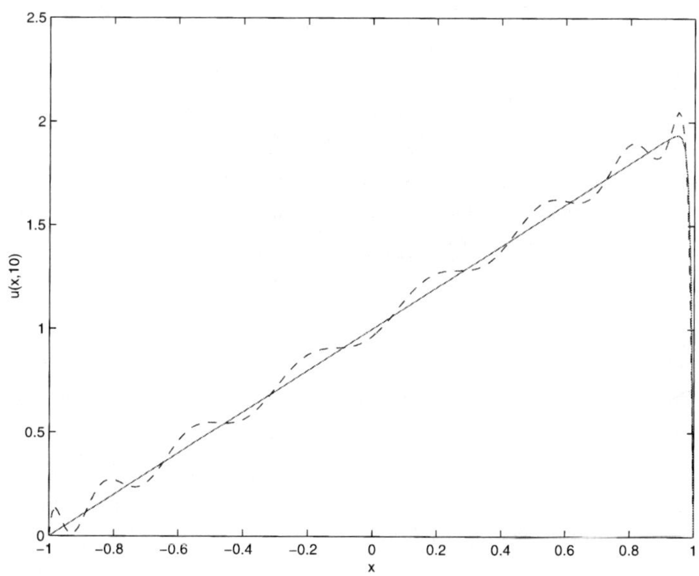

FIGURE 8. Numerical (dashed line) and exact (solid line) solution at $t = 10$, $\nu = 0.001$, $p = 32$.

of the methods arising naturally form the type of splitting considered. The numerical experiments are limited to some simple test problems in one space dimension and the methods will be further tested in more challenging applications.

References

1. U. M. Ascher, S. J. Ruuth, and B. T. R. Wetton, *Implicit-explicit methods for time-dependent partial differential equations*, SIAM J. Numer. Anal. **32** (1995), 797–823.
2. C. Canuto, M. Y. Hussaini, A. Quarteroni, and T. A. Zang, *Spectral methods in fluid dynamics*, Springer Ser. Comput. Phys., Springer-Verlag, New York, 1988.
3. E. Celledoni, A. Marthinsen, and B. Owren, *Commutator-free Lie group methods*, FCGS **19** (2003), 341–352.
4. E. Celledoni and I. Moret, *A Krylov projection method for systems of* ODE*s*, Volterra centennial (Tempe, AZ, 1996), Appl. Numer. Math. **24** (1997), 365–378.
5. S. M. Cox and P. C. Matthews, *Exponential time differencing for stiff systems*, J. Comput. Phys. **176** (2002), 430–455.
6. P. F. Fischer, G. W. Kruse, and F. Loth, *Spectral element method for transitional flows in complex geometries*, Proceedings of the Fifth International Conference on Spectral and High Order Methods (ICOSAHOM-01) (Uppsala), J. Sci. Comput., vol. 17, 2002, no. 1-4, pp. 81–98.
7. A. K. Kassam and L. N. Trefethen, *Fourth-order time stepping for stiff* PDE*s*, SIAM J. Sci. Comput., 2002 (submitted).
8. E. Hairer and G. Wanner, *Solving ordinary differential equations.* II. *Stiff and differential-algebraic problems*, 2nd ed., Springer Ser. Comput. Math., vol. 14. Springer-Verlag, Berlin, 1996.
9. M. Hochbruck and C. Lubich, *On Krylov subspace approximations of the matrix exponential operator*, SIAM J. Numer. Anal. **34** (1997), 1911–1925.
10. M. Hochbruck, C. Lubich, and H. Selhofer, *Exponential integrators for large systems of differential equations*, SIAM J. Sci. Comput. **19** (1998), 1552–1574.
11. S. Krogstad, *Generalized integrating factor methods for stiff* PDE*s*, J. Comput. Phys., 2003 (submitted).
12. S. Krogstad, RKMK-*related methods for stiff nonlinear* PDE*s*, Innovative Time Integrators for PDEs proceedings, Appl. Numer. Math., 2002, (submitted).
13. Y. Maday, A. T. Patera, and E. M. Rønquist, *An operator integration factor splitting method for time dependent problems: Application to incompressible fluid flows*, J. Sci. Comput. **5** (1990), 263–292.
14. L. N. Trefethen, *Spectral methods in* `Matlab`, Software, Environ. Tools, SIAM, Philadelphia, PA, 2000.

Centre for Advanced Study, Drammensveien 78, Oslo, Norway.
Current address: SINTEF Applied Math., Sem Sælandsvei 5, 7465 Trondheim, Norway.
E-mail address: `elenac@math.ntnu.no`

Centre de Recherches Mathématiques
CRM Proceedings and Lecture Notes
Volume **39**, 2005

Second Order Linear ODEs: Two Non-Liouvillian Approaches

E. S. Cheb-Terrab

ABSTRACT. There exists sound literature and algorithms for computing Liouvillian solutions for linear ODEs with rational coefficients. On the other hand, Liouvillian solutions are not easily computable when the ODE has non-rational coefficients. Also, special function solutions which are not Liouvillian appear frequently in mathematical physics problems but there are not so general algorithms for computing these solutions even in the rational coefficients case. This paper summarizes two related set of algorithms recently developed and implemented as computer algebra libraries in the Maple system. These libraries are at the base of Maple's ability for tackling this type of ODE problem.

Introduction

Given a second order linear ODE with rational coefficients $A(x)$, $B(x)$,

$$y'' + A(x)y' + B(x)y = 0 \tag{0.1}$$

the problem under consideration is that of systematically computing solutions for it even when the solution admits no Liouvillian form[1]. We are also interested in identifying ODE families where $A(x)$ and $B(x)$ are not rational in x and solutions, Liouvillian or not, can be systematically computed.

With that in mind the first thing to note is that non-Liouvillian solutions which can be represented symbolically (not as infinite sums) are representable using special functions, e.g. Bessel, Hermite or Legendre functions [**8**]. In turn these and most of the special functions frequently appearing in mathematical physics problems happen to be particular cases of the ${}_pF_q$ hypergeometric function for p equal to 0, 1 or 2 and q equal to 0 or 1 (see [**1**]). For example, the Bessel functions can be expressed in terms of ${}_0F_1$, all cylindrical functions as well as the Hermite, Laguerre, Whittaker and error family of functions can be expressed in terms of ${}_1F_1$ and all

2000 *Mathematics Subject Classification.* Primary: 34A05; Secondary: 68W30.
This work was developed in the framework of the MITACS project and supported by the Centre of Experimental and Constructive Mathematics, Simon Fraser University, Canada.
This is the final form of the paper.

[1]Functions that can be expressed in terms of exponentials, integrals, and algebraic functions, are called Liouvillian functions. The typical example is $\exp\left(\int R(x)dx\right)$ where $R(x)$ is a rational or an algebraic function representing the roots of a polynomial.

Chebyshev, Gegenbauer, Jacobi, Legendre and some others can be expressed in terms of ${}_2F_1$.

For our problem, one natural approach is then to directly attempt the computation of hypergeometric function solutions of these ${}_0F_1$, ${}_1F_1$ and ${}_2F_1$ types, since in this way we cover at once solutions involving all the related special functions. This approach was developed in collaboration with L. Chan [**2**] and is formulated in Section 1 as an *equivalence approach* to the ${}_pF_q$ differential equations.

Another approach for this problem, exploring a different idea, consists of studying the general form of the symmetries of first order non-linear Riccati type equations, with arbitrary coefficients $a(x), b(x), c(x)$

$$y' + a(x)y^2 + b(x)y + c(x) = 0. \tag{0.2}$$

Linear equations of the form (0.1) can systematically be mapped into Riccati ones. The key observation here is that, after imposing some restrictions on $\{a(x), b(x), c(x)\}$, there are varied symmetry groups whose infinitesimals can be systematically computed when (0.2) belongs to the corresponding symmetry class. This fact leads to solutions (Liouvillian or not) to (0.1), for varied families where the coefficients $A(x)$, $B(x)$ are not rational. This approach was developed in collaboration with T. Kolokolnikov [**3**] and is discussed in Section 2.

1. Computing ${}_2F_1$, ${}_1F_1$ and ${}_0F_1$ hypergeometric solutions

To compute ${}_pF_q$ solutions to (0.1) the idea is to formulate an equivalence approach to the ${}_2F_1$, ${}_1F_1$ and ${}_0F_1$ underlying hypergeometric differential equations, that is, to determine whether a given linear ODE can be obtained from one of these ${}_pF_q$ ODEs

$$\begin{aligned} (x^2-x)y'' + \big((a+b+1)x - c\big)y' + bay &= 0, \text{ admits } {}_2F_1 \text{ solutions},\\ xy'' + (c-x)y' - ay &= 0, \text{ admits } {}_1F_1 \text{ solutions},\\ xy'' + cy' - y &= 0, \text{ admits } {}_0F_1 \text{ solutions} \end{aligned} \tag{1.1}$$

where $\{a, b, c\}$ are arbitrary constants, by means of a transformation of a certain type. If so, the solution to the given linear ODE is obtained by applying the same transformation to the solution of the corresponding ${}_pF_q$ ODE above.

The first ODE in (1.1) has 3 regular singularities at 0, 1 and ∞. The second ODE in (1.1), also known as the confluent hypergeometric equation; has a regular singularity at 0 and an irregular one at ∞. The third order also has one regular and one irregular singularity but we considered the case separately in order to automatically obtain solutions expressed in terms of simpler (Bessel) functions. As we shall see, the structure of the singularities of these equations is a key for resolving the equivalence.

The approach of course also requires determining the values of the hypergeometric parameters a, b, c for which the equivalence exists, and it is clear that its chances of success depend crucially on how general is the class of transformations being considered. For instance, one can verify that for linear transformations[2] with arbitrary $F(x)$, $P(x)$

$$x \rightarrow F(x), \quad y \rightarrow P(x)y \tag{1.2}$$

[2]The problem of equivalence under transformations $\{x \rightarrow F(x), y(x) \rightarrow P(x)y + Q(x)\}$ for linear ODEs can always be mapped into one with $Q(x) = 0$, see [**5**].

the problem is too general in that to solve it requires solving first the given ODE, turning the approach of no practical use.

The transformations considered in this work are of the form

$$x \longrightarrow \frac{\alpha x^k + \beta}{\gamma x^k + \delta}, \quad y \rightarrow P(x)y \tag{1.3}$$

with $P(x)$ arbitrary and $\{\alpha, \beta, \gamma, \delta, k\}$ constants with respect to x. These transformations, which do not conform a class in the strict sense, can obtained by sequentially composing three different transformations each of which do constitute a class. The sequence starts with linear fractional—also called Möbius—transformations

$$x \longrightarrow \frac{\alpha x + \beta}{\gamma x + \delta} \tag{1.4}$$

followed by power transformations

$$x \longrightarrow x^k \tag{1.5}$$

followed by linear homogeneous transformations of the dependent variable

$$y \rightarrow Py. \tag{1.6}$$

1.1. Transformations $y \rightarrow P(x)y$ of the dependent variable. The first thing to note at this point is that transformations of the form (1.6) can easily be factored out of the problem: if two equations of the form (0.1), respectively with coefficients $\{A(x), B(x)\}$ and $\{C(x), D(x)\}$, can be obtained from each other by means of (1.6), the transformation relating them is computable directly from these coefficients. For that purpose first rewrite both equations in normal form, for instance for (0.1) use

$$y = ue^{-\int A/2\,dx} \tag{1.7}$$

to obtain

$$u'' = \left(\frac{A'}{2} + \frac{A^2}{4} - B\right)u \tag{1.8}$$

and the transformation relating the two hypothetical ODEs, when it exists, is given by $y = ue^{\int (C-A)/2\,dx}$. In what follows we will refer to

$$I(x) = \frac{A'}{2} + \frac{A^2}{4} - B, \tag{1.9}$$

the coefficient of u in (1.8), as *the invariant* [**7**], regardless of the fact that this object is only an absolute invariant under (1.6) and not under (1.4) or (1.5).

1.2. Transformations $x \rightarrow F(x)$ of the independent variable. By changing $x \rightarrow F(x)$ in (0.1), the invariant I_1 of the changed ODE can be expressed in terms of the invariant I_0 of (0.1) by

$$I_1(x) = F'^2 I_0\big(F(x)\big) + S(F') \tag{1.10}$$

where $S(x)$ is the Schwarzian [**10**]

$$S(F') = \frac{3F''^2}{4F'^2} - \frac{F'''}{2F'}. \tag{1.11}$$

The form of $S(F')$ is particularly simple when $F(x)$ is a Möbius transformation (1.4), in which case $S(F') = 0$, and also when $F(x)$ is a power transformation; this is of relevance in formulating a strategy for resolving the equivalence.

Regarding the Möbius transformations, perhaps the most important property for our purposes is that these transformations do not change the structure of the singularities of (0.1). Möbius transformations only move the location of the poles. For example, the ${}_0F_1$ hypergeometric equation

$$xy'' + cy' - y = 0 \tag{1.12}$$

has one regular singularity at the origin and one irregular at infinity and the transformed ODE, obtained from (1.12) by means of (1.4)

$$y'' + \frac{(2\alpha\gamma x + 2\gamma\beta - \gamma c\beta + c\alpha\delta)}{(\alpha x + \beta)(\gamma x + \delta)} y' - \frac{(\alpha\delta - \gamma\beta)^2}{(\gamma x + \delta)^3(\alpha x + \beta)} y = 0 \tag{1.13}$$

has also one regular and one irregular singularity, respectively located at $-\beta/\alpha$ and $-\delta/\gamma$. In the case of the ${}_2F_1$ differential equation (see (1.1)), under (1.4) the three regular singularities move from $\{0, 1, \infty\}$ to $\{-\delta/\gamma, -\beta/\alpha, (\delta - \beta)/(\alpha - \gamma)\}$.

So, from the singularities of an ODE, not only one can tell with respect to which of the three differential equations (1.1) could the equivalence under (1.4) be resolved, but also one can extract the values of the parameters $\{\alpha, \beta, \gamma, \delta\}$ entering the transformation (1.4).

Reversing the line of reasoning, through Mobius transformations one can formulate a complete classification of singularities of the linear ODEs "equivalent" to the ${}_pF_q$ equations (1.1). This classification is based on how the invariant of each of these equations is transformed under (1.4).

After transforming the ${}_2F_1$ differential equation, the invariant of the resulting equation has the form

$$I_{{}_2F_1} = \frac{\omega_2 x^2 + 2\omega_1 x + \omega_0}{(\sigma_1 x + \sigma_2)^2(\sigma_3 x + \sigma_4)^2(\sigma_5 x + \sigma_6)^2} \tag{1.14}$$

where all $\{\omega_i, \sigma_j\}$ can be expressed in terms of the parameters $\{a, b, c\}$ entering the hypergeometric equation (see (1.1)) and the parameters $\{\alpha, \beta, \delta, \gamma\}$ entering the transformation (1.4). The invariant of the transformed ${}_1F_1$ differential equation has the form

$$I_{{}_1F_1} = \frac{\omega_2 x^2 + 2\omega_1 x + \omega_0}{(\sigma_3 x + \sigma_4)^2(\sigma_5 x + \sigma_6)^4} \tag{1.15}$$

and that of the transformed ${}_0F_1$ equation has the form

$$I_{{}_0F_1} = \frac{\omega_1 x + \omega_0}{(\sigma_3 x + \sigma_4)^2(\sigma_5 x + \sigma_6)^3}. \tag{1.16}$$

These transformed invariants are all of the form

$$I_{{}_pF_q} = \frac{\prod_{i=1}^{m}(a_i x + b_i)}{\prod_{i=1}^{n}(c_i x + d_i^{q_i}}. \tag{1.17}$$

Cancellations between factors in the numerator and denominators of (1.17) may also happen and, independent of that, some coefficients $\{a_i, c_i\}$ can be zero[3]. So the degrees with respect to x of the numerators and denominators of (1.14), (1.15) and (1.16) can be lower than the maximum implicit by these equations; in this way the problem splits into cases.

---

[3]Provided that, in (1.4), $\alpha\delta - \gamma\beta \neq 0$ and also that in (0.1) the invariant remains finite, i.e. its denominator is not zero.

TABLE 1. Classification of linear ODEs equivalent to ${}_pF_q$ ODEs under Möbius

Class	Cases	Number of cases
${}_2F_1$	$[<=2*,[2*,2*,2*]],[<=2*,[2*,2*]]$	14
${}_1F_1$	$[2*,[2*,4]],[<=2,[6]],[<=2,[4]],[2*,[2*]],[2,[0]]$	13
${}_0F_1$	$[1*,[2*,3]],[<=1,[5]],[<=1,[3]],[1*,[2*]],[1,[0]]$	9

Taking possible cancellations into account, from the structure of the invariants (1.14), (1.15) and (1.16), all the possible different cases for each of the ${}_2F_1$, ${}_1F_1$, ${}_0F_1$ classes were determined. With this classification in hands, from the knowledge of the degrees with respect to x of the numerator and denominator of the invariant (1.17) of a given ODE, one can tell whether or not it can be obtained from the ${}_2F_1$, ${}_1F_1$ or ${}_0F_1$ equations (1.1) using (1.4).

These observations can be summarized in a classification table as follows, using the symbol $[\leq p, [q_1*, q_2*, \ldots, q_n*]]$, where p is the degree in x of the numerator of (1.17) and q_i are the powers of the factors entering the denominator of it. The symbol $\leq$, when present, refers to the value of p (can be less or equal to). The symbol $*$, when present, means there can be factors canceling between numerator and denominator, so that the actual value of the related q_i can be lower (provided p is also lower by the same amount). For example,

$$[\leq 2*, [2*, 2*]] \tag{1.18}$$

represents the following possible seven different "lists of values" (herein referred as cases) for the degrees of the numerator and denominator of the invariant

$$\begin{aligned} &[2*,[2*,2*]] = [2,[2,2]],[1,[1,2]],[0,[1,1]],[0,[0,2]],\\ &[1*,[2*,2*]] = [1,[2,2]],[0,[1,2]],\\ &\quad [0,[2,2]]. \end{aligned} \tag{1.19}$$

With this notation, the classification of all the possible cases equivalent to the ${}_2F_1$, ${}_1F_1$ and ${}_0F_1$ equations under Möbius transformations is given by

1.3. Transformations $x \to x^k$ of the independent variable. Using the results of the previous subsections it is possible to resolve the equivalence of a given linear ODE (0.1) and the hypergeometric equations (1.1) under compositions of transformations (1.6) of the dependent variable $y(x)$ and Möbius transformations (1.4) of the independent variable x. In this section a worth additional level of generalization is obtained by composing those two transformations with transformations $x \to x^k$ of the independent variable.

The first thing to note regarding power transformations is that, unlike Möbius transformations, they *do not preserve* the structure of singularities. The change in the invariant due to $x \to x^k$, however, has a simple and tractable structure; the Schwarzian (1.11) is given by:

$$S(F') = \frac{k^2-1}{4x^2}. \tag{1.20}$$

So, the changed invariant I_1 shown in (1.10) is expressed in terms of I_0 by

$$x^2 I_1(x) + \tfrac{1}{4} = ((x^k)^2 I_0(x^k) + \tfrac{1}{4})k^2. \tag{1.21}$$

This naturally suggests the introduction of a "shifted" invariant $J(x)$

$$J_i(x) = x^2 I_i(x) + \tfrac{1}{4} \tag{1.22}$$

for which the transformation rule under $x \to x^k$ has the simple form

$$J_1(x) = k^2 J_0(x^k). \tag{1.23}$$

The equivalence of two linear ODEs A and B under $x \to x^k$ can then be formulated as follows: given $J_{1A}(x)$, and $J_{2A}(x)$ compute k_A, k_B entering (1.23) such that the degrees with respect to x of $J_{0A}(x)$ and $J_{0B}(x)$ are minimized. This approach is systematic in that from the knowledge of $x \to x_A^k$ and $x \to x_B^k$ respectively leading to J_{0A}, J_{0B} with minimized degrees on can tell whether or not they are related through power transformations, and if so the mapping relating them is just $x \to x^{k_A - k_B}$.

In the framework of this work, where power transformations of x are to be composed with Möbius and linear transformations of the dependent variable $y(x)$ to match the structure of singularities of the hypergeometric equations (1.1) the computation of k minimizing the degrees of J_1 in (1.23) is formulated as follows: given the set

$$A := \frac{p_i}{q_i}, \quad i = 1 \text{ to } m \tag{1.24}$$

of (possibly rational) numbers entering as powers in the numerator and denominator of J_1, compute the smallest rational number $\widetilde{k}$ such that multiplying by it each element of A, all of them become integers. Then $k = 1/\widetilde{k}$.

1.4. Summary of this `hyper3` approach. Summarizing the material discussed in the previous subsections, an itemized description of the algorithm (so called `hyper3`) to resolve the equivalence proposed in the introduction is as follows

(1) Rewrite the given equation (0.1) we want to solve, in normal form

$$y'' = I(x)y \tag{1.25}$$

where $I(x)$ is the invariant (1.9).

(2) Compute $J_1(x)$, the shifted invariant (1.9), and use transformations $x \to x^k$ to reduce to the integer minimal values the powers entering the numerator and denominator; i.e., compute $J_0(x)$.

(3) By analyzing the structure of the singularities of $J_0(x)$, use the classification of singularities developed (Table 1) to tell whether an equivalence under Möbius transformations to any of the ${}_2F_1$, ${}_1F_1$ or ${}_0F_1$ equations (1.1) is possible; if so, compute the parameters $\{\alpha, \beta, \gamma, \delta\}$ directly from the singularities of $J_0(x)$.

(4) Compose the three transformations to obtain a transformation of the form

$$x \to \frac{\alpha x^k + \beta}{\gamma x^k + \delta}, \quad y \to P(x)y$$

mapping the ${}_pF_q$ equation involved into the ODE we are solving, then apply this transformation to the known solution of the ${}_pF_q$ equation resulting in the desired ODE solution.

Example. Consider the second order linear ODE

$$y'' = \frac{((2\nu - 2\mu)x^2 - 3x^4 - 2\mu - 2\nu - 1)}{x^5 - xy'} + \frac{2\nu(\nu/2 + \mu + 1)}{x^6 - x^2 y}. \tag{1.26}$$

This equation has five regular singular points: $\{0, 1, -1, i, -i\}$. Following the steps mentioned in the Summary, we rewrite the equation in normal form and compute the value of k leading to an equation with minimal values for the powers entering the invariant. The value of k found is $k = 2$ and the resulting equation in normal form is:

$$y'' = \frac{((\mu^2 + 2\nu^2 - 2\mu - 4)x^2 + (2\mu^2 - 2\nu^2)x + 2\mu + \mu^2)}{4x^2(x^4 - 2x^2 + 1)} y. \tag{1.27}$$

To construct the solution to (1.26) we also keep track of the power transformation of x and the linear transformation for $y(x)$ mapping (1.27) into (1.26), given by

$$\left\{ x \to x^2, y \to \sqrt{x} e^{\int (3x^4 + (-2\nu + 2\mu)x^2 + 1 + 2\nu + 2\mu)/(2(x^5 - x)\, dx} y \right\}. \tag{1.28}$$

In step three we check the singularities of (1.27): the equation has now only three regular singular points, at $\{0, 1, -1\}$. Using the notation of Section 1.2, the degrees of the numerator and the factors in the denominator are $[2, [2, 2, 2]]$. The equation then matches the classification table (see Table 1 in Section 1.2). So we proceed with constructing the Möbius transformation mapping the ${}_2F_1$ equation into (1.27) and in doing so computing the values of the hypergeometric parameters $\{a, b, c\}$ such that the equivalence under Möbius exists, obtaining:

$$x \to \frac{2x}{x - 1}, \quad \left\{ a = \frac{\nu}{2}, b = \frac{\nu}{2} - \mu, c = -\mu \right\}. \tag{1.29}$$

Composing these transformations and using the values above for $\{a, b, c\}$ we obtain the solution of (1.26)

$$\begin{aligned} y(x) = {} & C_1 \frac{(x+1)^{(\nu+1)/2}}{x^{\mu}/2(x-1)^{(\nu-\mu-1)/2}} \, {}_2F_1\left(\frac{\nu}{2}, \frac{\nu}{2} - \mu; -\mu; \frac{2x}{x-1}\right) \\ & + C_2 \frac{(x+1)^{(\nu+1)/2} x^{(\mu/2+1)}}{(x-1)^{(\nu+\mu+1)/2}} \, {}_2F_1\left(\frac{\nu}{2} + 1, \frac{\nu}{2} + \mu + 1; 2 + \mu; \frac{2x}{x-1}\right) \end{aligned} \tag{1.30}$$

where C_1 and C_2 are arbitrary constants.

2. Riccati ODEs with computable symmetries

As is known, linear ODEs (0.1) can always be rewritten as non-linear first order equations of Riccati type [**5**]. For instance

$$y'' = I(x) y \tag{2.1}$$

is mapped, using $y \to 1/\exp\left(\int yG\, dx\right)$, where $G(x)$ is arbitrary, into

$$y' = Gy^2 - \frac{G'}{G} y - \frac{I}{G}. \tag{2.2}$$

In this section we show a way to compute solutions for some families of linear ODE with possibly non-rational coefficients by showing how to systematically compute symmetries [**9**] for a corresponding first order Riccati equation. This approach works whenever the problem has some particular restrictions, though the ODE families are reasonably general[4].

We note that two generic Riccati equations with arbitrary $c_i \equiv c_i(x)$

$$y' = c_2 y^2 + c_1 y + c_0 \tag{2.3}$$

can always be mapped between themselves through a linear transformations

$$x \to F(x), \quad y \to P(x)y + Q(x) \tag{2.4}$$

where $\{F, P, Q\}$ are arbitrary. The point symmetries related to these linear transformations are also linear and have the form[5]

$$[\xi = h_1(x), \eta = h_2(x)y(x) + h_3(x)] \tag{2.5}$$

where $h_i(x)$ are functions related to $\{F, P, Q\}$. Therefore and since all Riccati equations are connected through transformations (2.4), all of these equations admit linear symmetries of the form (2.5). The idea is to compute Riccati ODE solutions by means of explicitly computing these linear symmetries.

Without loss of generality the symmetry (2.5) can be rewritten as

$$\left[\frac{p}{f}, -\frac{p'y + q'}{f}\right]. \tag{2.6}$$

Computing the differential invariants [**9**] associated to this symmetry we obtain a general invariant ODE family in terms of $\{f, p, q\}$

$$y' = \frac{f H(py + q)}{p^2} - \frac{p'y + q'}{p} \tag{2.7}$$

where H is an arbitrary function of its argument. The specialization of this equation into a Riccati type one is obtained taking H quadratic and depends on two arbitrary constants a, b

$$y' = f y^2 + \frac{((a + 2q)f - p')}{p} y + \frac{((a + q)q + b)f - q'p}{p^2}. \tag{2.8}$$

Five different tractable cases of this ODE family, where symmetries of the form (2.6) are computable directly by quadratures from the ODE coefficients and its derivatives, are as follows.

CASE $p' = 0$. In this case (2.6) is of the form

$$\left[\frac{1}{f}, -\frac{q'}{f}\right] \tag{2.9}$$

that is, both infinitesimals $[\xi, \eta]$ depend only on x. Constructing the PDE satisfied by these infinitesimals and a generic Riccati equation (2.3), this PDE splits into three equations

$$\xi' c_2 + \xi c_2' = 0, \quad \xi' c_1 + \xi c_1' + 2\eta c_2 = 0, \quad \eta' - \xi' c_0 - \xi c_0' - \eta c_1 = 0. \tag{2.10}$$

The conditions on the equation coefficients $c_i(x)$ such that (2.3) admits symmetries of this form can be obtained by running a differential elimination process

[4] For a more detailed discussion of the material of this section see [**3**].

[5] Herein $[\xi, \eta]$ are the infinitesimals entering the symmetry generator $\xi\partial/\partial x + \eta\partial/\partial y$.

on the system above (2.10), ranking the $c_i(x)$ higher than $\{\xi, \eta\}$—these conditions are also discussed in [**4**]. The system (2.10) has for solution

$$\xi = \frac{C_1}{c_2}, \quad \eta = \frac{C_1(c_2' c_1 - c_2 c_1')}{2c_2^3} \tag{2.11}$$

where C_1 is an arbitrary constant. With the symmetry in hands, an ODE solution can be constructed using standard procedures, e.g. using it to compute an integrating factor. In fact, as shown in [**4**], not only those of Riccati type by actually all first order ODEs having a symmetry of the form $[\xi(x), \eta(x)]$ can be integrated straightforwardly since a symmetry of that form is always computable by quadratures, without solving any auxiliary differential equations.

CASE $q' = 0$. In this case, the symmetry (2.6) has the form

$$\left[\frac{p}{f}, -\frac{p'y}{f}\right] \tag{2.12}$$

that is, the infinitesimals are of the form $[\xi(x), G(x)y]$. The determining PDE system for these infinitesimals and a generic Riccati equation (2.3) is given by

$$\begin{gathered}
\left(\frac{pf'}{f^2} - 2\frac{p'}{f}\right) f^2 c_2 + f(2p'c_2 - pc_2') = 0, \\
-p''f + p'(f' + fc_1) + \left(\frac{pf'}{f^2} - 2\frac{p'}{f}\right) f^2 c_1 - pf{c_1}' = 0, \\
\left(\frac{pf'}{f^2} - 2\frac{p'}{f}\right) f^2 c_0 - pfc_0' = 0.
\end{gathered} \tag{2.13}$$

This system is also solvable in the general case

$$f = C_2 c_2, \quad p = C_1 \frac{\sqrt{c_2}}{\sqrt{c_0}} \tag{2.14}$$

where C_1, C_2 are arbitrary constants. Running a differential elimination process on the system (2.13) ranking $\{f, p\}$ higher than the coefficients c_i one obtains the set of conditions satisfied by these coefficients such that the equation belongs to this family.

CASE $p = q$. In this case the symmetry has the form

$$\left[\frac{q}{f}, -\frac{q'y + q'}{f}\right]. \tag{2.15}$$

A direct computation shows that the transformation $y \to y - 1$ takes this symmetry into one of the form $[\xi(x), G(x)y]$ discussed in the previous section. So a method for this case consists of shifting the dependent variable by one followed by testing the conditions on the $c_i(x)$ mentioned in the previous section.

CASE $p = f$. In this case the symmetry has the form

$$\left[1, -\frac{f'y + q'}{f}\right]. \tag{2.16}$$

A direct computation shows that the transformation $y \to y/f(x)$, where from (2.8) f is the coefficient of y^2, takes this symmetry into $[1, -q']$, that is, of the form $[\xi(x), \eta(x)]$ discussed in the previous sections and so also computable by quadratures, without solving auxiliary differential equations.

Finally, the case $f = q$ also admits solution without solving auxiliary differential equations, but the problem requires more detailed manipulations—for details see [**3**].

Conclusions

In this presentation we discussed two algorithms for second order linear ODEs: one, so called `hyper3`, for computing non-Liouvillian solutions resolving an equivalence to the ${}_2F_1$, ${}_1F_1$ and ${}_0F_1$ equations; the other one, based on the symmetries of a Riccati equation, for computing solutions (Liouvillian or not) for linear ODEs when the coefficients are not necessarily rational.

The `hyper3` algorithm is the most successful one of the entire set of linear ODE algorithms of the Maple system. From the 363 linear ODE examples of Kamke's book [**6**] having rational coefficients, `hyper3` alone solves 331 (91 %), followed by Kovacic's algorithm solving 181 (50 %). Moreover, from these 181 examples admitting Liouvillian solutions, `hyper3` solves 163 (90 %). The fact that around one half of Kamke's examples only admit special function solutions of non-Liouvillian form also illustrates the relevance of this type of solution in the general framework of linear ODE problems.

Despite the simplicity of the approach, it is worth mentioning that till the end of 2001, when the routines for this algorithm were developed, no equivalent or similar algorithms were available in any of the Axiom, Maple, Mathematica, MuPAD or Reduce computer algebra systems (CAS). These CAS used to fail systematically whenever the linear ODE solution admitted no Liouvillian form, except for few textbook and sufficiently similar examples.

The approach for computing solutions for Riccati equations exploring properties of the linear symmetries admitted by these equations is also part of a more general algorithm [**3**]: linear symmetries for first order ODEs, when they exist, can be computed systematically (restrictions apply only for Riccati equations) and, as discussed in [**3**], this approach also consists of the most general one we are aware of for solving this type of problem.

Both algorithms can be extended in natural ways. `Hyper3` can be applied to equations with four regular singular points and also to linear ODEs of order three and higher. Regarding the symmetry algorithm, higher order linear ODEs can be reduced in order by one to a "Riccati like" non-linear equation in the same way second order linear equations can, and so the ideas discussed in Section 2 can be applied as well.

References

1. M. Abramowitz and I. A. Stegun, eds., *Handbook of mathematical functions*, Dover Publ., Inc., New York 1966.
2. L. Chan and E. S Cheb-Terrab, *Hypergeometric solutions for linear* ODE*s: an equivalence approach*, Comput. Phys. Comm., 2003 (submitted).
3. E. S. Cheb-Terrab and T. Kolokolnikov, *First-order ordinary differential equations, symmetries and linear transformations*, Eur. J. Appl. Math. **14** (2003), no. 2, 231–246.
4. E. S Cheb-Terrab and A. D. Roche, *Symmetries and first order* ODE *patterns*, Comput. Phys. Comm. **113** (1998), 239–260.
5. E. L. Ince, *Ordinary differential equations*, Dover Publ., New York, 1944.
6. E. Kamke, *Differentialgleichungen reeller Funktionen*, Chelsea Publ. Co., New York, N.Y., 1947 (German).
7. P. Olver, *Equivalence, invariants and symmetry*, Cambridge Univ. Press, Cambridge, 1995.

8. J. B. Seaborn, *Hypergeometric functions and their applications*, Text Appl. Math., vol. 8, Springer-Verlag, 1991.
9. H. Stephani, *Differential equations: their solution using symmetries*, Cambridge Univ. Press, Cambridge, 1989.
10. E. W. Weisstein, CRC *concise encyclopedia of mathematics*, CRC Press, Boca Raton, FL, 1999.

CECM-MITACS, Department of Mathematics, Simon Fraser University, Vancouver, Canada.

Current address: Maplesoft, Waterloo Maple Inc.

E-mail address: ecterrab@cecm.sfu.ca

Centre de Recherches Mathématiques
CRM Proceedings and Lecture Notes
Volume **39**, 2005

On Rational Solutions of the Fourth Painlevé Equation and Its Hamiltonian

P. A. Clarkson

ABSTRACT. There are two classes of rational solutions of the fourth Painlevé equation (P_{IV}), both of which are expressed in terms of logarithmic derivatives of the special polynomials that are defined through coupled second order, bilinear differential-difference equations that are equivalent to the Toda equation. These special polynomials, which were introduced by Noumi and Yamada [Nagoya Math. J. **153** (1999), 53–86], are generalizations of the standard Hermite polynomials and generalizations of polynomials introduced by Okamoto [Math. Ann. **275** (1986), 221–255]. In this paper, using the Hamiltonian theory for P_{IV}, it is shown that these generalized Hermite polynomials and generalized Okamoto polynomials also satisfy fourth order, bilinear ordinary differential equations and systems of difference equations. Further, rational solutions of some ordinary differential equations which are solvable in terms of solutions of P_{IV} are also expressed in terms of these polynomials.

1. Introduction

In this paper our interest is in rational solutions of the fourth Painlevé equation (P_{IV})

$$w'' = \frac{(w')^2}{2w} + \frac{3}{2}w^3 + 4zw^2 + 2(z^2 - \alpha)w + \frac{\beta}{w}, \tag{1.1}$$

where $' \equiv \mathrm{d}/\mathrm{d}z$ and α and β are arbitrary constants, and associated special polynomials.

Lukashevich [**28**], Gromak [**21**], and Murata [**29**] (see also [**5, 22, 33**]), proved the following theorem concerning rational solutions of P_{IV}.

2000 *Mathematics Subject Classification.* 33E17, 34M35.

I thank Mark Ablowitz, Carl Bender, Chris Cosgrove, Andy Hone, Arieh Iserles, Alexander Its, Nalini Joshi, Elizabeth Mansfield and Marta Mazzocco for their helpful comments and illuminating discussions. This research arose whilst involved with the "Digital Library of Mathematical Functions" project (see `http://dlmf.nist.gov`), in particular I thank Frank Olver and Dan Lozier for the opportunity to participate in this project. This paper was partially written whilst visiting INRIA, Sophia Antipolis, France; in particular I thank Evelyne Hubert for her hospitality and the British Council for their financial support through Alliance Franco–British grant no. PN 01.934.

This is the final form of the paper.

Theorem 1.1. $\mathrm{P_{IV}}$ *has rational solutions if and only if either*

$$\alpha = m, \quad \beta = -2(2n - m + 1)^2, \tag{1.2}$$

or

$$\alpha = m, \quad \beta = -2\left(2n - m + \tfrac{1}{3}\right)^2, \tag{1.3}$$

with $m, n \in \mathbb{Z}$. *Further the rational solutions for these parameter values are unique.*

Some simple rational solutions of $\mathrm{P_{IV}}$ are

$$w_1(z; \pm 2, -2) = \pm \tfrac{1}{z}, \quad w_2(z; 0, -2) = -2z, \quad w_3\left(z; 0, -\tfrac{2}{9}\right) = -\tfrac{2}{3}z. \tag{1.4}$$

It is known that there are three families of unique rational solutions of $\mathrm{P_{IV}}$, which have the solutions (1.4) as the simplest members. These are summarized in the following theorem (see [**5, 29, 33**] for further details).

Theorem 1.2. *There are three families of rational solutions of* $\mathrm{P_{IV}}$, *which have the forms*

$$w_1(z; \alpha_1, \beta_1) = P_{1,N-1}(z)/Q_{1,N}(z), \tag{1.5a}$$

$$w_2(z; \alpha_2, \beta_2) = -2z + P_{2,N-1}(z)/Q_{2,N}(z), \tag{1.5b}$$

$$w_3(z; \alpha_3, \beta_3) = -\tfrac{2}{3}z + P_{3,N-1}(z)/Q_{3,N}(z) \tag{1.5c}$$

where $P_{j,n}(z)$ *and* $Q_{j,n}(z)$, $j = 1, 2, 3$, *are polynomials of degree* N, *and*

$$(\alpha_1, \beta_1) = \left(\pm m, -2(1 + 2n + m)^2\right), \quad n \leq -1, \ m \geq -2n, \tag{1.6a}$$

$$(\alpha_2, \beta_2) = \left(m, -2(1 + 2n + m)^2\right), \quad n \geq 0, \ m \geq -n, \tag{1.6b}$$

$$(\alpha_3, \beta_3) = \left(m, \tfrac{2}{9}(1 + 6n - 3m)^2\right), \tag{1.6c}$$

with $m, n \in \mathbb{Z}$.

The three families given in this theorem are known as the "$-1/z$ hierarchy", the "$-2z$ hierarchy" and the "$-\frac{2}{3}z$ hierarchy", respectively (see [**5**] where the terminology was introduced). The "$-1/z$ hierarchy" and the "$-2z$ hierarchy" form the set of rational solutions of $\mathrm{P_{IV}}$ with parameter values given by (1.2) and the "$-\frac{2}{3}z$ hierarchy" forms the set with parameter values given by (1.3). The rational solutions of $\mathrm{P_{IV}}$ with parameter values given by (1.2) lie at the vertexes of the "Weyl chambers" and those with parameter values given by (1.3) lie at the centres of the "Weyl chamber" [**33**].

The two classes of rational solutions of $\mathrm{P_{IV}}$ with parameter values given by (1.2) and (1.3), can be expressed in terms of logarithmic derivatives of two sets of special polynomials which are defined through coupled second order, bilinear differential-difference equations that are equivalent to the Toda equation. These special polynomials were introduced by Noumi and Yamada [**30**] and are known as the *generalized Hermite polynomials* $H_{m,n}$, which are generalizations of the standard Hermite polynomials, and the *generalized Okamoto polynomials* $Q_{m,n}$, which are generalizations of polynomials introduced by Okamoto [**31**].

This paper is organised as follows. In §2 we discuss the derivation and properties of the generalized Hermite polynomials $H_{m,n}$ and the generalized Okamoto polynomials $Q_{m,n}$. Further it is shown that the hierarchies of rational solutions of $\mathrm{P_{IV}}$ can also be generated using Schlesinger transformations for $\mathrm{P_{IV}}$ given by Fokas, Mugan and Ablowitz [**18**]. In §3 we discuss the rational solutions of the the Hamiltonian system associated with $\mathrm{P_{IV}}$. In particular, it is shown that $H_{m,n}$ and

$Q_{m,n}$ also satisfy fourth order, bilinear ordinary differential equations and systems of difference equations as well as differential-difference equations.

2. Special polynomials associated with rational solutions of P_{IV}

2.1. Okamoto polynomials. In a comprehensive study of the fourth Painlevé equation P_{IV}, Okamoto [**31**] (see also [**19, 27, 30**]) defined two sets of polynomials analogous to the Yablonskii–Vorob'ev polynomials which are associated with rational solutions of P_{II} [**34, 35**] (see also [**14, 19, 31**]). The polynomials given by Okamoto are defined in Theorems 2.1 and 2.2 below, where they have been scaled compared to Okamoto's original definition, where the polynomials are monic, so that they are for the standard P_{IV}.

Theorem 2.1. *Suppose $Q_n(z)$ satisfies the recursion relation*

$$Q_{n+1}Q_{n-1} = \tfrac{9}{2}[Q_nQ_n'' - (Q_n')^2] + [2z^2 + 3(2n-1)]Q_n^2, \tag{2.1}$$

with $Q_0(z) = Q_1(z) = 1$. Then

$$w_n = w(z;\alpha_n,\beta_n) = -\frac{2}{3}z + \frac{\mathrm{d}}{\mathrm{d}z}\left\{\ln\left[\frac{Q_{n+1}(z)}{Q_n(z)}\right]\right\}, \tag{2.2}$$

for $n \geq 0$, satisfies P_{IV} with $(\alpha_n,\beta_n) = (2n, -2/9)$.

Reference [**31**] are defined:

Theorem 2.2. *Suppose $Q_In(z)$ satisfies the recursion relation*

$$R_{n+1}R_{n-1} = \tfrac{9}{2}[R_nR_n'' - (R_n')^2] + 2(z^2 + 3n)R_n^2 \tag{2.3}$$

with $R_0(z) = 1$ and $R_1(z) = \sqrt{2}\,z$. Then

$$\widehat{w}_n = w(z;\widehat{\alpha}_n,\hat{\beta}_n) = -\frac{2}{3}z + \frac{\mathrm{d}}{\mathrm{d}z}\left\{\ln\left[\frac{R_{n+1}}{R_n}\right]\right\}, \tag{2.4}$$

for $n \geq 0$, satisfies P_{IV} with $(\widehat{\alpha}_n,\hat{\beta}_n) = (2n+1, -8/9)$.

The polynomials $Q_n(z)$ are polynomials of degree $n(n-1)$, in fact they are monic polynomials in $\zeta = \sqrt{2}\,z$ with integer coefficients, which is the form in which Okamoto [**31**] originally defined these polynomials. Further the polynomials $Q_n(z)$ are even polynomials, i.e. monic polynomials in $\zeta^2 = 2z^2$ of degree $\frac{1}{2}n(n-1)$. The polynomials $Q_In(z)$ are polynomials of degree n^2, in fact they are monic polynomials in $\zeta = 2z$ with integer coefficients, which is the form in which Okamoto [**31**] originally defined these polynomials. In [**12**] plots of the locations of the roots, in the complex plane, for the Okamoto polynomials $Q_n(z) = 0$, defined by (2.1), and $Q_In(z) = 0$, defined by (2.3), are given. These both take the form of two "triangles" with the polynomials $Q_In(z)$ having an additional row of roots on a straight line, the real axis, between the two "triangles". The term "triangles" is used since the roots lie on arcs, rather than straight lines and so are only approximately triangular.

2.2. Generalized Hermite polynomials and generalized Okamoto polynomials. Noumi and Yamada [**30**] generalized the results of Okamoto [**31**] and introduced the *generalized Hermite polynomials* $H_{m,n}(z)$, which are defined in Theorem 2.3 and *generalized Okamoto polynomials* $Q_{m,n}(z)$, which are defined in Theorem 2.4. Noumi and Yamada [**30**] expressed both the generalized Hermite polynomials and the generalized Okamoto polynomials in terms of Schur functions related to the so-called modified Kadomtsev–Petviashvili (mKP) hierarchy. Kajiwara and Ohta [**27**] also expressed rational solutions of P_{IV} in terms of Schur functions by expressing the solutions in the form of determinants.

Theorem 2.3. *Suppose $H_{m,n}(z)$ satisfies the recurrence relations*

$$2mH_{m+1,n}H_{m-1,n} = H_{m,n}H''_{m,n} - (H'_{m,n})^2 + 2mH^2_{m,n}, \tag{2.5a}$$

$$2nH_{m,n+1}H_{m,n-1} = -H_{m,n}H''_{m,n} + (H'_{m,n})^2 + 2nH^2_{m,n}, \tag{2.5b}$$

with $H_{0,0} = H_{1,0} = H_{0,1} = 1$ and $H_{1,1} = 2z$, then

$$w^{(\mathrm{I})}_{m,n} = -\frac{\mathrm{d}}{\mathrm{d}z}\left\{\ln\left(\frac{H_{m,n+1}}{H_{m,n}}\right)\right\} \equiv -2m\frac{H_{m+1,n}H_{m-1,n+1}}{H_{m,n+1}H_{m,n}}, \tag{2.6a}$$

$$w^{(\mathrm{II})}_{m,n} = \frac{\mathrm{d}}{\mathrm{d}z}\left\{\ln\left(\frac{H_{m+1,n}}{H_{m,n}}\right)\right\} \equiv 2n\frac{H_{m,n+1}H_{m+1,n-1}}{H_{m+1,n}H_{m,n}}, \tag{2.6b}$$

where $w^{(\mathrm{J})}_{m,n} = w(z;\alpha^{(\mathrm{J})}_{m,n},\beta^{(\mathrm{J})}_{m,n})$ for $\mathrm{J} = \mathrm{I},\mathrm{II}$, is a solution of P_{IV}, respectively for

$$\alpha^{(\mathrm{I})}_{m,n} = -(m+2n+1), \quad \beta^{(\mathrm{I})}_{m,n} = -2m^2, \tag{2.7a}$$

$$\alpha^{(\mathrm{II})}_{m,n} = 2m+n+1, \quad \beta^{(\mathrm{II})}_{m,n} = -2n^2. \tag{2.7b}$$

Each generalized Hermite polynomial $H_{m,n}(z)$ is of degree mn with integer coefficients [**30**]. In fact $H_{m,n}(\frac{1}{2}\zeta)$ is a monic polynomial in ζ with integer coefficients. The polynomials $H_{m,n}(z)$ possess the symmetry $H_{m,n}(\mathrm{i}z) = \mathrm{i}^{mn}H_{n,m}(z)$ (note that mn is the degree of $H_{m,n}$). Hence, if $x_j + \mathrm{i}y_k$ is a (complex) root of $H_{m,n}(z)$, then $y_j + \mathrm{i}x_j$ is a root of $H_{n,m}(z)$. Further the relationship between the rational solutions defined by (2.6) is $w^{(\mathrm{I})}_{m,n}(z) = -\mathrm{i}w^{(\mathrm{II})}_{n,m}(\mathrm{i}z)$. This also follows since if $w(z;\alpha,\beta)$ is a solution of P_{IV} then so is $\overline{w}(z;\overline{\alpha},\bar{\beta}) = -\mathrm{i}w(\mathrm{i}z;\alpha,\beta)$, with $(\overline{\alpha},\bar{\beta}) = (-\alpha,\beta)$. Also $H_{n,1}(z) = H_n(z)$ and $H_{1,n}(z) = \mathrm{i}^{-n}H_n(\mathrm{i}z)$, where $H_n(z)$ is the standard Hermite polynomial.

The rational solutions of P_{IV} defined by (2.6) include all the solutions in the "$-1/z$" and "$-2z$" hierarchies, as is easily verified by comparing the parameters in (2.7) with those in (1.6a) and (1.6b). Further they are the set of rational solutions of P_{IV} with parameter values given by (1.2). The rational solutions of P_{IV} generated by the generalized Hermite polynomials $H_{m,n}(z)$ are special cases of the special function solutions, often called *one-parameter families of solutions*, which are expressible in terms of Whittaker functions $M_{\kappa,\mu}(\xi)$ and $W_{\kappa,\mu}(\xi)$, or equivalently parabolic cylinder functions $D_\nu(\xi)$ (cf. [**12**]).

The two representations of the rational solutions $w^{(\mathrm{I})}_{m,n}$ and $w^{(\mathrm{II})}_{m,n}$ in (2.6) imply that

$$H'_{m,n+1}H_{m,n} - H_{m,n+1}H'_{m,n} = 2mH_{m+1,n}H_{m-1,n+1}, \tag{2.8a}$$

$$H'_{m+1,n}H_{m,n} - H_{m+1,n}H'_{m,n} = 2nH_{m,n+1}H_{m+1,n-1}. \tag{2.8b}$$

These also follow from [**30**, Prop. 4.2], where the third relation

$$H'_{m+1,n+1}H_{m,n} - H_{m+1,n+1}H'_{m,n} = 2(m+n+1)H_{m+1,n}H_{m,n+1}, \tag{2.9}$$

is additionally given.

Plots of the locations of the roots of the polynomials $H_{m,n}(z)$ for various choices of m and n, are given in [**12**]. These plots, which are invariant under reflections in the real and imaginary z-axes, take the form of $m \times n$ "rectangles", though these are only approximate rectangles as can be seen by looking at the actual values of the roots.

Theorem 2.4. *Suppose $Q_{m,n}(z)$ satisfies the recurrence relations*

$$Q_{m+1,n}Q_{m-1,n} = \tfrac{9}{2}[Q_{m,n}Q''_{m,n} - (Q'_{m,n})^2] \tag{2.10a}$$
$$\qquad + [2z^2 + 3(2m+n-1)]Q^2_{m,n}, \tag{2.10b}$$
$$Q_{m,n+1}Q_{m,n-1} = \tfrac{9}{2}[Q_{m,n}Q''_{m,n} - (Q'_{m,n})^2] \tag{2.10c}$$
$$\qquad + [2z^2 + 3(1-m-2n)]Q^2_{m,n}, \tag{2.10d}$$

with $Q_{0,0} = Q_{1,0} = Q_{0,1} = 1$ and $Q_{1,1} = \sqrt{2}\, z$, then

$$\widetilde{w}^{(\mathrm{I})}_{m,n} = -\tfrac{2}{3}z - \frac{\mathrm{d}}{\mathrm{d}z}\left\{\ln\left(\frac{Q_{m,n+1}}{Q_{m,n}}\right)\right\} \equiv -\frac{1}{3}\sqrt{2}\,\frac{Q_{m+1,n}Q_{m-1,n+1}}{Q_{m,n+1}Q_{m,n}} \tag{2.11a}$$

$$\widetilde{w}^{(\mathrm{II})}_{m,n} = -\tfrac{2}{3}z + \frac{\mathrm{d}}{\mathrm{d}z}\left\{\ln\left(\frac{Q_{m+1,n}}{Q_{m,n}}\right)\right\} \equiv -\frac{1}{3}\sqrt{2}\frac{Q_{m,n+1}Q_{m+1,n-1}}{Q_{m+1,n}Q_{m,n}}, \tag{2.11b}$$

where $\widetilde{w}^{(\mathrm{J})}_{m,n} = w(z;\widetilde{\alpha}^{(\mathrm{J})}_{m,n},\widetilde{\beta}^{(\mathrm{J})}_{m,n})$ for $\mathrm{J} = \mathrm{I},\mathrm{II}$, are solutions of $\mathrm{P_{IV}}$, respectively for

$$\widetilde{\alpha}^{(\mathrm{I})}_{m,n} = -(m+2n), \quad \widetilde{\beta}^{(\mathrm{I})}_{m,n} = -2\left(m-\tfrac{1}{3}\right)^2, \tag{2.12a}$$
$$\widetilde{\alpha}^{(\mathrm{II})}_{m,n} = 2m+n, \qquad \widetilde{\beta}^{(\mathrm{II})}_{m,n} = -2\left(n-\tfrac{1}{3}\right)^2. \tag{2.12b}$$

Each polynomial $Q_{m,n}(z)$ is of degree $d_{m,n} = m^2+n^2+mn-m-n$ with integer coefficients [**30**]. Further $Q_{m,n}(z)$ is a monic polynomial in $\zeta = \sqrt{2}\, z$ with integer coefficients. The polynomials $Q_{m,n}(z)$ possess the symmetry $Q_{m,n}(\mathrm{i}z) = \mathrm{i}^{d_{m,n}}Q_{n,m}(z)$, where $d_{m,n}$ is the degree of $Q_{m,n}$. Hence, if $x_j + \mathrm{i}y_k$ is a (complex) root of $Q_{m,n}(z)$, then $y_j + \mathrm{i}x_j$ is a root of $Q_{n,m}(z)$. Further the relationship between the rational solutions defined by (2.11) is $\widetilde{w}^{(\mathrm{I})}_{m,n}(z) = -\mathrm{i}\widetilde{w}^{(\mathrm{II})}_{n,m}(\mathrm{i}z)$. The original Okamoto polynomials defined in Theorems 2.1 and 2.2 are respectively given by $Q_m(z) = Q_{m,0}(z)$ and $Q_Im(z) = Q_{m,1}(z)$.

The rational solutions of $\mathrm{P_{IV}}$ defined by (2.11) include all the solutions in the "$-\frac{2}{3}z$" hierarchy, as is easily verified by comparing the parameters in (2.12) with those in (1.6c). Further they are the set of rational solutions of $\mathrm{P_{IV}}$ with parameter values given by (1.3).

The two representations of the rational solutions $\widetilde{w}^{(\mathrm{I})}_{m,n}$ and $\widetilde{w}^{(\mathrm{II})}_{m,n}$ in (2.11) imply that

$$Q'_{m,n+1}Q_{m,n} - Q_{m,n+1}Q'_{m,n} + \tfrac{2}{3}zQ_{m,n+1}Q_{m,n} = \tfrac{1}{3}\sqrt{2}\,Q_{m+1,n}Q_{m-1,n+1}, \tag{2.13a}$$

$$Q'_{m+1,n}Q_{m,n} - Q_{m+1,n}Q'_{m,n} - \tfrac{2}{3}zQ_{m,n+1}Q_{m,n} = -\tfrac{1}{3}\sqrt{2}\,Q_{m,n+1}Q_{m+1,n-1}. \tag{2.13b}$$

These also follow from [**30**, Prop. 4.2], where the third relation

$$Q'_{m+1,n+1}Q_{m,n} - Q_{m+1,n+1}Q'_{m,n} = \sqrt{2}\,(3m+3n+1)Q_{m+1,n}Q_{m,n+1}. \tag{2.14}$$

is additionally given.

Plots of the locations of the roots of the polynomials $Q_{m,n}(z)$, for various m and n, are given in [**12**]. The roots of the polynomial $Q_{m,n}(z)$ takes the form of $m \times n$ "rectangles" with an "equilateral triangle", which have either $m-1$ or $n-1$

roots on each of its sides. These are only approximate rectangles and equilateral triangles as can be seen by looking at the actual values of the roots. We remark that as for the generalized Hermite polynomials above, the plots are invariant under reflections in the real and imaginary z-axes.

An important, well-known property of classical orthogonal polynomials, such as the Hermite, Laguerre or Legendre polynomials whose roots all lie on the real line (cf. [**1**]), is that the roots of successive polynomials interlace. Thus for a set of classical polynomials $\varphi_n(z)$, for $n = 0, 1, 2, \ldots$, if $z_{n,m}$ and $z_{n,m+1}$ are two successive roots of $\varphi_n(z)$, i.e. $\varphi_n(z_{n,m}) = 0$ and $\varphi_n(z_{n,m+1}) = 0$, then $\varphi_{n-1}(\zeta_{n-1}) = 0$ and $\varphi_{n+1}(\zeta_{n+1}) = 0$ for some ζ_{n-1} and ζ_{n+1} such that $z_{n,m} < \zeta_{n-1}, \zeta_{n+1} < z_{n,m+1}$. Further the derivatives $\varphi_n'(z)$ and $\varphi_{n+1}'(z)$ also have roots in the interval $(z_{n,m}, z_{n,m+1})$, that is $\varphi_n'(\xi_n) = 0$ and $\varphi_{n+1}'(\xi_{n+1}) = 0$ for some ξ_n and ξ_{n+1} such that $z_{n,m} < \xi_n, \xi_{n+1} < z_{n,m+1}$.

An interesting open question is whether there are analogous results for the polynomials $H_{m,n}(z)$ and $Q_{m,n}(z)$. Obviously there are significant differences since the polynomials $H_{m,n}(z)$ and $Q_{m,n}(z)$ have complex roots, whereas the classical polynomial $\varphi_n(z)$ has real roots. The pattern of the roots of $H_{m,n}(z)$ and $Q_{m,n}(z)$ given in [**12**] are highly symmetric and structured, suggesting that they have interesting properties. An particularly intriguing question is whether there there is any "interlacing of roots" (in the complex plane), analogous to that for classical polynomials (on the real line); though we do not expect any specific relationship between the roots of the polynomials $H_{m,n}(z)$ and $Q_{m,n}(z)$ with roots of any classical polynomial. Further it is necessary to define what is meant by "interlacing of roots in the complex plane".

To illustrate some possible "interlacing of roots", the roots of the generalized Hermite polynomials $H_{6,6}(z)$ (•), $H_{6,7}(z)$ (■), $H_{7,6}(z)$ (♦) and $H_{7,7}(z)$ (○) are simultaneously plotted in Figure 1 and the roots of the generalized Okamoto polynomials $Q_{5,5}(z)$ (•), $Q_{5,6}(z)$ (■), $Q_{6,5}(z)$ (♦) and $Q_{6,6}(z)$ (○), are simultaneously plotted in Figure 2. These figures give some experimental evidence that there is structure to

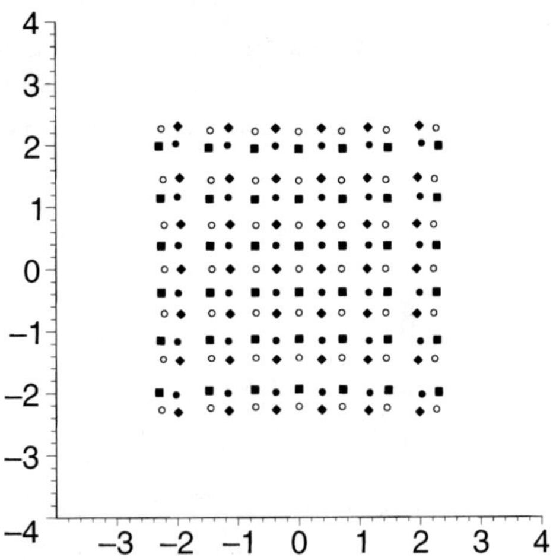

FIGURE 1. Roots of the generalized Hermite polynomials $H_{6,6}$ (•), $H_{6,7}$ (■), $H_{7,6}$ (♦) and $H_{7,7}$ (○).

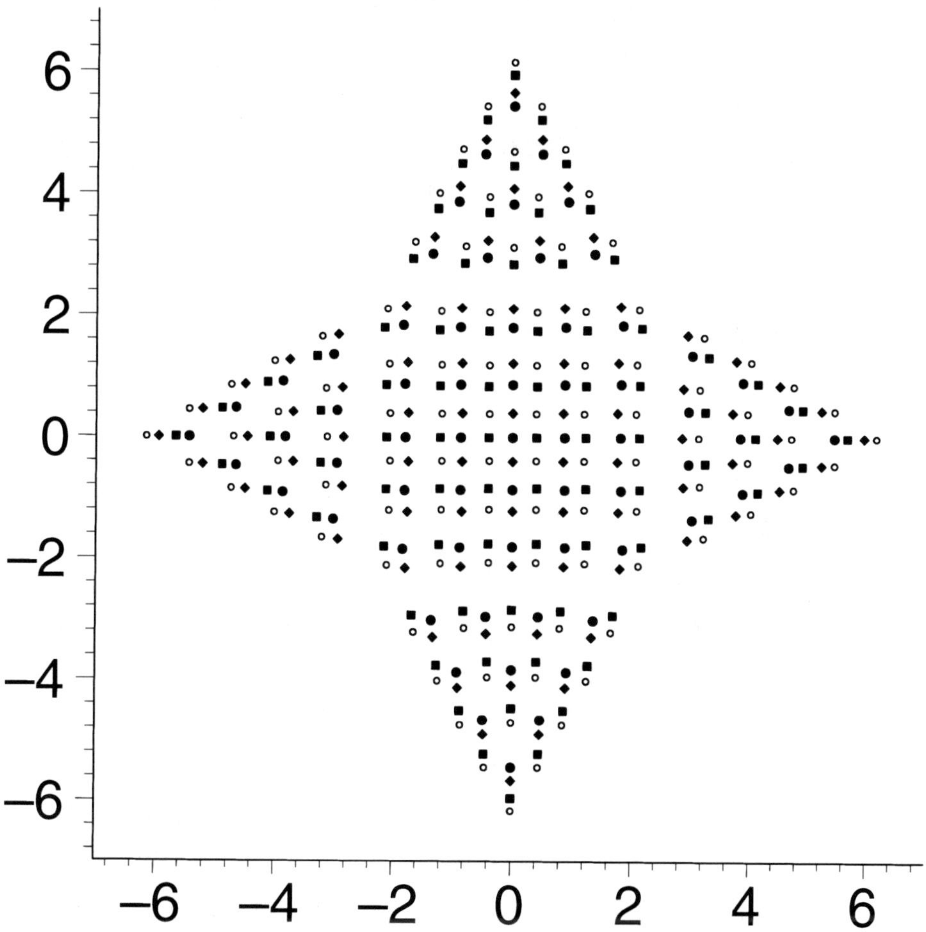

FIGURE 2. Roots of the generalized Okamoto polynomials $Q_{5,5}$ (•), $Q_{5,6}$ (■), $Q_{6,5}$ (♦) and $Q_{6,6}$ (○).

the relative positions of the roots. Similar structures are observed for the roots of the Yablonskii–Vorob'ev polynomials polynomials which are associated with rational solutions of P_{II} [**14**] and the roots of the polynomials which are associated with rational and algebraic solutions of P_{III} [**11**]. We feel that this "interlacing of roots" for the polynomials $H_{m,n}(z)$ and $Q_{m,n}(z)$ warrants further analytical and numerical studies. We shall not pursue these questions any further here.

2.3. Schlesinger transformations for P_{IV}. Fokas, Mugan and Ablowitz [**18**], using a Schlesinger transformation formulation associated with the isomonodromy formulation of P_{IV}, deduced the following transformations $\mathcal{R}_1$—$\mathcal{R}_4$ for P_{IV}.

$$\mathcal{R}_1 : w_1 = \frac{(w' + \sqrt{-2\beta})^2 + (4\alpha + 4 - 2\sqrt{-2\beta})w^2 - w^2(w+2z)^2}{2w(w^2 + 2zw - w' - \sqrt{-2\beta})}, \tag{2.15}$$

$$\mathcal{R}_2 : w_2 = \frac{(w' - \sqrt{-2\beta})^2 + (4\alpha - 4 - 2\sqrt{-2\beta})w^2 - w^2(w+2z)^2}{2w(w^2 + 2zw + w' - \sqrt{-2\beta})}, \tag{2.16}$$

TABLE 1. The effect of the Schlesinger transformations $\mathcal{R}_1$–$\mathcal{R}_5$ and $\mathcal{R}_7$ on the parameters.

	α	β	θ_0	θ_∞
$\mathcal{R}_1$	$\alpha+1$	$-\frac{1}{2}(2-\sqrt{-2\beta})^2$	θ_0-1	θ_∞
$\mathcal{R}_2$	$\alpha-1$	$-\frac{1}{2}(2+\sqrt{-2\beta})^2$	θ_0+1	θ_∞
$\mathcal{R}_3$	$\alpha+1$	$-\frac{1}{2}(2+\sqrt{-2\beta})^2$	θ_0+1	$\theta_\infty+1$
$\mathcal{R}_4$	$\alpha-1$	$-\frac{1}{2}(2-\sqrt{-2\beta})^2$	θ_0-1	$\theta_\infty-1$
$\mathcal{R}_5\equiv\mathcal{R}_1\mathcal{R}_3$	$\alpha+2$	β	θ_0	$\theta_\infty+1$
$(4+\sqrt{-2\beta})^2(4-\sqrt{-2\beta})^2\mathcal{R}_7$ $\equiv\mathcal{R}_2\mathcal{R}_4$	$\alpha-2$	β	θ_0	$\theta_\infty-1$

$$\mathcal{R}_3 : w_3 = \frac{(w'-\sqrt{-2\beta})^2-(4\alpha+4+2\sqrt{-2\beta})w^2-w^2(w+2z)^2}{2w(w^2+2zw-w'+\sqrt{-2\beta})}, \tag{2.17}$$

$$\mathcal{R}_4 : w_4 = \frac{(w'+\sqrt{-2\beta})^2+(4\alpha-4+2\sqrt{-2\beta})w^2-w^2(w+2z)^2}{2w(w^2+2zw+w'+\sqrt{-2\beta})} \tag{2.18}$$

where $w\equiv w(z;\alpha,\beta)$, $w_j\equiv w(z;\alpha_j,\beta_j)$. Fokas, Mugan and Ablowitz [**18**] defined the composite transformations $\mathcal{R}_5=\mathcal{R}_1\mathcal{R}_3$ and $\mathcal{R}_7=\mathcal{R}_2\mathcal{R}_4$ given by

$$\mathcal{R}_5 : w_5 = \frac{(w'-w^2-2zw)^2+2\beta}{2w[w'-w^2-2zw+2(\alpha+1)]}, \tag{2.19}$$

$$\mathcal{R}_7 : w_7 = \frac{(w'+w^2+2zw)^2+2\beta}{2w[w'+w^2+2zw-2(\alpha-1)]}, \tag{2.20}$$

respectively. We remark that $\mathcal{R}_5$ is the transformation $\mathcal{T}_+$ defined by Murata [**29**]. The effect of these Schlesinger transformations on the parameters α, β, θ_0 and θ_∞ is given in Table 1.

Applying the Schlesinger transformations $\mathcal{R}_1$—$\mathcal{R}_5$ and $\mathcal{R}_7$ to the rational solutions $w^{(\mathrm{I})}_{m,n}$, $w^{(\mathrm{II})}_{m,n}$, $\widetilde{w}^{(\mathrm{I})}_{m,n}$ and $\widetilde{w}^{(\mathrm{II})}_{m,n}$ of $\mathrm{P_{IV}}$, defined by (2.6) and (2.11), increments the indices m and n by ±1. These are summarized as follows, where $v_{m,n}$ is either $v_{m,n}$ or $\widetilde{w}_{m,n}$.

$$\begin{aligned} v^{(\mathrm{I})}_{m+1,n} &= \mathcal{R}_2(v^{(\mathrm{I})}_{m,n}), & v^{(\mathrm{I})}_{m-1,n} &= \mathcal{R}_1(v^{(\mathrm{I})}_{m,n}),\\ v^{(\mathrm{I})}_{m,n+1} &= \mathcal{R}_7(v^{(\mathrm{I})}_{m,n}), & v^{(\mathrm{I})}_{m,n-1} &= \mathcal{R}_5(v^{(\mathrm{I})}_{m,n}),\\ v^{(\mathrm{II})}_{m+1,n} &= \mathcal{R}_5(v^{(\mathrm{II})}_{m,n}), & v^{(\mathrm{II})}_{m-1,n} &= \mathcal{R}_7(v^{(\mathrm{II})}_{m,n}),\\ v^{(\mathrm{II})}_{m,n+1} &= \mathcal{R}_1(v^{(\mathrm{II})}_{m,n}), & v^{(\mathrm{II})}_{m,n-1} &= \mathcal{R}_2(v^{(\mathrm{II})}_{m,n}). \end{aligned}$$

We remark that the Schlesinger transformations $\mathcal{R}_1$—$\mathcal{R}_5$ and $\mathcal{R}_7$ are Bäcklund transformations of $\mathrm{P_{IV}}$ which involve $(w')^2$. The standard Bäcklund transformation for $\mathrm{P_{IV}}$ has the form

$$\widehat{w}(z;\widehat{\alpha},\widehat{\beta}) = \frac{A(w,z)w'+B(w,z)}{C(w,z)w'+D(w,z)}, \tag{2.21}$$

where $A(w,z)$, $B(w,z)$, $C(w,z)$ and $D(w,z)$ are polynomial in w and locally analytic in z, with $AD-BC\neq0$ [**21, 28**]. In [**5**] it is shown that the Schlesinger transformations $\mathcal{R}_1$—$\mathcal{R}_5$ and $\mathcal{R}_7$ can be written in terms of the standard Bäcklund transformations for $\mathrm{P_{IV}}$. Second-order difference equations which are linear the highest coefficient, in particular the discrete Painlevé equations, can be derived from Bäcklund transformations of the form (2.21) [**17**], whereas Bäcklund transformations which involve $(w')^2$ usually give rise to difference equations that are of second-degree in the highest coefficient [**15**].

3. The Hamiltonian representation of $\mathrm{P_{IV}}$

3.1. Rational solutions for the $\mathrm{P_{IV}}$ Hamiltonian. The Hamiltonian for $\mathrm{P_{IV}}$ is [**31**]

$$\mathrm{H_{IV}}(q,p,z;\theta_0,\theta_\infty) = 2qp^2 - (q^2+2zq+2\theta_0)p + \theta_\infty q \tag{3.1}$$

then from Hamilton's equation we have

$$q' = \frac{\partial \mathrm{H_{IV}}}{\partial p} = 4qp - q^2 - 2zq - 2\theta_0, \tag{3.2a}$$

$$p' = -\frac{\partial \mathrm{H_{IV}}}{\partial q} = -2p^2 + 2pq + 2zp - \theta_\infty. \tag{3.2b}$$

Eliminating p in (3.2), then $q = w$ satisfies $\mathrm{P_{IV}}$ with $(\alpha,\beta) = (1-\theta_0+2\theta_\infty, -2\theta_0^2)$, and eliminating q in (3.2), then $w = -2p$ satisfies $\mathrm{P_{IV}}$ with $(\alpha,\beta) = (-1+2\theta_0-\theta_\infty, -2\theta_\infty^2)$.

The Hamiltonian function $\sigma(z;\theta_0,\theta_\infty) = \mathrm{H_{IV}}(q,p,z;\theta_0,\theta_\infty)$ satisfies

$$(\sigma'')^2 - 4(z\sigma'-\sigma)^2 + 4\sigma'(\sigma'+2\theta_0)(\sigma'+2\theta_\infty) = 0. \tag{3.3}$$

This equation is equivalent to equation SD-I.c in the classification of second-order, second-degree ordinary differential equations with the Painlevé property due to Cosgrove and Scoufis [**16**], an equation first derived and solved by Chazy [**10**] and rederived by Bureau [**7–9**]. It was also derived by Jimbo and Miwa [**26**] and Okamoto [**31**] in a Hamiltonian description of $\mathrm{P_{IV}}$. Conversely, if σ is a solution of (3.3), then

$$q = -\frac{\sigma''-2z\sigma'+2\sigma}{2(\sigma'+2\theta_\infty)}, \qquad p = \frac{\sigma''+2z\sigma'-2\sigma}{2(\sigma'+2\theta_0)}, \tag{3.4}$$

are solutions of (3.2).

Due to the relationship between the Hamiltonian function σ and the associated τ-functions given by [**31**]

$$\frac{\mathrm{d}}{\mathrm{d}z}\ln\tau(z;\theta_0,\theta_\infty) = \sigma(z;\theta_0,\theta_\infty) \tag{3.5}$$

then rational solutions of (3.3) have the form

$$h_{m,n} = \frac{\mathrm{d}}{\mathrm{d}z}\ln H_{m,n}, \qquad \theta_0 = -n,\ \theta_\infty = m, \tag{3.6a}$$

$$\sigma_{m,n} = \frac{4}{27}z^3 - \frac{2}{3}(m-n)z + \frac{\mathrm{d}}{\mathrm{d}z}\ln Q_{m,n}, \tag{3.6b}$$

$$\theta_0 = -n+\frac{1}{3},\ \theta_\infty = m-\frac{1}{3},$$

where $H_{m,n}$ are the generalized Hermite polynomials and $Q_{m,n}$ the generalized Okamoto polynomials. Consequently $h_{m,n}$ and $\sigma_{m,n}$ satisfy the respective differential equations

$$(h''_{m,n})^2 - 4(zh'_{m,n} - h_{m,n})^2 + 4h'_{m,n}(h'_{m,n} - 2n)(h'_{m,n} + 2m) = 0, \tag{3.7a}$$

$$(\sigma''_{m,n})^2 - 4(z\sigma'_{m,n} - \sigma_{m,n})^2 + 4\sigma'_{m,n}(\sigma'_{m,n} - 2n + \tfrac{2}{3})(\sigma'_{m,n} + 2m - \tfrac{2}{3}) = 0. \tag{3.7b}$$

We note that the relationship between the rational solutions of $\mathrm{P_{IV}}$ and the Hamiltonian functions $h_{m,n}$ and $\sigma_{m,n}$ is

$$w^{(\mathrm{I})}_{m,n} = -\frac{\mathrm{d}}{\mathrm{d}z}\left\{\ln\left(\frac{H_{m,n+1}}{H_{m,n}}\right)\right\} = h_{m,n} - h_{m,n+1}, \tag{3.8a}$$

$$w^{(\mathrm{II})}_{m,n} = \frac{\mathrm{d}}{\mathrm{d}z}\left\{\ln\left(\frac{H_{m+1,n}}{H_{m,n}}\right)\right\} = h_{m+1,n} - h_{m,n}, \tag{3.8b}$$

$$\widetilde{w}^{(\mathrm{I})}_{m,n} = -\frac{2}{3}z - \frac{\mathrm{d}}{\mathrm{d}z}\left\{\ln\left(\frac{Q_{m,n+1}}{Q_{m,n}}\right)\right\} = \sigma_{m,n} - \sigma_{m,n+1}, \tag{3.8c}$$

$$\widetilde{w}^{(\mathrm{II})}_{m,n} = -\frac{2}{3}z + \frac{\mathrm{d}}{\mathrm{d}z}\left\{\ln\left(\frac{Q_{m+1,n}}{Q_{m,n}}\right)\right\} = \sigma_{m+1,n} - \sigma_{m,n}. \tag{3.8d}$$

3.2. Differential equations for $H_{m,n}$ and $Q_{m,n}$. Here we derive ordinary differential equations for $H_{m,n}$ and $Q_{m,n}$ using the Hamiltonian representation of $\mathrm{P_{IV}}$. Differentiating (3.4) with respect to z yields

$$\sigma''' + 6(\sigma')^2 - 4(z^2 + 2\theta_0 + 2\theta_\infty)\sigma' + 4z\sigma + 8\theta_0\theta_\infty = 0. \tag{3.9}$$

Then substituting $\theta_\infty = m$, (3.6a) into (3.9) yields the following fourth-order, bilinear equation for $H_{m,n}$

$$\begin{aligned} H_{m,n}H''''_{m,n} &- 4H'_{m,n}H'''_{m,n} + 3(H''_{m,n})^2 \\ &- 4(z^2 + 2n - 2m)[H_{m,n}H''_{m,n} - (H'_{m,n})^2] + 4zH_{m,n}H'_{m,n} - 8mnH^2_{m,n} = 0. \end{aligned} \tag{3.10}$$

Similarly, substituting (3.6b)) into (3.9) yields the following fourth-order, bilinear equation for $Q_{m,n}$

$$\begin{aligned} Q_{m,n}Q''''_{m,n} &- 4Q'_{m,n}Q'''_{m,n} + 3(Q''_{m,n})^2 + \tfrac{4}{3}z^2[Q_{m,n}Q''_{m,n} - (Q'_{m,n})^2] \\ &+ 4zQ_{m,n}Q'_{m,n} - \tfrac{8}{3}(m^2 + n^2 + mn - m - n)Q^2_{m,n} = 0. \end{aligned} \tag{3.11}$$

It is anticipated that these ordinary differential equations (3.10) and (3.11) will be useful for the derivation of properties of $H_{m,n}$ and $Q_{m,n}$. For example, if we seek a monic polynomial solution in the form $F(z) = z^r + a_1 z^{r-1} + \cdots + a_{r-1}z + a_r$, since (3.10) and (3.11) are homogeneous then we can assume that the coefficient of z^r is unity, then it is easy to see that necessarily $r = mn$ for (3.10) and $r = m^2 + n^2 + mn - m - n$ for (3.11), which is a very simple proof of the degrees of $H_{m,n}$ and $Q_{m,n}$. Also it is easy to see that $a_{2j+1} = 0$ and to obtain recurrence relations for the coefficients a_{2j}.

3.3. Difference equations for $H_{m,n}$ and $Q_{m,n}$. Here we illustrate how to derive pure difference equations for $H_{m,n}$ and $Q_{m,n}$, again using the Hamiltonian representation of $\mathrm{P_{IV}}$.

First we derive difference equations satisfied by $H_{m,n}$. Setting $q = q_{m,n}$, $p = -\frac{1}{2}v_{m,n}$, $\theta_0 = -n$ and $\theta_\infty = m$ in (3.2) yields

$$q'_{m,n} = -q^2_{m,n} - 2q_{m,n}v_{m,n} - 2zq_{m,n} + 2n, \tag{3.12a}$$

$$v'_{m,n} = v^2_{m,n} + 2v_{m,n}q_{m,n} + 2zv_{m,n} + 2m. \tag{3.12b}$$

Then $q_{m,n}$ satisfies P_{IV} with $(\alpha, \beta) = (2m + n + 1, -2n^2)$ and $v_{m,n}$ satisfies P_{IV} with $(\alpha, \beta) = (-m - 2n - 1, -2m^2)$. Therefore

$$q_{m,n} = w^{(II)}_{m,n} = h_{m+1,n} - h_{m,n}, \quad v_{m,n} = w^{(I)}_{m,n} = h_{m,n+1} - h_{m,n},$$

and so

$$q_{m,n} = \frac{\mathrm{d}}{\mathrm{d}z}\left\{\ln\left(\frac{H_{m+1,n}}{H_{m,n}}\right)\right\} = 2n\frac{H_{m,n+1}H_{m+1,n-1}}{H_{m+1,n}H_{m,n}}, \tag{3.13a}$$

$$v_{m,n} = -\frac{\mathrm{d}}{\mathrm{d}z}\left\{\ln\left(\frac{H_{m,n+1}}{H_{m,n}}\right)\right\} = -2m\frac{H_{m+1,n}H_{m-1,n+1}}{H_{m,n+1}\,H_{m,n}}. \tag{3.13b}$$

Substituting the logarithmic derivative representation of the rational solution in the left-hand side of the system (3.12) and the other representation in the right-hand side, then using the fact that $H_{m,n}$ satisfies (2.5) yields difference equations for $H_{m,n}$. Since $H_{m,n}$ satisfies two second differential-difference equations in (2.5) and there are two second derivatives to be substituted, there are four possibilities for each of the two equations (3.13), which results in eight difference equations.

Next we derive difference equations satisfied by $Q_{m,n}$. Setting $q = \widetilde{q}_{m,n}$, $p = -\frac{1}{2}\widetilde{v}_{m,n}$, $\theta_0 = -n + \frac{1}{3}$ and $\theta_\infty = m - \frac{1}{3}$ in (3.2) yields

$$\widetilde{q}'_{m,n} = -\widetilde{q}^2_{m,n} - 2\widetilde{q}_{m,n}\widetilde{v}_{m,n} - 2z\widetilde{q}_{m,n} + 2n - \tfrac{2}{3}, \tag{3.14a}$$

$$\widetilde{v}'_{m,n} = \widetilde{v}^2_{m,n} + 2\widetilde{v}_{m,n}\widetilde{q}_{m,n} + 2z\widetilde{v}_{m,n} + 2m - \tfrac{2}{3}. \tag{3.14b}$$

Then $\widetilde{q}_{m,n}$ satisfies P_{IV} with $(\alpha, \beta) = \left(2m + n, -2(n - \frac{1}{3})^2\right)$ and $\widetilde{v}_{m,n}$ satisfies P_{IV} with $(\alpha, \beta) = \left(-m - 2n, -2(m - \frac{1}{3})^2\right)$. Therefore

$$\widetilde{q}_{m,n} = \widetilde{w}^{(II)}_{m,n} = \sigma_{m+1,n} - \sigma_{m,n}, \quad \widetilde{v}_{m,n} = \widetilde{w}^{(I)}_{m,n} = \sigma_{m,n+1} - \sigma_{m,n},$$

and so

$$\widetilde{q}_{m,n} = -\tfrac{2}{3}z + \frac{\mathrm{d}}{\mathrm{d}z}\left\{\ln\left(\frac{Q_{m+1,n}}{Q_{m,n}}\right)\right\} = -\tfrac{1}{3}\sqrt{2}\frac{Q_{m,n+1}Q_{m+1,n-1}}{Q_{m+1,n}Q_{m,n}}, \tag{3.15a}$$

$$\widetilde{v}_{m,n} = -\tfrac{2}{3}z - \frac{\mathrm{d}}{\mathrm{d}z}\left\{\ln\left(\frac{Q_{m,n+1}}{Q_{m,n}}\right)\right\} = -\tfrac{1}{3}\sqrt{2}\frac{Q_{m+1,n}Q_{m-1,n+1}}{Q_{m,n+1}Q_{m,n}}. \tag{3.15b}$$

Substituting the logarithmic derivative representation of the rational solution in the left-hand side of the system (3.14) and the other representation in the right-hand side, then using the fact that $Q_{m,n}$ satisfies (2.10) yields difference equations for $Q_{m,n}$. Again since $Q_{m,n}$ satisfies two second differential-difference equations in (2.10) and there are two second derivatives to be substituted, there are four possibilities for each of the two equations (3.15), which results in eight difference equations.

3.4. Okamoto's ρ and ρ' transformations. Okamoto [**32**] gives two transformations associated with the Hamiltonian formulation of P_{IV} from which we can also derive difference equations for $H_{m,n}$ and $Q_{m,n}$. These are the $\boldsymbol{\rho}$ transformation, which we shall denote $\mathcal{R}_\infty$, and the $\boldsymbol{\rho}'$ transformation, which we shall denote $\mathcal{R}_0$. The rationale for these names is that the $\mathcal{R}_\infty$ transformation increases θ_∞ by 1 and leaves θ_0 invariant whilst the $\mathcal{R}_0$ transformation increases θ_0 by 1 and leaves θ_∞ invariant.

Suppose (q_0, p_0) satisfy the Hamiltonian system

$$q_0' = 4q_0p_0 - q_0^2 - 2zq_0 - 2\Theta_0, \tag{3.16a}$$

$$p_0' = -2p_0^2 + 2p_0q_0 + 2zp_0 - \Theta_\infty. \tag{3.16b}$$

Eliminating p_0 in (3.16), then q_0 satisfies P_{IV} with $(\alpha,\beta) = (1-\Theta_0+2\Theta_\infty, -2\Theta_0^2)$, and eliminating q_0 in (3.16), then $p_0 = -\frac{1}{2}w$, where w satisfies P_{IV} with $(\alpha,\beta) = (-1+2\Theta_0-\Theta_\infty, -2\Theta_\infty^2)$. We remark that for the generalized Hermite polynomials $H_{m,n}$, the parameters are $\Theta_0 = \Theta_\infty = 0$, and for the generalized Okamoto polynomials $Q_{m,n}$, the parameters are $\Theta_0 = \frac{1}{3}$ and $\Theta_\infty = -\frac{1}{3}$.

The $\mathcal{R}_\infty$ transformation is defined by

$$q_{j+1} = -q_j + 2p_j - 2z - \frac{2(q_j - 2p_j + 2z)(\Theta_\infty + j + 1)}{q_j(q_j - 2p_j + 2z) + 2(\Theta_0 - \Theta_\infty - j - 1)}, \tag{3.17a}$$

$$p_{j+1} = -\tfrac{1}{2}q_j - \Theta_0 - \Theta_\infty - j - 1/q_j - 2p_j + 2z, \tag{3.17b}$$

with inverse transformation

$$q_{j-1} = -2p_j + 2p_j(\Theta_0 - \Theta_\infty - j)/q_jp_j - \Theta_\infty - j, \tag{3.17c}$$

$$p_{j-1} = \tfrac{1}{2}q_j - p_j + z + \frac{p_j(\Theta_0 - \Theta_\infty - j)}{q_jp_j - \Theta_\infty - j} + \frac{\Theta_\infty + j}{2p_j} \tag{3.17d}$$

where $(q_j, p_j) = \mathcal{R}_\infty^j(q_0, p_0)$, i.e. applying the $\mathcal{R}_\infty$ transformation j times, and the mapping of the parameters is $\mathcal{R}_\infty^j(\Theta_0, \Theta_\infty) = (\Theta_0, \Theta_\infty + j)$.

The transformations $\mathcal{R}_\infty$ and $\mathcal{R}_0$ can be combined as follows. Suppose $(q_{m,n}, p_{m,n})$ satisfy the Hamiltonian system

$$q_{m,n}' = 4q_{m,n}p_{m,n} - q_{m,n}^2 - 2zq_{m,n} - 2(\Theta_0 - n), \tag{3.18a}$$

$$p_{m,n}' = -2p_{m,n}^2 + 2p_{m,n}q_{m,n} + 2zp_{m,n} - (\Theta_\infty + m). \tag{3.18b}$$

The Hamiltonian $\sigma_{m,n} = \mathrm{H}_{IV}(q_{m,n}, p_{m,n}, z)$ is

$$\sigma_{m,n} = 2q_{m,n}p_{m,n}^2 - [q_{m,n}^2 + 2zq_{m,n} + 2(\Theta_0 - n)]p_{m,n} + (\Theta_\infty + m)q_{m,n}, \tag{3.19}$$

and furthermore

$$\sigma_{m+1,n} = \sigma_{m,n} + q_{m,n}, \qquad \sigma_{m,n+1} = \sigma_{m,n} + 2p_{m,n}.$$

Consequently, from the $\mathcal{R}_\infty$ transformation (3.17) and the $\mathcal{R}_0$ transformation (3.17) it can be shown that $q_{m,n}$ and $p_{m,n}$ satisfy the following discrete equations

$$\begin{aligned} q_{m+1,n} &= -q_{m,n} + 2p_{m,n} - 2z \\ &\quad - \frac{2(q_{m,n} - 2p_{m,n} + 2z)(\Theta_\infty + m + 1)}{q_{m,n}(q_{m,n} - 2p_{m,n} + 2z) + 2(\Theta_0 - \Theta_\infty - m - n - 1)}, \\ p_{m+1,n} &= -\frac{1}{2}q_{m,n} - \frac{\Theta_0 - \Theta_\infty - m - n - 1}{q_{m,n} - 2p_{m,n} + 2z}, \end{aligned}$$

$$
\begin{aligned}
q_{m-1,n} &= -2p_{m,n} + \frac{2p_{m,n}(\Theta_0 - \Theta_\infty - m - n)}{q_{m,n}p_{m,n} - \Theta_\infty - m}, \\
p_{m-1,n} &= \frac{1}{2}q_{m,n} - p_{m,n} + z + \frac{p_{m,n}(\Theta_0 - \Theta_\infty - m - n)}{q_{m,n}p_{m,n} - \Theta_\infty - m} + \frac{\Theta_\infty + m}{2p_{m,n}}, \\
q_{m,n+1} &= -2p_{m,n} - \frac{2(\Theta_0 - \Theta_\infty - m - n - 1)}{q_{m,n} - 2p_{m,n} + 2z}, \\
p_{m,n+1} &= \frac{1}{2}q_{m,n} - p_{m,n} + z \\
&\qquad - \frac{(q_{m,n} - 2p_{m,n} + 2z)(\Theta_0 - n - 1)}{2p_{m,n}(q_{m,n} - 2p_{m,n} + 2z) + 2(\Theta_0 - \Theta_\infty - m - n - 1)}, \\
q_{m,n-1} &= 2p_{m,n} - q_{m,n} - 2z - \frac{q_{m,n}(\Theta_0 - \Theta_\infty - m - n)}{q_{m,n}p_{m,n} - \Theta_0 + n} - \frac{2(\Theta_0 - n)}{q_{m,n}}, \\
p_{m,n-1} &= -\frac{1}{2}q_{m,n} - \frac{q_{m,n}(\Theta_0 - \Theta_\infty - m - n)}{2(q_{m,n}p_{m,n} - \Theta_0 + n)}.
\end{aligned}
$$

Substituting (3.13) and (3.15), with $p_{m,n} = -\frac{1}{2}v_{m,n}$ into these difference equations yields eight difference equations for $H_{m,n}$ and $Q_{m,n}$, respectively, which are equivalent to the difference equations derived above.

4. Rational solutions of ordinary differential equations related to $\mathrm{P_{IV}}$

In this section we express rational solutions of some ordinary differential equations which are related to $\mathrm{P_{IV}}$ in terms of the generalized Hermite polynomials $H_{m,n}$ and the Okamoto polynomials $Q_{m,n}$.

Example 4.1. The equation

$$V'''' + VV'' + (V')^2 + 4zV' + 8V = 32z^2, \tag{4.1}$$

arises from the generalized scaling reduction

$$u(x,t) = \frac{\sqrt{3}}{24t}V(z) - \frac{(x - 3\kappa t^2)^2}{24t^2}, \qquad z = \frac{\sqrt[4]{3}(x + \kappa t^2)}{2\sqrt{t}}, \tag{4.2}$$

with κ an arbitrary constant, of the Boussinesq equation,

$$u_{tt} + 3(u^2)_{xx} + u_{xxxx} = 0, \tag{4.3}$$

ref. [**13**]. Equation (4.1) is solvable in terms of $\mathrm{P_{IV}}$ since if

$$V = -6(w' + w^2 + 2zw) - 4z^2 + 4(\alpha - 1), \tag{4.4}$$

where $w = w(z; \alpha, \beta)$ satisfies $\mathrm{P_{IV}}$ then V satisfies (4.1); in fact there is a one-to-one relationship between solutions of (4.1) and those of $\mathrm{P_{IV}}$ (see [**13**] for further details). Further, by differentiating (3.4) three times with respect to z, it is straightforward to show that if σ satisfies (3.4) then

$$V = -4z^2 + 2(\theta_0 + \theta_\infty) + 12\sigma', \tag{4.5}$$

satisfies (4.1). Hence, from (3.6), then rational solutions of (4.1) have the form

$$V_{m,n} = -4z^2 + 8(m - n) + 12\frac{\mathrm{d}^2}{\mathrm{d}z^2}\ln H_{m,n}, \tag{4.6a}$$

$$\widetilde{V}_{m,n} = \frac{4}{3}z^2 + 12\frac{\mathrm{d}^2}{\mathrm{d}z^2}\ln Q_{m,n}. \tag{4.6b}$$

Example 4.2. The equation

$$(W'')^2 = 4(zW' - W)^2 - 4(W')^3 + 4\kappa^2 W', \tag{4.7}$$

is a special case of (3.4), the "σ-equation" associated with the P_{IV} Hamiltonian, and arises from the scaling reduction

$$u(x,t) = t^{-1/2} R(z) \exp\{i\theta(z)\}, \quad z = x/t^{1/2}, \tag{4.8}$$

of the defocusing nonlinear Schrödinger equation

$$iu_t = u_{xx} - 2|u|^2 u \tag{4.9}$$

(see [**6, 20**] for details). Essentially the relationship between $W(z)$, the solution of (4.7), and $R(z)$, in the scaling reduction (4.8), is $W(z) = R^2(z)$, possibly after a transformation of variables. Equation (4.7) can be solved in terms of P_{IV} since if

$$W = \frac{(w')^2}{8w} - \frac{1}{8}w^3 - \frac{1}{2}zw^2 + \frac{1}{2}(\alpha + 1 - z^2)w + \frac{1}{3}(\alpha+1)z + \frac{\beta}{4w}, \tag{4.10}$$

where $w = w(z;\alpha,\beta)$ satisfies P_{IV} with the constraint $\beta = -\frac{2}{9}(\alpha+1)^2$, then (4.10) satisfies (4.7) provided that

$$\kappa^2 = \tfrac{1}{3}(\alpha+1)^2 - \tfrac{1}{2}\beta = \tfrac{4}{9}(\alpha+1)^2.$$

Therefore from 3.6, rational solutions of 4.7 have the form

$$W_n = H'_{n,n}/H_{n,n}, \qquad \kappa = \pm 2n, \tag{4.11a}$$

$$\widetilde{W}_n = \tfrac{4}{27}z^3 + Q'_{n,n}/Q_{n,n}, \quad \kappa = \pm 2(n - \tfrac{1}{3}). \tag{4.11b}$$

generalized Okamoto polynomials. Hence the rational solutions of (4.7) have a "square pattern".

Hirota and Nakamura [**23**] (see also [**6, 24, 25**]) show that the de-focusing nonlinear Schrödinger equation (4.9) has rational solutions expressed in terms of rational solutions of P_{IV}, which decay algebraically as $|x| \to \infty$, in the form

$$u_n(x,t) = g_n(x,t)/f_n(x,t), \tag{4.12}$$

where $g_n(x,t)$ is a polynomial of degree $n^2 - 1$ in x and $f_n(x,t)$ is a polynomial of degree n^2 in x. In terms of the bi-Hermite polynomials $H_{m,n}(z)$, it can be shown that

$$\begin{aligned} g_n(x,t) &= nt^{(n^2-1)/2} e^{\pi i/4} H_{n-1,n+1}(z), \\ f_n(x,t) &= t^{n^2/2} e^{-\pi i/4} H_{n,n}(z), \end{aligned} \qquad z = \frac{x e^{\pi i/4}}{2t^{1/2}}.$$

Thus we obtain the algebraically decaying rational solution of the de-focusing nonlinear Schrödinger equation (4.9) given by

$$u_n(x,t) = \frac{in}{t^{1/2}} \left[\frac{H_{n-1,n+1}(z)}{H_{n,n}(z)} \right]_{z = xe^{\pi i/4}/(2t^{1/2})}. \tag{4.13}$$

References

1. M. Abramowitz and I. A. Stegun, *Handbook of mathematical functions*, 10th ed., Dover, New York, 1972.
2. V. E. Adler, *Nonlinear chains and Painlevé equations*, Phys. D **73** (1994), 335–351.
3. H. Airault, *Rational solutions of Painlevé equations*, Stud. Appl. Math. **61** (1979), 31–53.
4. H. Airault, H. P. McKean, and J. Moser, *Rational and elliptic solutions of the* KdV *equation and related many-body problems*, Comm. Pure Appl. Math. **30** (1977), 95–148.

5. A. P. Bassom, P. A. Clarkson, and A. C. Hicks, *Bäcklund transformations and solution hierarchies for the fourth Painlevé equation*, Stud. Appl. Math. **95** (1995), 1–71.
6. M. Boiti and F. Pempinelli, *Nonlinear Schrödinger equation, Bäcklund transformations and Painlevé transcendents*, Nuovo Cimento B **59** (1980), 40–58.
7. F. Bureau, *Differential equations with fixed critical points*. II, Ann. Mat. Pura Appl. (4) **66** (1964), 1–116.
8. ———, *Differential equations with fixed critical points*, Ann. Mat. Pura Appl. (4) **64** (1964), 229–364.
9. ———, *Équations différentielles du second ordre en Y et du second degré en $\ddot{Y}$ dont l'intégrale générale est à points critiques fixes*, Ann. Mat. Pura Appl. (4) **91** (1972), 163–281.
10. J. Chazy, *Sur les équations différentielles du troisième ordre et d'ordre supérieur dont l'intégrale générale a ses points critiques fixes*, Acta Math. **34** (1911), 317–385.
11. P. A. Clarkson, *The third Painlevé equation and associated special polynomials*, J. Phys. A **36** (2003) (to appear).
12. ———, *The fourth Painlevé equation and associated special polynomials*, J. Math. Phys. **44** (2003), no. 11, 5350–5374.
13. P. A. Clarkson and M. D. Kruskal, *New similarity solutions of the Boussinesq equation*, J. Math. Phys. **30** (1989), 2201–2213.
14. P. A. Clarkson and E. L. Mansfield, *The second Painlevé equation, its hierarchy and associated special polynomials*, Nonlinearity **16** (2003), R1–R26.
15. P. A. Clarkson, E. L. Mansfield, and H. N. Webster, *On the relation between the continuous and discrete Painlevé equations*, Theoret. and Math. Phys. **122** (2000), 1–16.
16. C. M. Cosgrove and G. Scoufis, *Painlevé classification of a class of differential equations of the second order and second degree*, Stud. Appl. Math. **88** (1993), 25–87.
17. A. S. Fokas, B. Grammaticos, and A. Ramani, *From continuous to discrete Painlevé equations*, J. Math. Anal. Appl. **180** (1993), 342–360.
18. A. S. Fokas, U. Mugan, and M. J. Ablowitz, *A method of linearisation for Painlevé equations: Painlevé* IV, V, Phys. D **30** (1988), 247–283.
19. S. Fukutani, K. Okamoto, and H. Umemura, *Special polynomials and the Hirota bilinear relations of the second and fourth Painlevé equations*, Nagoya Math. J. **159** (2000), 179–200.
20. L. Gagnon, B. Grammaticos, A. Ramani, and P. Winternitz, *Lie symmetries of a generalised nonlinear Schrödinger equation*. III. *Reductions to third-order ordinary differential equations*, J. Phys. A **22** (1989), 499–509.
21. V. I. Gromak, *Theory of the fourth Painlevé equations*, Differential Equations **23** (1987), 506–513.
22. V. I. Gromak, I. Laine, and S. Shimomura, *Painlevé differential equations in the complex plane*, de Gruyter Stud. Math., vol. 28, Walter de Gruyter & Co., Berlin, 2002.
23. R. Hirota and A. Nakamura, *A new example of explode-decay solitary waves in one-dimension*, J. Phys. Soc. Japan **54** (1985), 491–499.
24. A. N. W. Hone, *Integrable systems and their finite-dimensional reductions*, Ph.D thesis, University of Edinburgh, 1996.
25. ———, *Crum transformation and rational solutions of the non-focusing nonlinear Schrödinger equation*, J. Phys. A **30** (1997), 7473–7483.
26. M. Jimbo and T. Miwa, *Monodromy preserving deformations of linear ordinary differential equations with rational coefficients*. II, Phys. D **2** (1981), 407–448.
27. K. Kajiwara and Y. Ohta, *Determinant structure of the rational solutions for the Painlevé* IV *equation*, J. Phys. A **31** (1998), 2431–2446.
28. N. A. Lukashevich, *Theory of the fourth Painlevé equation*, Differential Equations **3** (1967), 395–399.
29. Y. Murata, *Rational solutions of the second and the fourth Painlevé equations*, Funkcial. Ekvac. **28** (1985), 1–32.
30. M. Noumi and Y. Yamada, *Symmetries in the fourth Painlevé equation and Okamoto polynomials*, Nagoya Math. J. **153** (1999), 53–86.
31. K. Okamoto, *Studies on the Painlevé equations* III. *Second and fourth Painlevé equations,* P_{II} *and* P_{IV}, Math. Ann. **275** (1986), 221–255.
32. ———, *Algebraic relations among six adjacent τ-functions related to the fourth Painlevé equation*, Kyushu J. Math. **50** (1996), 513–532.

33. H. Umemura and H. Watanabe, *Solutions of the second and fourth Painlevé equations* I, Nagoya Math. J. **148** (1997), 151–198.
34. A. P. Vorob'ev, *On rational solutions of the second Painlevé equation*, Differential Equations **1** (1965), 58–59.
35. A. I. Yablonskii, *On rational solutions of the second Painlevé equations*, Vestnik Akad. Nauk BSSR Ser. Fiz. Tkh. Nauk. **3** (1959), 30–35 (Russian).

Institute of Mathematics & Statistics, University of Kent, Canterbury, CT2 7NF, UK.

E-mail address: P.A.Clarkson@kent.ac.uk

Centre de Recherches Mathématiques
CRM Proceedings and Lecture Notes
Volume **39**, 2005

Comparison of Symmetry Preserving Difference Schemes with Standard Numerical Methods

C. Cyr–Gagnon

ABSTRACT. The application of Lie group theory to difference equations is much more recent than its application to differential ones. For instance, Sophus Lie classified second-order ordinary differential equations according to their symmetries in 1891. A similar classification of second-order difference schemes was only performed in the year 2000. Here we present some new results on the symmetry classification of third-order difference schemes. We also present some numerical computations showing that symmetry preserving discretizations of differential equations have smaller errors than standard integration schemes. Work done in collaboration with A. Bourlioux and P. Winternitz.

1. Introduction

At the end of the XIXth century, the Lie group theory was perceived as a transformation theory of differential equation solutions. S. Lie classified second-order ordinary differential equations according to their symmetries in 1891. And it is only hundred years later that this theory has been developed into a powerful tool to classify and to solve differential equations. However, the applications of this theory to difference equations are much more recent. For example, a classification of second-order ordinary difference equations according to Lie groups leaving invariant solutions was performed in the year 2000 by V. Dorodnitsyn, R. Kozlov, and P. Winternitz [**1**]. In this article, they transformed the procedure for the continuous case into a similar one for the discrete case. We do a similar work for the third-order invariant schemes and we compare the symmetry preserving schemes with standard ones.

2. Third-order invariant schemes

2.1. Generalities. We present the procedure for finding third-order symmetry preserving schemes. We have new results for this classification; we found several third-order schemes. We adapted the method to determine second-order invariant

2000 *Mathematics Subject Classification.* Primary: 65L12; Secondary: 34A26.
Key words and phrases. Difference equation, Lie algebra, symmetry group, invariant scheme, prolongation.
This is the final form of the paper.

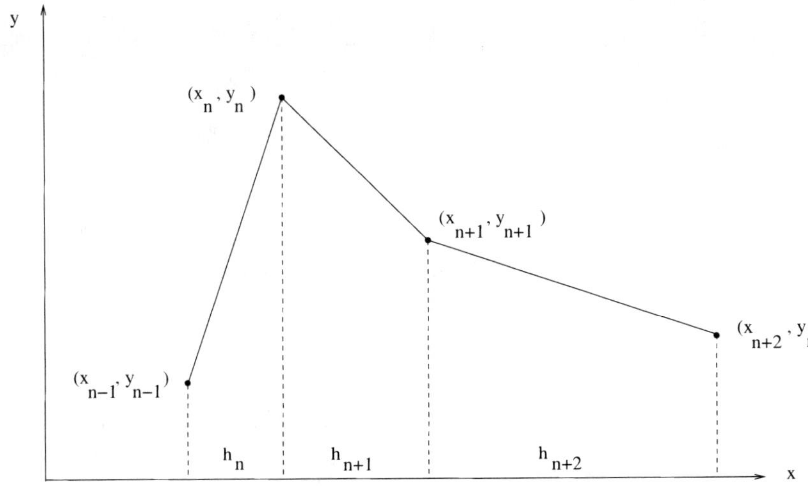

FIGURE 2.1. 4-point stencil.

schemes elaborated in *Lie group classification of second-order ordinary difference equations* [**1**].

We consider a 4-point stencil. We have an 8-dimensional subspace of the space of independent and dependent variables.

$$\{(x_{n-1}, x_n, x_{n+1}, x_{n+2}; y_{n-1}, y_n, y_{n+1}, y_{n+2})\}$$

The scheme can be represented by two equations

$$E(x_{n-1}, x_n, x_{n+1}, x_{n+2}; y_{n-1}, y_n, y_{n+1}, y_{n+2}) = 0, \tag{2.1}$$

$$F(x_{n-1}, x_n, x_{n+1}, x_{n+2}; y_{n-1}, y_n, y_{n+1}, y_{n+2}) = 0 \tag{2.2}$$

such that $\det[d(E,F)/d(x_{n-1}, y_{n-1})] \neq 0$ and $\det[d(E,F)/d(x_{n+2}, y_{n+2})] \neq 0$.

The second equation gives the lattice on which the first equation is considered.

From those four points, we can create the discrete derivative formulas. We take

$$h_n = x_n - x_{n-1}, \quad h_{n+1} = x_{n+1} - x_n, \quad h_{n+2} = x_{n+2} - x_{n+1}. \tag{2.3}$$

The first derivatives are

$$y_x^{(n)} = \frac{y_n - y_{n-1}}{h_n}, \quad y_x^{(n+1)} = \frac{y_{n+1} - y_n}{h_{n+1}}, \quad y_x^{(n+2)} = \frac{y_{n+2} - y_{n+1}}{h_{n+2}}. \tag{2.4}$$

The second derivatives are

$$\begin{aligned} y_{xx}^{(n+1)} &= \frac{2}{h_{n+1} + h_n}\left(\frac{y_{n+1} - y_n}{h_{n+1}} - \frac{y_n - y_{n-1}}{h_n}\right), \\ y_{xx}^{(n+2)} &= \frac{2}{h_{n+2} + h_{n+1}}\left(\frac{y_{n+2} - y_{n+1}}{h_{n+2}} - \frac{y_{n+1} - y_n}{h_{n+1}}\right). \end{aligned} \tag{2.5}$$

The third derivative is

$$y_{xxx}^{(n+2)} = \frac{3}{h_{n+2} + h_{n+1} + h_n}(y_{xx}^{(n+2)} - y_{xx}^{(n+1)}). \tag{2.6}$$

There are two conditions on the model. First, in the continuous limit as h_n, h_{n+1} and h_{n+2} go to 0, the equation (2.1) must go to an ordinary differential equation

$$E(x, y, y', y'', y''') = 0 \tag{2.7}$$

and the equation (2.2) must go to the identity 0=0.

Secondly, the difference system should be invariant under the same symmetry group as the preceding ODE (2.7).

The form of the vector fields is the same as for the second and third-order continuous case,

$$X_\alpha = \xi_\alpha(x,y)\frac{\partial}{\partial x} + \eta_\alpha(x,y)\frac{\partial}{\partial y}, \quad \alpha = 1, \dots, m \tag{2.8}$$

where m is the dimension of the Lie algebra.

Here we have to prolong the action of the vector fields to an 8-dimensional space.

The third discrete prolongation formula for vector fields is

$$\mathrm{pr}^{(3)} X_\alpha = \sum_{i=n-1}^{n+2} \left(\xi_\alpha(x_i, y_i)\frac{\partial}{\partial x_i} + \eta_\alpha(x_i, y_i)\frac{\partial}{\partial y_i} \right). \tag{2.9}$$

To obtain the invariant scheme, we require that

$$\mathrm{pr}^{(3)} X_\alpha \; I(x_{n-1}, x_n, x_{n+1}, x_{n+2}; y_{n-1}, y_n, y_{n+1}, y_{n+2}) = 0, \quad \alpha = 1, \dots, m. \tag{2.10}$$

By solving these equations, using the method of characteristics, we obtain a set of elementary invariants $\{I_1, \dots, I_r\}$. We determine r as follows:

$$r = \dim M - \operatorname{rank} Z$$

where $M \subset \{(x_{n-1}, x_n, x_{n+1}, x_{n+2}; y_{n-1}, y_n, y_{n+1}, y_{n+2})\}$ and the matrix Z is

$$Z = \begin{pmatrix} \xi_{1,n-1} & \eta_{1,n-1} & \xi_{1,n} & \eta_{1,n} & \xi_{1,n+1} & \eta_{1,n+1} & \xi_{1,n+2} & \eta_{1,n+2} \\ \vdots & \vdots & \vdots & \vdots & \vdots & \vdots & \vdots & \vdots \\ \xi_{m,n-1} & \eta_{m,n-1} & \xi_{m,n} & \eta_{m,n} & \xi_{m,n+1} & \eta_{m,n+1} & \xi_{m,n+2} & \eta_{m,n+2} \end{pmatrix}$$

Finally, the symmetry preserving difference scheme is given by two functions of the invariants.

$$E(I_1, \dots, I_r) = 0, \quad F(I_1, \dots, I_r) = 0. \tag{2.11}$$

Example 2.1. We consider the following symmetry algebra of dimension 4,

$$X_1 = \frac{\partial}{\partial x}, \quad X_2 = \frac{\partial}{\partial y}, \quad X_3 = y\frac{\partial}{\partial y}, \quad X_4 = y^2\frac{\partial}{\partial y}.$$

We already know [**2**] that the ordinary differential equation invariant under this group is

$$\frac{y'''}{y'} - \frac{3}{2}\frac{y''^2}{y'^2} = C.$$

Let's find the invariant scheme. We demand

$$\mathrm{pr}^{(3)} X_1 E = \mathrm{pr}^{(3)} X_2 E = \mathrm{pr}^{(3)} X_3 E = 0$$

and we obtain the following invariants:

$$I_1 = x_n - x_{n-1}, \quad I_2 = x_{n+1} - x_n, \quad I_3 = x_{n+2} - x_{n+1},$$
$$I_4 = \frac{y_n - y_{n-1}}{y_{n+1} - y_n}, \quad I_5 = \frac{y_{n+1} - y_n}{y_{n+2} - y_{n+1}}.$$

Then, we have to satisfy the condition

$$\begin{aligned}
&\mathrm{pr}^{(3)}\, X_4 E(I_1, \dots, I_5) = 0 \\
&\iff \sum_{i=1}^{5} \mathrm{pr}^{(3)}\, X_4 I_i) \frac{\partial E}{\partial I_i} = 0 \\
&\iff \frac{(y_n - y_{n-1})(y_{n+1} - y_{n-1})}{y_{n+1} - y_n} \frac{\partial E}{\partial I_4} + \frac{(y_{n+1} - y_n)(y_{n+2} - y_n)}{y_{n+2} - y_n} \frac{\partial E}{\partial I_5} = 0.
\end{aligned}$$

Because $\partial E/\partial I_1$, $\partial E/\partial I_2$ and $\partial E/\partial I_3$ do not figure in this equation, we already get the invariants,

$$I_1 = x_n - x_{n-1}, \quad I_2 = x_{n+1} - x_n, \quad I_3 = x_{n+2} - x_{n+1}.$$

We divide the equation by $(y_{n+1} - y_n)$ and we obtain,

$$I_4(1 + I_4) \frac{\partial E}{\partial I_4} + (1 + I_5) \frac{\partial E}{\partial I_5} = 0. \tag{2.12}$$

By using the method of characteristics, we solve the PDE (2.12) to obtain,

$$\frac{dI_4}{I_4(1 + I_4)} = \frac{dI_5}{1 + I_5}$$

and the invariant is

$$\phi = \frac{y_{n+2} - y_n}{y_{n+2} - y_{n+1}} \frac{y_{n+1} - y_{n-1}}{y_n - y_{n-1}}.$$

So, a basis of the invariants is

$$\{I_1, I_2, I_3, \phi\}$$

and there are 4 invariants, it really is the dimension of M, here 8, minus the rank of Z, here 4.

Let's consider the step h constant. It is possible because I_1, I_2 and I_3 are invariants.

So, our aim is to transform ϕ to obtain an invariant scheme which goes to the invariant EDO when $h \to 0$.

To do that, we use Taylor series development,

$$\begin{aligned}
y_n &= y(x), \\
y_{n+1} &= y + hy' + \frac{h^2}{2} y'' + \frac{h^3}{6} y''' + \cdots, \\
y_{n-1} &= y - hy' + \frac{h^2}{2} y'' - \frac{h^3}{6} y''' + \cdots, \\
y_{n+2} &= y + 2hy' + 2h^2 y'' + \frac{4h^3}{3} y''' + \cdots.
\end{aligned}$$

We find

$$\phi = 4 - 2h^2 \left(\frac{y'''}{y'} - \frac{3}{2} \left(\frac{y''}{y'} \right)^2 \right) + O(h^3) \iff \frac{4 - \phi}{2h^2} = \frac{y'''}{y'} - \frac{3}{2} \left(\frac{y''}{y'} \right)^2 + O(h).$$

The invariant model can be written as

$$\frac{4-\phi}{2h^2} = C \tag{2.13}$$

with a constant step h.

3. Numerical comparison

In this section, we compare symmetry preserving schemes with centered-difference ones.

3.1. First example. Let us consider the following second-order linear equation

$$y'' = y \tag{3.1}$$

and its symmetry group generated by the Lie algebra

$$X_1 = \frac{\partial}{\partial x}, \quad X_2 = y\frac{\partial}{\partial y}, \quad X_3 = e^x\frac{\partial}{\partial y}, \quad X_4 = e^{-x}\frac{\partial}{\partial y},$$
$$X_5 = e^{2x}\left(\frac{\partial}{\partial x} + y\frac{\partial}{\partial y}r\right), \quad X_6 = ye^x\left(\frac{\partial}{\partial x} + y\frac{\partial}{\partial y}\right),$$
$$X_7 = e^{-2x}\left(\frac{\partial}{\partial x} - y\frac{\partial}{\partial y}\right), \quad X_8 = ye^{-x}\left(\frac{\partial}{\partial x} - y\frac{\partial}{\partial y}\right).$$

For the subalgebra $\{X_1, \dots, X_6\}$, the second discrete prolongation does not allow any invariant. However, it allows an invariant manifold

$$\begin{aligned} I = y_n e^{-x_n}(e^{-2x_{n+1}} - e^{-2x_{n-1}}) + y_{n+1}e^{-x_{n+1}}(e^{-2x_{n-1}} - e^{-2x_n}) \\ + y_{n-1}e^{-x_{n-1}}(e^{-2x_n} - e^{-2x_{n+1}}) = 0 \end{aligned} \tag{3.2}$$

and the expression

$$S = \frac{e^{-2x_n} - e^{-2x_{n-1}}}{e^{-2x_{n+1}} - e^{-2x_n}}$$

is an invariant over the manifold (3.2).

Thus, a symmetry preserving scheme can be

$$I = 0, \tag{3.3}$$
$$S = C, \quad \text{where } C \text{ is a constant.} \tag{3.4}$$

First, we study the lattice given by (3.4).

We set

$$h_- = x_n - x_{n-1}, \quad h_+ = x_{n+1} - x_n.$$

So, we have

$$\frac{e^{-2x_n} - e^{-2(x_n-h_-)}}{e^{-2(x_n+h_+)} - e^{-2x_n}} = C \iff 1 - e^{2h_-} = c(c^{-2h_+} - 1). \tag{3.5}$$

It allows us to write h_+ as a function of h_-,

$$h_+ = -\frac{1}{2}\ln\left(\frac{1-e^{2h_-}}{C} + 1\right).$$

But for h to be positive, we must have

$$0 < \frac{1-e^{2h_-}}{C} + 1 < 1 \iff -1 < \frac{1-e^{2h_-}}{C} < 0. \tag{3.6}$$

Because $h_- > 0$, we have $e^{2h_-} > 1$ and $C > 0$. So, the condition is

$$-1 < \frac{1 - e^{2h_-}}{C} \iff h_- > \frac{\ln(C+1)}{2}. \tag{3.7}$$

Moreover, there is a difficulty because the relation of h_+ in function of h_- is decreasing. So, there exists an h_- such that $h_- < \ln(C+1)/2$. Therefore, we have to make a good choice of C and the initial h to be able to cover the whole interval of integration.

Our aim is to compare the symmetry preserving scheme with the standard centered-difference scheme. To do this, we should take a constant h. However, we must verify if the lattice (3.4) allows an equidistant lattice.

We assume $h_- = h_+ = h$. The equation (3.5) gives

$$1 - e^{2h} = C(e^{-2h} - 1) \iff Ce^{-2h} + e^{2h} = C + 1 \iff h = \frac{\ln C}{2}. \tag{3.8}$$

Hence, we can say that it is possible to consider a constant $h = \ln C/2$ with the supplementary condition, $C > 1$.

Secondly, we study the difference equations of the schemes. The centered-difference scheme is

$$\frac{y_{n+1} - 2y_n + y_{n+1}}{h^2} = y_n.$$

We now simplify the equation (3.3) by considering h constant. We take $x_n = x_0 + nh$ where x_0 is the lower boundary of the interval of integration and n is the number of achieved steps.

We obtain

$$y_{n+1} - (e^{-h} + e^h)y_n + y_{n-1} = 0. \tag{3.9}$$

It is interesting to transform this scheme to a form similar to the centered-difference scheme.

We can rewrite the scheme (3.9) as

$$\frac{y_{n+1} - 2y_n + y_{n+1}}{h^2} = 2\left(\frac{\cosh h - 1}{h^2}\right)y_n.$$

Thus, we can observe that the difference is the coefficient of y_n in the right part.

The exact solution of the ODE (3.1) is

$$y(x) = c_1 e^x + c_2 e^{-x}$$

where c_1 and c_2 are arbitrary constants.

Moreover, we observe that the exact solution satisfies the symmetry preserving scheme. It follows that the local truncation error is 0. Thus, if we could work with an infinite arithmetic, the scheme would give the exact solution. But in the numerical context, the precison of the computer is not infinite. So, it could be interesting to study if the rounding errors increase such that the numerical solution differs from the exact solution.

We consider the initial-value problem $\{x_0, y_0, x_1, y_1\}$ and we choose the exact solution $y(x) = e^{-x} + e^x$ where x is in $[0, 2]$.

We take

$$x_0 = 0, \quad y_0 = 2, \qquad x_1 = h, \quad y_1 = e^{-h} + e^h.$$

The Figures (3.1), (3.2) and (3.3) show the graphics of absolute errors for different values of h with a logarithmic scale.

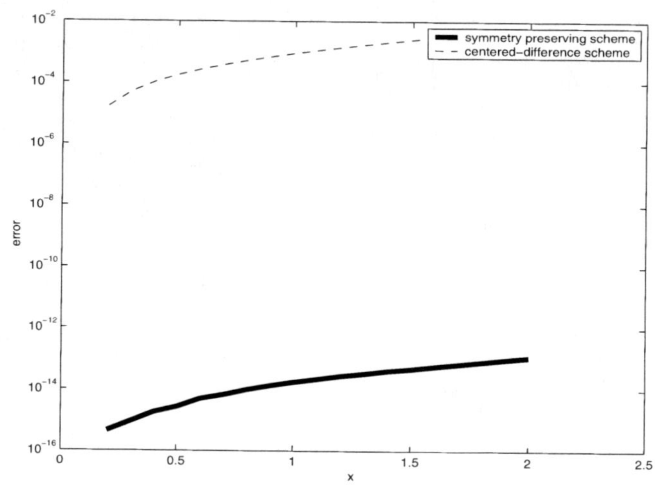

FIGURE 3.1. Absolute errors for $h = 0.1$.

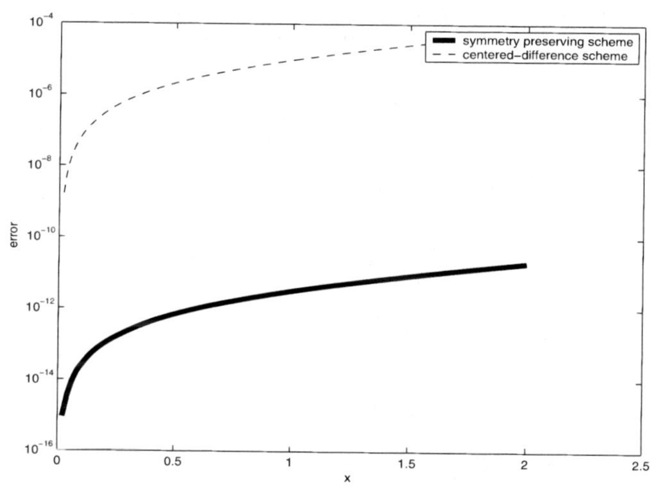

FIGURE 3.2. Absolute errors for $h = 0.01$.

First, we observe that the error for the symmetry preserving scheme is not really zero. There is an accumulation of round off errors. Indeed, when h is divided by 10, the error for this scheme is multiplied by 100. Then, we remark that the error for the centered-difference scheme is larger than the one for the scheme (3.9). Thus, the scheme obtained by the Lie group theory seems to have a certain advantage as far as precision is concerned. We also note that the more h decreases, the more the two curves get closer. This behavior is reasonable because when the number of steps increases, the error of discretization decreases and round off errors of the invariant scheme increase. Moreover, the rate of convergence is 2 for the centered-difference scheme, since when h is divided by 10, the error is divided by 100.

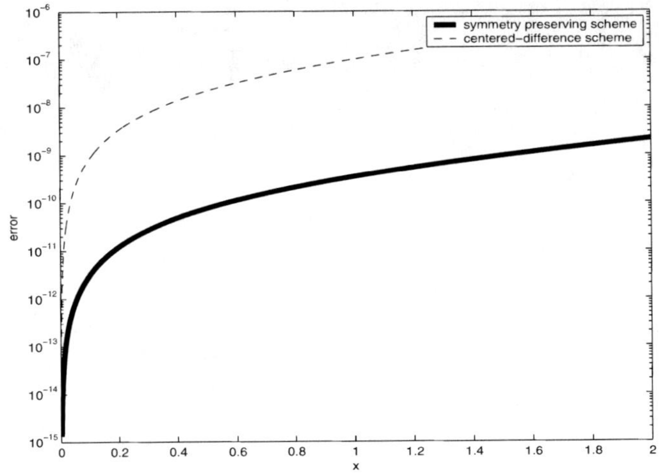

FIGURE 3.3. Absolute errors for $h = 0.001$.

3.2. Second example. We consider the second-order ordinary differential equation,

$$y''x^2 + 4xy' + 2y = (2xy + x^2y')^{k-2/k-1}, \quad \text{with } k \neq 0, \tfrac{1}{2}, 1, 2. \tag{3.10}$$

The Lie algebra of the symmetry group of this equation is

$$X_1 = \frac{\partial}{\partial x} - \frac{2y}{x}\frac{\partial}{\partial y}, \quad X_2 = \frac{1}{x^2}\frac{\partial}{\partial y}, \quad X_3 = x\frac{\partial}{\partial x} + (k-2)y\frac{\partial}{\partial y}$$

and we obtain three difference invariants:

$$I_1 = \frac{x_{n+1} - x_n}{x_n - x_{n-1}}, \quad I_2 = \frac{x_{n+1}^2 y_{n+1} - x_n^2 y_n}{(x_{n+1} - x_n)^k}, \quad I_3 = \frac{x_n^2 y_n - x_{n-1}^2 y_{n-1}}{(x_n - x_{n-1})^k}$$

The symmetry preserving scheme is

$$I_2 - I_3 = \left(\frac{I_2 + I_3}{2}\right)^{k-2/k-1}, \quad I_1 = 1$$

and in term of the variables (x_i, y_i), with $i = n-1, \ldots, n+1$, it becomes

$$\begin{aligned} & x_{n+1}^2 y_{n+1} - 2x_n^2 y_n + x_{n-1}^2 y_{n-1} \\ & \qquad = \left(\tfrac{1}{2}\right)^{k-2/k-1} h^{k/k-1} (x_{n+1}^2 y_{n+1} - x_{n-1}^2 y_{n-1})^{k-2/k-1}, \\ & x_{n+1} - x_n = x_n - x_{n-1} = h. \end{aligned} \tag{3.11}$$

The second equation represents an equidistant lattice with step h. We want to compare this scheme with the standard one

$$\frac{(y_{n+1} - 2y_n + y_{n-1})}{h^2} x_n^2 + 2x_n \frac{(y_{n+1} - y_{n-1})}{h} + 2y_n = \left(2x_n y_n + x_n^2 \frac{(y_{n+1} - y_{n-1})}{2h}\right)^{k-2/k-1}$$

where h is a constant step.

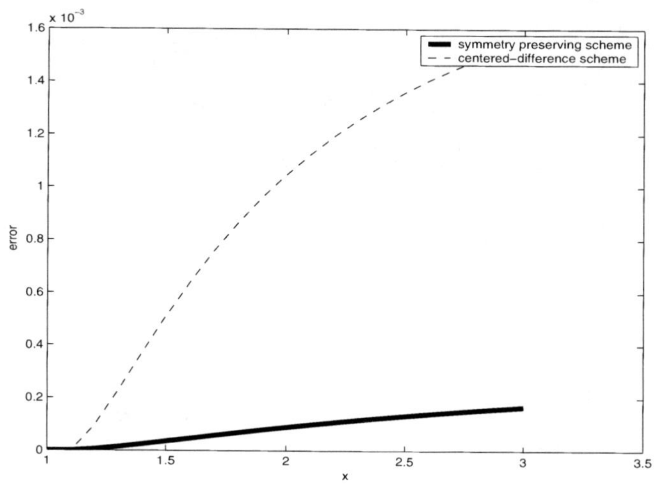

FIGURE 3.4. Absolute errors for $h = 0.1$.

The difficulty is that both schemes are non linear and implicit. Thus, we have to solve a nonlinear equation at each step. We used the function *fzero* in MATLAB.

The exact solution is of the form

$$y(x) = \left(\frac{1}{k-1}\right)^{k-1} \frac{1}{x^2 k}(x - x_0)^k + \frac{u_0}{x^2}.$$

For this example, we choose $k = 3$, $x_0 = 0$, $u_0 = 0$ and x in $[1, 3]$. So we have

$$y(x) = \frac{x}{12} + \frac{1}{x^2}, \quad x \neq 0.$$

For the initial-value problem $\{x_0, y_0, x_1, y_1\}$, we take

$$x_0 = 1, \quad y_0 = y(1) = \frac{13}{12},$$

$$x_1 = 1 + h, \quad y_1 = y(1+h) = \frac{1+h}{12} + \frac{1}{(1+h)^2}.$$

The figures (3.4), (3.5) and (3.6) show the graphics of absolute errors for different values of h.

First, we observe that the symmetry preserving scheme has smaller error than the centered scheme. Secondly, both methods have a rate of convergence of 2 because when h is divided by 10, the error is divided by 100. Finally, we know that the behavior of the error is

$$E \leq M\ h^k$$

where M is a constant and k is the rate of convergence. We obtained that k is the same for both methods. The symmetry preserving scheme is more precise because M is smaller.

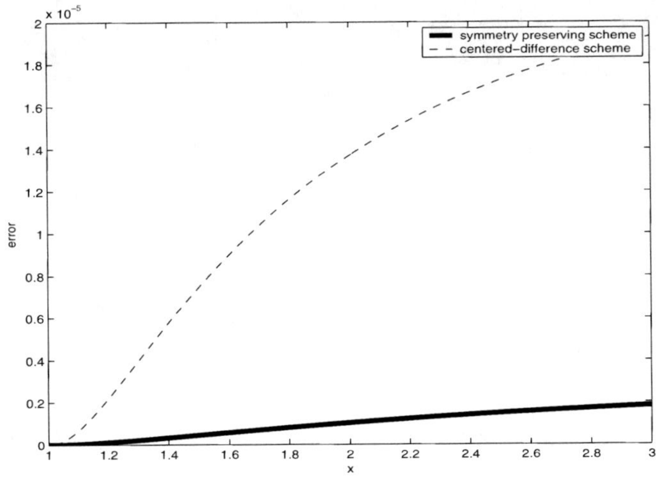

FIGURE 3.5. Absolute errors for $h = 0.01$.

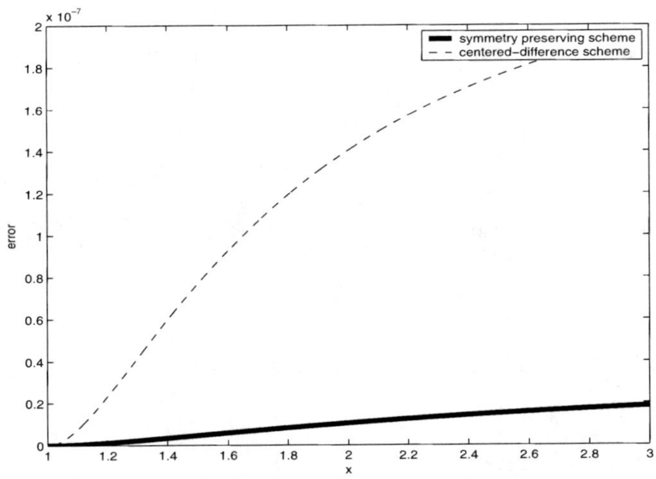

FIGURE 3.6. Absolute errors for $h = 0.001$.

3.3. Third example. We consider the following third-order ordinary equation:

$$\frac{y'''}{y'} - \frac{3}{2}\frac{y''^2}{y'^2} = \frac{-k^2}{2}. \tag{3.12}$$

The Lie symmetry algebra of this equation is

$$X_1 = \frac{\partial}{\partial x}, \quad X_2 = \frac{\partial}{\partial y}, \quad X_3 = y\frac{\partial}{\partial y}, \quad X_4 = y^2\frac{\partial}{\partial y}$$

and we obtain four invariants

$$I_1 = x_n - x_{n-1}, \quad I_2 = x_{n+1} - x_n, \quad I_3 = x_{n+2} - x_{n+1},$$
$$\phi = \frac{y_{n+2} - y_n}{y_{n+2} - y_{n+1}} \frac{y_{n+1} - y_{n-1}}{y_n - y_{n-1}}.$$

To compare the symmetry preserving scheme with a standard one, we choose a constant step h.

We obtain the following symmetry preserving scheme:

$$\frac{4-\phi}{2h^2} = \frac{-k^2}{2}.$$

The standard method is to use the equation

$$\frac{3\,(y_{n+2} - 3y_{n+1} + 3y_n - y_{n-1})}{h^2\,(y_{n+2} - y_{n-1})} - \frac{27\,(y_{n+2} - y_{n+1} - y_n + y_{n-1})^2}{8h^2\,(y_{n+2} - y_{n-1})^2} = \frac{-k^2}{2}.$$

We already observe that the symmetry preserving scheme will be easier to apply because it is linear and the centered-difference scheme is not.

The exact solution is of the form

$$y(x) = \frac{A}{k(e^{kx_0} - e^{kx})} + d.$$

For this example, we choose $k = 1$, $A = 1$, $d = 0$, $x_0 = 0$ and x ε [1,11]. So we have

$$y(x) = \frac{1}{1 - e^x}.$$

For the initial-value problem $\{x_0, y_0, x_1, y_1, x_2, y_2\}$, we take

$$x_0 = 1, \quad y_0 = (1-e)^{-1},$$
$$x_1 = 1 + h, \quad y_1 = (1 - e^{1+h})^{-1},$$
$$x_2 = 1 + 2h, \quad y_2 = (1 - e^{1+2h})^{-1}.$$

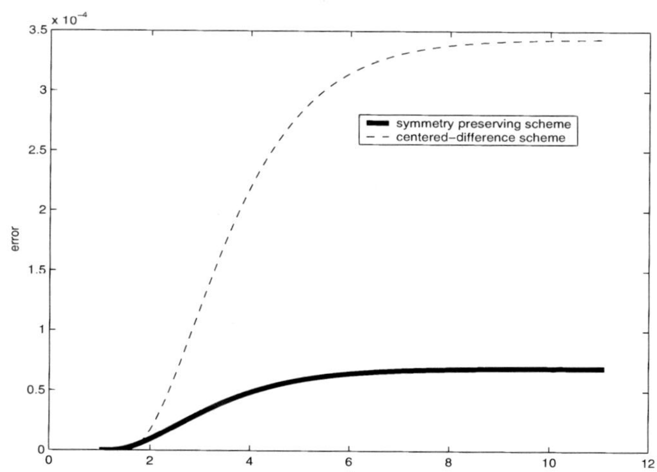

FIGURE 3.7. Absolute errors for $h = 0.1$.

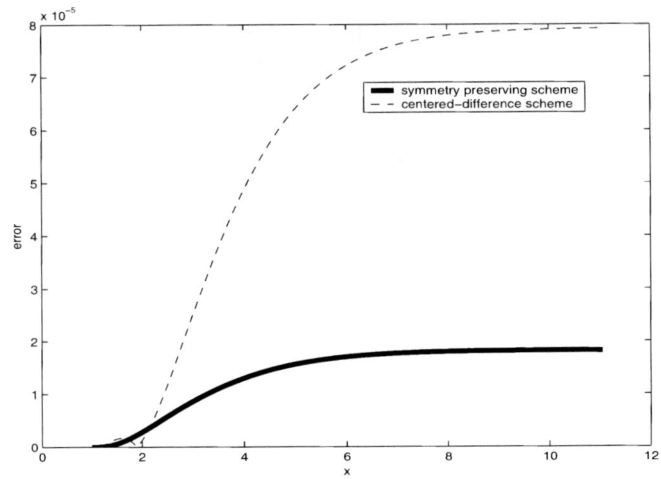

FIGURE 3.8. Absolute errors for $h = 0.05$.

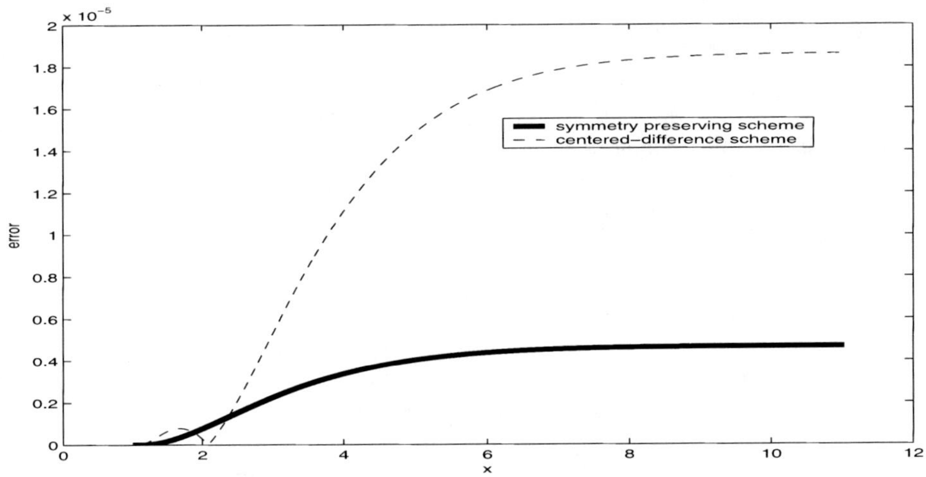

FIGURE 3.9. Absolute errors for $h = 0.025$.

The figures (3.7), (3.8) and (3.9) show the graphics of absolute errors for different values of h.
We observe once again that the invariant scheme is more precise because it has smaller error of discretization. Both methods also have a rate of convergence of 2 because when h is divided by 2, the error is divided by 4. Finally, knowing that the behavior of the error is $E \leq M\ h^k$ and that both methods has the same rate of convergence, we can conclude that the invariant scheme is more precise because M is smaller.

4. Conclusion

The main conclusions to be drawn from the research reported here are the followings:

(1) It is possible to discretize ordinary differential equations while preserving most, if not all of their Lie point symmetries.
(2) The corresponding invariant difference schemes can be used in numerical calculations and actually compare favorably to standard schemes.

References

1. V. Dorodnitsyn, R. Kozlov, and P. Winternitz, *Lie group classification of second-order ordinary difference equations*, J. Math. Phys. **41** (2000), no. 1, 480–504.
2. Omri Gat, *Symmetry algebras of third-order ordinary differential equations*, J. Math. Phys. **33** (1992), no. 9, 2966–2971.
3. A. Gonzalez–Lopez, N. Kamran, and P. J. Olver, *Lie algebras of vector fields in the real plane*, Proc. London Math. Soc. (3) **64** (1992), 339–368.
4. James E. Humphreys, *Introduction to Lie algebras and representation theory*, Springer-Verlag New York Inc., 1972.
5. D. Levi, S Tremblay, and P. Winternitz, *Lie point symmetries of difference equations and lattices*, J. Phys. **33** (2000), 8507–8523.
6. Sophus Lie, *Vorlesungen ber continuierliche Gruppen mit geometrischen und anderen Anwendungen*, Dr. G. Scheffers, Teubner, Leipzig, 1893.
7. P. J. Olver, *Applications of Lie groups to differential equations*, Springer-Verlag New York Inc., 1986.
8. P. Winternitz, *Lie groups and solutions of nonlinear partial differential equations*, CRM-1841, 1993.

DÉPARTEMENT DE MATHÉMATIQUES ET DE STATISTIQUE, UNIVERSITÉ DE MONTRÉAL, C.P. 6128, SUCC. CENTRE-VILLE, MONTRÉAL, QUÉ. H3C 3J7, CANADA.

Current address: Cégep de Saint-Laurent, 625 avenue Sainte-Croix, Saint-Laurent, Qué. H4L 3X7, Canada.

E-mail address: cgagnon.netrover.com

E-mail address: ccyrgagnon@cegep-st-laurent.qc.ca

Centre de Recherches Mathématiques
CRM Proceedings and Lecture Notes
Volume **39**, 2005

Symbolic Computation of Polynomial Conserved Densities, Generalized Symmetries, and Recursion Operators for Nonlinear Differential-Difference Equations

W. Hereman, J. A. Sanders, J. Sayers, and J. P. Wang

This paper is dedicated to Ryan Sayers (1982–2003)

ABSTRACT. Algorithms for the symbolic computation of polynomial conserved densities, fluxes, generalized symmetries, and recursion operators for systems of nonlinear differential-difference equations are presented. In the algorithms we use discrete versions of the Fréchet and variational derivatives and the Euler and homotopy operators.

The algorithms are illustrated for prototypical nonlinear polynomial lattices, including the Kac–van Moerbeke (Volterra) and Toda lattices. Results are shown for the modified Volterra and Ablowitz–Ladik lattices.

1. Introduction

The study of complete integrability of nonlinear differential-difference equations (DDEs) largely parallels that of PDEs [**1, 3, 24**]. Indeed, as in the continuous case, the existence of large numbers of generalized symmetries and conserved densities is a good indicator for complete integrability. Albeit useful, such predictors do not provide proof of complete integrability. Based on the first few densities and symmetries, quite often one can explicitly construct a recursion operator which maps symmetries of the equation into new symmetries. The existence of a recursion operator, which allows one to generate commuting symmetries, then confirms complete integrability.

There is a vast body of work on the complete integrability of DDEs. Consult e.g. [**3, 13**] for various approaches and additional references. In this article we describe algorithms to symbolically compute polynomial conservation laws, generalized symmetries, and recursion operators for DDEs. The design of these algorithms

---

2000 *Mathematics Subject Classification.* Primary: 37J35, 37K10; Secondary: 35Q58, 37K05.

This material is based upon work supported by the National Science Foundation under Grant no. CCR-9901929.

M. Hickman and B. Deconinck are gratefully acknowledged for valuable discussions. D. Baldwin is thanked for proof reading the manuscript.

This is the final form of the paper.

relies on related work for PDEs [**6, 11, 12**] and work by Oevel *et al* [**19**] and Zhang *et al* [**25**].

In our algorithms we use discrete versions of the Fréchet and variational derivatives, as well as discrete Euler and homotopy operators. We draw on the close analogy between the continuous and discrete (in space) cases. As shown in [**14, 16**], this analogy can be completely formalized and both theories can be formulated in terms of complexes. The same applies for the formulation in terms of Lie algebra complexes in [**24**]. This allows one to translate by analogy the existing algorithms immediately (although complications arise when there is explicit dependence on the lattice point, n, in the discrete case). One of the more useful tools following from the abstract theory is the homotopy operator, which is based on scaling vector fields, and goes back to Poincaré in the continuous case. This operator allows one to directly integrate differential forms and can be straightforwardly implemented in computer algebra systems (CAS), since it reduces to integration over one scaling parameter. The discrete analogue as described in [**14, 16**] does the corresponding job in the discrete case and we use it to compute fluxes. In this paper we do not explicitly use the abstract framework, yet, it has been a motivating force for the development of our algorithms.

In contrast to the symmetry approach in [**3**], our algorithms rely on specific assumptions. For example, we will use the dilation invariance of DDEs in the construction of densities, symmetries, and recursion operators. At the cost of generality, our algorithms can be implemented in major CAS. Our Mathematica package `InvariantsSymmetries.m` [**7**] computes densities and symmetries, and therefore aids in automated testing of complete integrability of semi-discrete lattices. Mathematica code to compute recursion operators of DDEs is under development.

The paper is organized as follows. In Section 2, we present key definitions, necessary tools, and prototypical examples which will be used throughout the paper. The examples include the Kac–van Moerbeke (KvM) [**15**], Toda [**23**] and Ablowitz–Ladik (AL) lattices [**2**]. The discrete higher Euler operators (variational derivatives) and the discrete homotopy operator are introduced in Section 3. These operators are applied in the construction of densities and fluxes in Section 5. The algorithm for higher-order symmetries is outlined in Section 6. The algorithms for scalar and matrix recursion operators are given in Section 7 and 8. The paper concludes with two more examples in Section 9.

2. Key definitions and prototypical examples

Definition 2.1. A nonlinear (autonomous) DDE is an equation of the form

$$\dot{\mathbf{u}}_n = \mathbf{F}(\ldots, \mathbf{u}_{n-1}, \mathbf{u}_n, \mathbf{u}_{n+1}, \ldots), \tag{2.1}$$

where $\mathbf{u}_n$ and $\mathbf{F}$ are vector-valued functions with N components. The integer n corresponds to discretization in space; the dot denotes one derivative with respect to the continuous time variable (t).

For simplicity, we denote the components of $\mathbf{u}_n$ by $(u_n, v_n, w_n, \ldots)$ and write $\mathbf{F}(\mathbf{u}_n)$, although $\mathbf{F}$ typically depends on $\mathbf{u}_n$ and a finite number of its forward and backward shifts. We assume that $\mathbf{F}$ is polynomial with constant coefficients. If present, parameters are denoted by lower-case Greek letters. No restrictions are imposed on the forward or backward shifts or the degree of nonlinearity in $\mathbf{F}$.

Example 2.2. The AL lattice [**2**],

$$\begin{aligned}\dot{u}_n &= (u_{n+1} - 2u_n + u_{n-1}) + u_n v_n (u_{n+1} + u_{n-1}),\\ \dot{v}_n &= -(v_{n+1} - 2v_n + v_{n-1}) - u_n v_n (v_{n+1} + v_{n-1}),\end{aligned} \tag{2.2}$$

is a completely integrable discretization of the nonlinear Schrödinger equation.

Definition 2.3. A DDE is said to be dilation invariant if it is invariant under a scaling (dilation) symmetry.

Example 2.4. The KvM lattice [**15**],

$$\dot{u}_n = u_n (u_{n+1} - u_{n-1}), \tag{2.3}$$

is invariant under $(t, u_n) \to (\lambda^{-1} t, \lambda u_n)$, where λ is an arbitrary scaling parameter.

Example 2.5. The Toda lattice [**23**] in polynomial form [**8**],

$$\dot{u}_n = v_{n-1} - v_n, \quad \dot{v}_n = v_n (u_n - u_{n+1}), \tag{2.4}$$

is invariant under the scaling symmetry

$$(t, u_n, v_n) \to (\lambda^{-1} t, \lambda^1 u_n, \lambda^2 v_n). \tag{2.5}$$

Definition 2.6. We define the weight, w, of a variable as the exponent in the scaling parameter λ which multiplies the variable.

As a result of definition 2.6, t is always replaced by $\frac{t}{\lambda}$ and $w(\mathrm{d}/\mathrm{d}t) = w(\mathrm{D}_t) = 1$. In view of (2.5), we have $w(u_n) = 1$, and $w(v_n) = 2$ for the Toda lattice.

Weights of dependent variables are nonnegative, rational, and independent of n. For example, $w(u_{n-1}) = w(u_n) = w(u_{n+1})$, etc.

Definition 2.7. The rank of a monomial is defined as the total weight of the monomial. An expression is uniform in rank if all its terms have the same rank.

Example 2.8. In the first equation of (2.4), all the monomials have rank 2; in the second equation all the monomials have rank 3. Conversely, requiring uniformity in rank for each equation in (2.4) allows one to compute the weights of the dependent variables (and thus the scaling symmetry) with simple linear algebra. Indeed,

$$w(u_n) + 1 = w(v_n), \quad w(v_n) + 1 = w(u_n) + w(v_n), \tag{2.6}$$

yields

$$w(u_n) = 1, \quad w(v_n) = 2, \tag{2.7}$$

which is consistent with (2.5).

Dilation symmetries, which are special Lie-point symmetries, are common to many lattice equations. Lattices described by polynomial DDEs that do not admit a dilation symmetry can be made scaling invariant by extending the set of dependent variables with auxiliary parameters with appropriate scales.

Example 2.9. The AL lattice is not dilation invariant. We therefore introduce an auxiliary parameter α. Replacing (2.2) by

$$\begin{aligned}\dot{u}_n &= \alpha(u_{n+1} - 2u_n + u_{n-1}) + u_n v_n (u_{n+1} + u_{n-1}),\\ \dot{v}_n &= -\alpha(v_{n+1} - 2v_n + v_{n-1}) - u_n v_n (v_{n+1} + v_{n-1}),\end{aligned} \tag{2.8}$$

and requiring uniformity in rank then gives

$$\begin{aligned} w(u_n)+1 &= w(\alpha)+w(u_n) = 2w(u_n)+w(v_n), \\ w(v_n)+1 &= w(\alpha)+w(v_n) = 2w(v_n)+w(u_n). \end{aligned} \tag{2.9}$$

Obviously, $w(u_n) + w(v_n) = w(\alpha) = 1$. Several scales are possible. The choice $w(u_n) = w(v_n) = \frac{1}{2}, w(\alpha) = 1$, corresponds to the scaling symmetry

$$(t, u_n, v_n, \alpha) \to (\lambda^{-1}t, \lambda^{1/2}u_n, \lambda^{1/2}v_n, \lambda\alpha). \tag{2.10}$$

Definition 2.10. A scalar function $\rho_n(\mathbf{u}_n)$ is a conserved density of (2.1) if there exists a scalar function $J_n(\mathbf{u}_n)$, called the associated flux, such that [**20**]

$$\mathrm{D}_t\rho_n + \Delta J_n = 0 \tag{2.11}$$

is satisfied on the solutions of (2.1).

In (2.11), we used the (forward) difference operator, $\Delta = \mathrm{D} - \mathrm{I}$, defined by

$$\Delta J_n = (\mathrm{D}-\mathrm{I})J_n = J_{n+1} - J_n, \tag{2.12}$$

where D denotes the up-shift (forward or right-shift) operator, $\mathrm{D}J_n = J_{n+1}$, and I is the identity operator. The operator Δ takes the role of a spatial derivative on the shifted variables as many DDEs arise from discretization of a PDE in $(1+1)$ variables. Most, but not all, densities are polynomial in $\mathbf{u}_n$.

Example 2.11. The first three density-flux pairs [**12**] for (2.3) are

$$\rho_n^{(0)} = \ln(u_n), \qquad J_n^{(0)} = u_n + u_{n-1}, \tag{2.13}$$

$$\rho_n^{(1)} = u_n, \qquad J_n^{(1)} = u_n u_{n-1}, \tag{2.14}$$

$$\rho_n^{(2)} = \tfrac{1}{2}u_n^2 + u_n u_{n+1}, \qquad J_n^{(2)} = u_{n-1}u_n(u_n + u_{n+1}). \tag{2.15}$$

Example 2.12. The first four density-flux pairs [**8**] for (2.4) are

$$\rho_n^{(0)} = \ln(v_n), \qquad J_n^{(0)} = u_n, \tag{2.16}$$

$$\rho_n^{(1)} = u_n, \qquad J_n^{(1)} = v_{n-1}, \tag{2.17}$$

$$\rho_n^{(2)} = \tfrac{1}{2}u_n^2 + v_n, \qquad J_n^{(2)} = u_n v_{n-1}, \tag{2.18}$$

$$\rho_n^{(3)} = \tfrac{1}{3}{u_n}^3 + u_n(v_{n-1} + v_n), \qquad J_n^{(3)} = u_{n-1}u_n v_{n-1} + {v_{n-1}}^2. \tag{2.19}$$

The densities in (2.16)–(2.19) are uniform of ranks 0 through 3. The corresponding fluxes are also uniform in rank with ranks 1 through 4. In general, if in (2.11) rank $\rho_n = R$ then rank $J_n = R + 1$, since $w(\mathrm{D}_t) = 1$. All the pieces in (2.11) are uniform in rank. This comes as no surprise since the conservation law (2.11) holds on solutions of (2.1), hence it 'inherits' the dilation symmetry of (2.1).

In Section 5 we will give an algorithm to compute polynomial conserved densities and fluxes and use (2.4) to illustrate the steps. Non-polynomial densities (which are usually easy to find by hand) can be computed with the method given in [**13**].

Definition 2.13. Compositions of D and it's inverse D^{-1} define an equivalence relation ($\equiv$) on monomials in the components of $\mathbf{u}_n$. All shifted monomials are equivalent.

Example 2.14. For example, $u_{n-1}v_{n+1} \equiv u_n v_{n+2} \equiv u_{n+1}v_{n+3} \equiv u_{n+2}v_{n+4}$.

Factors in a monomial in u_n and its shifts are ordered by $u_{n+j} \prec u_{n+k}$ if $j < k$.

Definition 2.15. The main representative of an equivalence class of monomials in u_n and its shifts is the monomial with u_n in the first position.

Example 2.16. For example, $u_n u_{n+2}$ (not $u_{n-2}u_n$) is the main representative of the class $\{\ldots, u_{n-2}u_n, u_{n-1}u_{n+1}, u_n u_{n+2}, u_{n+1}u_{n+3}, \ldots\}$.

For monomials involving u_n, v_n, w_n, etc. and their shifts, we lexicographically order the variables, i.e. $u_n \prec v_n \prec w_n$, etc.

Example 2.17. For example, $u_n v_{n+2}$ (not $u_{n-2}v_n$) is the main representative of the class $\{\ldots, u_{n-2}v_n, u_{n-1}v_{n+1}, u_n v_{n+2}, \ldots\}$.

Definition 2.18. A vector function $\mathbf{G}(\mathbf{u}_n)$ is called a generalized symmetry of (2.1) if the infinitesimal transformation $\mathbf{u}_n \to \mathbf{u}_n + \varepsilon\mathbf{G}$ leaves (2.1) invariant up to order ε. Consequently, $\mathbf{G}$ must satisfy [**20**]

$$\mathrm{D}_t\mathbf{G} = \mathbf{F}'(\mathbf{u}_n)[\mathbf{G}] \tag{2.20}$$

on solutions of (2.1). $\mathbf{F}'(\mathbf{u}_n)[\mathbf{G}]$ is the Fréchet derivative of $\mathbf{F}$ in the direction of $\mathbf{G}$.

For the scalar case ($N = 1$), the Fréchet derivative in the direction of G is

$$F'(u_n)[G] = \frac{\partial}{\partial\varepsilon}F(u_n + \varepsilon G)|_{\varepsilon=0} = \sum_k \frac{\partial F}{\partial u_{n+k}}\mathrm{D}^k G, \tag{2.21}$$

which defines the Fréchet derivative operator

$$F'(u_n) = \sum_k \frac{\partial F}{\partial u_{n+k}}\mathrm{D}^k. \tag{2.22}$$

In the vector case with components u_n and v_n, the Fréchet derivative operator is

$$\mathbf{F}'(\mathbf{u}_n) = \begin{pmatrix} \sum_k \frac{\partial F_1}{\partial u_{n+k}}\mathrm{D}^k & \sum_k \frac{\partial F_1}{\partial v_{n+k}}\mathrm{D}^k \\ \sum_k \frac{\partial F_2}{\partial u_{n+k}}\mathrm{D}^k & \sum_k \frac{\partial F_2}{\partial v_{n+k}}\mathrm{D}^k \end{pmatrix}. \tag{2.23}$$

Applied to $\mathbf{G} = (G_1 \quad G_2)^{\mathrm{T}}$, where T is transpose, one obtains

$$F_i{}'(\mathbf{u}_n)[\mathbf{G}] = \sum_k \frac{\partial F_i}{\partial u_{n+k}}\mathrm{D}^k G_1 + \sum_k \frac{\partial F_i}{\partial v_{n+k}}\mathrm{D}^k G_2, \quad i = 1, 2. \tag{2.24}$$

In (2.21) and (2.24) summation is over all positive and negative shifts (including $k = 0$). For $k > 0$, $\mathrm{D}^k = \mathrm{D}\circ\mathrm{D}\circ\cdots\circ\mathrm{D}$ (k times). Similarly, for $k < 0$ the down-shift operator D^{-1} is applied repeatedly. The generalization of (2.23) to a system with N components should be obvious.

Example 2.19. The first two symmetries [**12**] of (2.3) are

$$G^{(1)} = u_n(u_{n+1} - u_{n-1}), \tag{2.25}$$

$$G^{(2)} = u_n u_{n+1}(u_n + u_{n+1} + u_{n+2}) - u_{n-1}u_n(u_{n-2} + u_{n-1} + u_n). \tag{2.26}$$

These symmetries are uniform in rank (rank 2 and 3, respectively). The symmetries of ranks 0 and 1 are both zero.

Example 2.20. The first two non-trivial symmetries [**12**] of (2.4) are

$$\mathbf{G}^{(1)} = \begin{pmatrix} v_{n-1} - v_n \\ v_n(u_n - u_{n+1}) \end{pmatrix}, \tag{2.27}$$

$$\mathbf{G}^{(2)} = \begin{pmatrix} v_n(u_n + u_{n+1}) - v_{n-1}(u_{n-1} + u_n) \\ v_n(u_{n+1}^2 - u_n^2 + v_{n+1} - v_{n-1}) \end{pmatrix}. \tag{2.28}$$

The above symmetries are uniform in rank. For example, rank $G_1^{(2)} = 3$ and rank $G_2^{(2)} = 4$. Symmetries of lower ranks are trivial.

An algorithm to compute polynomial generalized symmetries will be outlined in Section 6 and applied to (2.4).

Definition 2.21. A recursion operator $\mathcal{R}$ connects symmetries

$$\mathbf{G}^{(j+s)} = \mathcal{R}\mathbf{G}^{(j)}, \tag{2.29}$$

where $j = 1, 2, \ldots$, and s is the gap length. The symmetries are linked consecutively if $s = 1$. This happens in most, but not all, cases. For N-component systems, $\mathcal{R}$ is an $N \times N$ matrix operator.

The defining equation for $\mathcal{R}$ [**20, 24**] is

$$\mathrm{D}_t\mathcal{R} + [\mathcal{R}, \mathbf{F}'(\mathbf{u}_n)] = \frac{\partial \mathcal{R}}{\partial t} + \mathcal{R}'[\mathbf{F}] + \mathcal{R} \circ \mathbf{F}'(\mathbf{u}_n) - \mathbf{F}'(\mathbf{u}_n) \circ \mathcal{R} = 0, \tag{2.30}$$

where [,] denotes the commutator and $\circ$ the composition of operators. The operator $\mathbf{F}'(\mathbf{u}_n)$ was defined in (2.23). $\mathcal{R}'[\mathbf{F}]$ is the Fréchet derivative of $\mathcal{R}$ in the direction of $\mathbf{F}$. For the scalar case, the operator $\mathcal{R}$ is often of the form

$$\mathcal{R} = U(u_n)\mathcal{O}\big((\mathrm{D} - \mathrm{I})^{-1}, \mathrm{D}^{-1}, \mathrm{I}, \mathrm{D}\big)V(u_n), \tag{2.31}$$

and then

$$\mathcal{R}'[F] = \sum_k (\mathrm{D}^k F)\frac{\partial U}{\partial u_{n+k}}\mathcal{O}V + \sum_k U\mathcal{O}(\mathrm{D}^k F)\frac{\partial V}{\partial u_{n+k}}. \tag{2.32}$$

For the vector case, the elements of the $N \times N$ operator matrix $\mathcal{R}$ are often of the form $\mathcal{R}_{ij} = U_{ij}(\mathbf{u}_n)\mathcal{O}_{ij}\big((\mathrm{D} - \mathrm{I})^{-1}, \mathrm{D}^{-1}, \mathrm{I}, \mathrm{D}\big)V_{ij}(\mathbf{u}_n)$. For the two-component case

$$\begin{aligned} \mathcal{R}'[\mathbf{F}]_{ij} &= \sum_k (\mathrm{D}^k F_1)\frac{\partial U_{ij}}{\partial u_{n+k}}\mathcal{O}_{ij}V_{ij} + \sum_k (\mathrm{D}^k F_2)\frac{\partial U_{ij}}{\partial v_{n+k}}\mathcal{O}_{ij}V_{ij} \\ &\quad + \sum_k U_{ij}\mathcal{O}_{ij}(\mathrm{D}^k F_1)\frac{\partial V_{ij}}{\partial u_{n+k}} + \sum_k U_{ij}\mathcal{O}_{ij}(\mathrm{D}^k F_2)\frac{\partial V_{ij}}{\partial v_{n+k}}. \end{aligned} \tag{2.33}$$

Example 2.22. The KvM lattice (2.3) has recursion operator

$$\begin{aligned} \mathcal{R} &= u_n(\mathrm{I} + \mathrm{D})(u_n\mathrm{D} - \mathrm{D}^{-1}u_n)(\mathrm{D} - \mathrm{I})^{-1}\frac{1}{u_n}\mathrm{I} \\ &= u_n\mathrm{D}^{-1} + (u_n + u_{n+1})\mathrm{I} + u_n\mathrm{D} + u_n(u_{n+1} - u_{n-1})(\mathrm{D} - \mathrm{I})^{-1}\frac{1}{u_n}\mathrm{I}. \end{aligned} \tag{2.34}$$

Example 2.23. The Toda lattice (2.4) has recursion operator

$$\mathcal{R} = \begin{pmatrix} -u_n\mathrm{I} & -\mathrm{D}^{-1} - \mathrm{I} + (v_{n-1} - v_n)(\mathrm{D} - \mathrm{I})^{-1}\frac{1}{v_n}\mathrm{I} \\ -v_n\mathrm{I} - v_n\mathrm{D} & -u_{n+1}\mathrm{I} + v_n(u_n - u_{n+1})(\mathrm{D} - \mathrm{I})^{-1}\frac{1}{v_n}\mathrm{I} \end{pmatrix}. \tag{2.35}$$

In Section 7 we will give an algorithm for the computation of scalar recursion operators like (2.34). In Section 8 we cover the matrix case and show how (2.35) is computed. The algorithms complement those for recursion operators of PDEs presented in [**11**] and elsewhere in these proceedings [**4**].

We now introduce two powerful tools which will be used in the computation of densities and fluxes.

3. The discrete variational derivative (Euler operator)

Definition 3.1. A function $E_n(\mathbf{u}_n)$ is a total difference if there exists a function $J_n(\mathbf{u}_n)$ such that $E_n = \Delta J_n = (\mathrm{D} - \mathrm{I})J_n$.

Theorem 3.2. *A necessary and sufficient condition for a function* E_n *with positive shifts up to level* p_0, *to be a total difference is that*

$$\mathcal{L}^{(0)}_{\mathbf{u}_n}(E_n) = 0. \tag{3.1}$$

$\mathcal{L}^{(0)}_{\mathbf{u}_n}$ *is the discrete variational derivative (Euler operator)* [**3**] *defined by*

$$\mathcal{L}^{(0)}_{\mathbf{u}_n} = \frac{\partial}{\partial \mathbf{u}_n}\left(\sum_{k=0}^{p_0} \mathrm{D}^{-k}\right) = \frac{\partial}{\partial \mathbf{u}_n}(\mathrm{I} + \mathrm{D}^{-1} + \mathrm{D}^{-2} + \cdots + \mathrm{D}^{-p_0}). \tag{3.2}$$

A proof is given in e.g. [**13**].

Remark 3.3. To verify that an expression $E(u_{n-q}, \ldots, u_n, \ldots, u_{n+p})$ involving negative shifts is a total difference, one must first remove the negative shifts by replacing E_n by $\widetilde{E}_n = \mathrm{D}^q E_n$. Applied to $\widetilde{E}_n$, (3.2) terminates at $p_0 = p + q$.

We now introduce a tool to invert the total difference operator $\Delta = \mathrm{D} - \mathrm{I}$.

4. The discrete homotopy operator

We assume that $\mathcal{L}^{(0)}_{\mathbf{u}_n}(E_n) = 0$. So, $\exists J_n$ such that $E_n = \Delta J_n$. To compute $J_n = \Delta^{-1}(E_n)$ one must invert the operator $\Delta = \mathrm{D} - \mathrm{I}$. Working with the formal inverse, $\Delta^{-1} = \mathrm{D}^{-1} + \mathrm{D}^{-2} + \mathrm{D}^{-3} + \cdots$, is impractical, perhaps impossible. We therefore present the (discrete) homotopy operator which circumvents the formal infinite series. In analogy to the continuous case [**20**], we first introduce the discrete higher Euler operators.

Definition 4.1. The discrete higher Euler operators are defined by

$$\mathcal{L}^{(i)}_{\mathbf{u}_n} = \frac{\partial}{\partial \mathbf{u}_n}\left(\sum_{k=i}^{p_i} \binom{k}{i} \mathrm{D}^{-k}\right), \quad i \geq 1. \tag{4.1}$$

The higher Euler operators all terminate at some maximal shift p_i.

Example 4.2. For scalar component u_n, the first higher Euler operators are:

$$\mathcal{L}^{(1)}_{u_n} = \frac{\partial}{\partial u_n}(\mathrm{D}^{-1} + 2\mathrm{D}^{-2} + 3\mathrm{D}^{-3} + 4\mathrm{D}^{-4} + \cdots + p_1\,\mathrm{D}^{-p_1}), \tag{4.2}$$

$$\mathcal{L}^{(2)}_{u_n} = \frac{\partial}{\partial u_n}(\mathrm{D}^{-2} + 3\mathrm{D}^{-3} + 6\mathrm{D}^{-4} + 10\mathrm{D}^{-5} + \cdots + \tfrac{1}{2}p_2(p_2 - 1)\mathrm{D}^{-p_2}), \tag{4.3}$$

where the p_i are the highest shifts in the arguments of the operators.

Similar formulae hold for $\mathcal{L}^{(i)}_{v_n}$. Next, we introduce the homotopy operators. For simplicity we show the formulae for the two component case $\mathbf{u}_n = (u_n, v_n)$.

Definition 4.3. The homotopy operator is defined as,

$$\mathcal{H}_{\mathbf{u}_n} = \int_0^1 (f_{1,n}(\mathbf{u}_n)[\lambda \mathbf{u}_n] + f_{2,n}(\mathbf{u}_n)[\lambda \mathbf{u}_n]) \frac{d\lambda}{\lambda}, \tag{4.4}$$

where

$$f_{1,n}(\mathbf{u}_n) = \sum_{i=0}^{m_1-1} (\mathrm{D}-\mathrm{I})^i (u_n \mathcal{L}_{u_n}^{(i+1)}), \quad f_{2,n}(\mathbf{u}_n) = \sum_{i=0}^{m_2-1} (\mathrm{D}-\mathrm{I})^i (v_n \mathcal{L}_{v_n}^{(i+1)}), \tag{4.5}$$

where m_1 and m_2 are the highest shifts in u_n and v_n in the arguments of the operators. Furthermore, $f_{1,n}(\mathbf{u}_n)[\lambda \mathbf{u}_n]$ means that in $f_{1,n}(\mathbf{u}_n)$ one replaces $\mathbf{u}_n \to \lambda \mathbf{u}_n$, and, of course, $\mathbf{u}_{n+1} \to \lambda \mathbf{u}_{n+1}$, $\mathbf{u}_{n+2} \to \lambda \mathbf{u}_{n+2}$, etc. Same for $f_{2,n}(\mathbf{u}_n)[\lambda \mathbf{u}_n]$.

Theorem 4.4. *$J_n = \Delta^{-1}(E_n)$ can be computed as $J_n = \mathcal{H}_{\mathbf{u}_n}(E_n)$.*

A similar theorem and proof for continuous homotopy operators is given in [**20**, p. 372]. By constructing an appropriate discrete variational bicomplex one can prove the above theorem. See [**14, 16**] and elsewhere in these proceedings [**17**].

5. Computation of polynomial conserved densities and fluxes

As an example, we compute density-flux pair (2.19) for (2.4). Assuming that the weights (2.7) are computed and the rank of the density is selected ($R = 3$ here), the algorithm has three steps. See [**8, 10, 12**] for details about the algorithm and alternate ways to compute fluxes.

Step 1 (Construct the form of the density). List all monomials in u_n and v_n of rank 3 or less: $\mathcal{M} = \{{u_n}^3, {u_n}^2, u_n v_n, u_n, v_n\}$.

Next, for each monomial in $\mathcal{M}$, introduce the correct number of t-derivatives so that each term has weight 3. Thus, using (2.4),

$$\frac{\mathrm{d}^0}{\mathrm{d}t^0}({u_n}^3) = {u_n}^3, \qquad \frac{\mathrm{d}^0}{\mathrm{d}t^0}(u_n v_n) = u_n v_n,$$

$$\frac{\mathrm{d}}{\mathrm{d}t}({u_n}^2) = 2u_n \dot{u}_n = 2u_n v_{n-1} - 2u_n v_n, \qquad \frac{\mathrm{d}}{\mathrm{d}t}(v_n) = \dot{v}_n = u_n v_n - u_{n+1} v_n,$$

$$\frac{\mathrm{d}^2}{\mathrm{d}t^2}(u_n) = \frac{\mathrm{d}}{\mathrm{d}t}(\dot{u}_n) = \frac{\mathrm{d}}{\mathrm{d}t}(v_{n-1} - v_n) = u_{n-1}v_{n-1} - u_n v_{n-1} - u_n v_n + u_{n+1} v_n.$$

Gather all terms in $\mathcal{N} = \{{u_n}^3, u_n v_{n-1}, u_n v_n, u_{n-1} v_{n-1}, u_{n+1} v_n\}$. Identify members belonging to the same equivalence classes and replace them by their main representatives. For example, $u_n v_{n-1} \equiv u_{n+1} v_n$, so the latter is replaced by $u_n v_{n-1}$. So, $\mathcal{N}$ is replaced by $\mathcal{P} = \{{u_n}^3, u_n v_{n-1}, u_n v_n\}$, which has the building blocks of the density. Linear combination of the monomials in $\mathcal{P}$ with constant coefficients c_i gives the candidate density,

$$\rho_n = c_1 {u_n}^3 + c_2 u_n v_{n-1} + c_3 u_n v_n. \tag{5.1}$$

Step 2 (Determine the coefficients). Require that (2.11) holds. Compute $\mathrm{D}_t \rho_n$. Use (2.4) to remove $\dot{u}_n$ and $\dot{v}_n$ and their shifts. Thus,

$$\begin{aligned} E_n = \mathrm{D}_t \rho_n = {} & (3c_1 - c_2) u_n^2 v_{n-1} + (c_3 - 3c_1) u_n^2 v_n + (c_3 - c_2) v_{n-1} v_n \\ & + c_2 u_{n-1} u_n v_{n-1} + c_2 v_{n-1}^2 - c_3 u_n u_{n+1} v_n - c_3 v_n^2. \end{aligned} \tag{5.2}$$

TABLE 1. Computation of $f_{1,n}(\mathbf{u}_n)(-\widetilde{E}_n)$.

i	$\mathcal{L}_{u_n}^{(i+1)}(-\widetilde{E}_n)$	$(\mathrm{D}-\mathrm{I})^i\left(u_n\mathcal{L}_{u_n}^{(i+1)}(-\widetilde{E}_n)\right)$
0	$u_{n-1}v_{n-1}+u_{n+1}v_n$	$u_nu_{n-1}v_{n-1}+u_nu_{n+1}v_n$
1	$u_{n-1}v_{n-1}$	$u_{n+1}u_nv_n-u_nu_{n-1}v_{n-1}$

TABLE 2. Computation of $f_{2,n}(\mathbf{u}_n)(-\widetilde{E}_n)$.

i	$\mathcal{L}_{v_n}^{(i+1)}(-\widetilde{E}_n)$	$(\mathrm{D}-\mathrm{I})^i\left(v_n\mathcal{L}_{v_n}^{(i+1)}(-\widetilde{E}_n)\right)$
0	$u_nu_{n+1}+2v_n$	$v_nu_nu_{n+1}+2v_n^2$

Compute $\widetilde{E}_n=\mathrm{D}E_n$. First, apply (3.2) for component u_n to $\widetilde{E}_n$:

$$\begin{aligned}\mathcal{L}_{u_n}^{(0)}(\widetilde{E}_n)&=\frac{\partial}{\partial u_n}(\mathrm{I}+\mathrm{D}^{-1}+\mathrm{D}^{-2})(\widetilde{E}_n)\\&=2(3c_1-c_2)u_nv_{n-1}+2(c_3-3c_1)u_nv_n\\&\qquad+(c_2-c_3)u_{n-1}v_{n-1}+(c_2-c_3)u_{n+1}v_n.\end{aligned}\tag{5.3}$$

Second, apply (3.2) for component v_n to $\widetilde{E}_n$:

$$\begin{aligned}\mathcal{L}_{v_n}^{(0)}(\widetilde{E}_n)&=\frac{\partial}{\partial v_n}(\mathrm{I}+\mathrm{D}^{-1})(\widetilde{E}_n)\\&=(3c_1-c_2)u_{n+1}^2+(c_3-c_2)v_{n+1}+(c_2-c_3)u_nu_{n+1}\\&\qquad+2(c_2-c_3)v_n+(c_3-3c_1)u_n^2+(c_3-c_2)v_{n-1}.\end{aligned}\tag{5.4}$$

Both (5.3) and (5.4) must vanish identically. Solve the linear system

$$\mathcal{S}=\{3c_1-c_2=0,c_3-3c_1=0,c_2-c_3=0\}.\tag{5.5}$$

The solution is $3c_1=c_2=c_3$. Substituting $c_1=\frac{1}{3},c_2=c_3=1$, into (5.1)

$$\rho_n=\tfrac{1}{3}{u_n}^3+u_n(v_{n-1}+v_n).\tag{5.6}$$

Step 3 (Compute the flux). In view of (2.11), we will compute $J_n=\Delta^{-1}(-E_n)$ with the homotopy operator introduced in Section 4. Alternative methods are described in [**8, 13**].

Insert $c_1=\frac{1}{3}$, $c_2=c_3=1$ into (5.2) and compute

$$\widetilde{E}_n=\mathrm{D}E_n=u_nu_{n+1}v_n+v_n^2-u_{n+1}u_{n+2}v_{n+1}-v_{n+1}^2.\tag{5.7}$$

We apply formulae (4.5) to $-\widetilde{E}_n$. The pieces are listed in Tables 1 and 2.

Adding the terms in the right columns in Table 1 and Table 2,

$$f_{1,n}(\mathbf{u}_n)(-\widetilde{E}_n)=2u_nu_{n+1}v_n,\quad f_{2,n}(\mathbf{u}_n)(-\widetilde{E}_n)=u_nu_{n+1}v_n+2v_n^2.\tag{5.8}$$

Thus, the homotopy operator (4.4) gives

$$\tag{5.9} \begin{aligned} \widetilde{J}_n &= \int_0^1 \left(f_{1,n}(\mathbf{u}_n)(-\widetilde{E}_n)[\lambda \mathbf{u}_n] + f_{2,n}(\mathbf{u}_n)(-\widetilde{E}_n)[\lambda \mathbf{u}_n] \right) \frac{d\lambda}{\lambda} \\ &= \int_0^1 (3\lambda^2 u_n u_{n+1} v_n + 2\lambda v_n^2)\, d\lambda \\ &= u_n u_{n+1} v_n + v_n^2. \end{aligned}$$

After a backward shift, $J_n = \mathrm{D}^{-1}\widetilde{J}_n$, we obtain the final result:

$$\tag{5.10} \rho_n = \tfrac{1}{3}u_n^3 + u_n(v_{n-1} + v_n), \quad J_n = u_{n-1}u_n v_{n-1} + v_{n-1}^2.$$

6. Computation of polynomial generalized symmetries

As an example, we compute symmetry (2.28) of rank (3, 4) for (2.4). The two steps of the algorithm [**9**] are similar to those in Section 5.

Step 1 (Construct the form of the symmetry). Start by listing all monomials in u_n and v_n of ranks 3 and 4, or less:

$$\mathcal{M}_1 = \{u_n^3, u_n^2, u_n v_n, u_n, v_n\}, \quad \mathcal{M}_2 = \{u_n^4, u_n^3, u_n^2 v_n, u_n^2, u_n v_n, u_n, v_n^2, v_n\}.$$

As in Step 1 in Section 5, for each monomial in $\mathcal{M}_1$ and $\mathcal{M}_2$, introduce the necessary t-derivatives so that each term has rank 3 and 4, respectively. At the same time, use (2.4) to remove all t-derivatives. Doing so, based on $\mathcal{M}_1$ we obtain

$$\tag{6.1} \mathcal{N}_1 = \{u_n^3, u_{n-1}v_{n-1}, u_n v_{n-1}, u_n v_n, u_{n+1}v_n\}.$$

Similarly, based on $\mathcal{M}_2$, we find

$$\tag{6.2} \begin{aligned} \mathcal{N}_2 = \{&u_n^4, u_{n-1}^2 v_{n-1}, u_{n-1}u_n v_{n-1}, u_n^2 v_{n-1}, v_{n-2}v_{n-1}, v_{n-1}^2, u_n^2 v_n, \\ &u_n u_{n+1} v_n, u_{n+1}^2 v_n, v_{n-1}v_n, v_n^2, v_n v_{n+1}\}. \end{aligned}$$

In contrast to the strategy for densities, we do not introduce the main representatives, but linearly combine the monomials in $\mathcal{N}_1$ and $\mathcal{N}_2$ with constant coefficients c_i to get the form of the candidate symmetry:

$$\tag{6.3} \begin{aligned} G_1 &= c_1 u_n^3 + c_2 u_{n-1}v_{n-1} + c_3 u_n v_{n-1} + c_4 u_n v_n + c_5 u_{n+1} v_n, \\ G_2 &= c_6 u_n^4 + c_7 u_{n-1}^2 v_{n-1} + c_8 u_{n-1}u_n v_{n-1} + c_9 u_n^2 v_{n-1} + c_{10} v_{n-2}v_{n-1} \\ &\quad + c_{11} v_{n-1}^2 + c_{12} u_n^2 v_n + c_{13} u_n u_{n+1} v_n + c_{14} u_{n+1}^2 v_n + c_{15} v_{n-1} v_n \\ &\quad + c_{16} v_n^2 + c_{17} v_n v_{n+1}. \end{aligned}$$

To compute the coefficients c_i we require that (2.20) holds on solutions of (2.1). Compute $\mathrm{D}_t G_1$ and $\mathrm{D}_t G_2$ and use (2.4) to remove $\dot{u}_n, \dot{v}_n$, and their shifts. This gives the left hand sides of (2.20). Use (2.24) to get the right hand sides of (2.20):

$$\tag{6.4} \begin{aligned} F_1{}'(\mathbf{u}_n)[\mathbf{G}] &= \mathrm{D}^{-1}G_2 - \mathrm{I}G_2, \\ F_2{}'(\mathbf{u}_n)[\mathbf{G}] &= v_n \mathrm{I}G_1 - v_n \mathrm{D}G_1 + (u_n - u_{n+1})\mathrm{I}G_2. \end{aligned}$$

Substitute (6.3) into (6.4) and equate the corresponding left and right hand sides. Since all monomials in u_n, v_n and their shifts are independent, one obtains the linear system that determines the coefficients c_i. The solution is

$$\tag{6.5} \begin{aligned} &c_1 = c_6 = c_7 = c_8 = c_9 = c_{10} = c_{11} = c_{13} = c_{16} = 0, \\ &c_2 = c_3 = -c_4 = -c_5 = c_{12} = -c_{14} = c_{15} = -c_{17}. \end{aligned}$$

With the choice $c_{17} = 1$, symmetry (6.3) finally becomes

$$\begin{aligned} G_1 &= v_n(u_n + u_{n+1}) - v_{n-1}(u_{n-1} - u_n), \\ G_2 &= v_n(u_{n+1}^2 - u_n^2 + v_{n+1} - v_{n-1}). \end{aligned} \tag{6.6}$$

7. Computation of scalar recursion operators

Recursion operators can be constructed as $\mathcal{R} = H_2 {H_1}^{-1}$, where H_1 and H_2 are appropriate Hamiltonian operators [**19**]. We use a less general but direct method based on dilation invariance. Furthermore, we assume that the form of discrete recursion operators mimics its continuous counterpart [**5, 18, 21, 22, 24**]. With the appropriate ansatz for $\mathcal{R}$, defining equation (2.30) is then used to determine the coefficients.

In this section we show how to compute recursion operator (2.34) of (2.4). We observe that (2.34) in expanded form naturally splits into two pieces:

$$\mathcal{R} = \mathcal{R}_0 + \mathcal{R}_1, \tag{7.1}$$

where $\mathcal{R}_0$ contains only terms with shift operators $\mathrm{D}^{-1}, \mathrm{I}$, and D, and $\mathcal{R}_1$ has terms involving $(\mathrm{D} - \mathrm{I})^{-1}$.

Step 1 (Determine the rank of the recursion operator). In view of (2.29), and assuming that the symmetries are linked consecutively ($s = 1$), the recursion operator $\mathcal{R}$ has rank

$$R = \operatorname{rank} \mathcal{R} = \operatorname{rank} G^{(2)} - \operatorname{rank} G^{(1)} = 3 - 2 = 1, \tag{7.2}$$

where we used (2.25) and (2.26). If the assumption turns out to be correct then the recursion operator has rank 1. If the assumption fails because symmetries are not linked consecutively, then R must be adjusted and the subsequent steps must be repeated. See [**4**] for examples of PDEs for which that happens.

Step 2 (Determine the coefficients). We split this into two sub-steps.

(i) *Determine the form of* $\mathcal{R}_0$. The candidate $\mathcal{R}_0$ is a sum of terms involving $\mathrm{D}^{-1}, \mathrm{I}$, and D. The coefficients of these terms are linear combinations of u_{n-1}, u_n, and u_{n+1}, so that all terms have the correct rank.

$$\begin{aligned} \mathcal{R}_0 = (c_1 u_{n-1} + c_2 u_n + c_3 u_{n+1})\mathrm{D}^{-1} + (c_4 u_{n-1} + c_5 u_n + c_6 u_{n+1})\mathrm{I} \\ + (c_7 u_{n-1} + c_8 u_n + c_9 u_{n+1})\mathrm{D}, \end{aligned} \tag{7.3}$$

where the c_i are constant coefficients.

A few remarks are in place. First, in $\mathcal{R}_0$ we moved the operators D^{-1}, I, and D all the way to the right. Second, the maximum up-shift and down-shift operator that should be included can be determined by comparing two consecutive symmetries. Indeed, if the maximum up-shift in the first symmetry is u_{n+p}, and the maximum up-shift in the next symmetry is u_{n+p+r}, then $\mathcal{R}_0$ must have $\mathrm{D}, \mathrm{D}^2, \ldots, \mathrm{D}^r$. The same line of reasoning determines the minimum down-shift operator to be included. In our example, there is no need to include terms in D^{-2}, D^2, etc. Third, the coefficients of the operators can be restricted to linear combinations of the terms appearing in F. Hence, no terms in $u_{n\pm2}$, $u_{n\pm3}$ and so on occur in (7.3).

(ii) *Determine the form of* $\mathcal{R}_1$. As in the continuous case [**11**], $\mathcal{R}_1$ is a linear combination (with constant coefficients $\tilde{c}_{jk}$) of sums of all suitable products of symmetries and covariants (Fréchet derivatives of densities) sandwiching $(\mathrm{D}-\mathrm{I})^{-1}$, i.e.

$$\mathcal{R}_1 = \sum_j \sum_k \tilde{c}_{jk} G^{(j)} (\mathrm{D}-\mathrm{I})^{-1} \rho_n^{(k)\prime}. \tag{7.4}$$

For (2.3), $G^{(1)}$ in (2.25) and $\rho_n^{(0)} = \ln(u_n)$ in (2.13) are the only suitable pair. Indeed, using (2.22) we have $\rho_n^{(0)\prime} = \ln(u_n)' = 1/u_n \mathrm{I}$, which has rank -1. Combined with $G^{(1)}$ of rank 2, we have a term of rank 1. Other combinations of symmetries and covariants would exceed rank 1. Therefore,

$$\mathcal{R}_1 = \tilde{c}_{10} u_n (u_{n+1} - u_{n-1})(\mathrm{D}-\mathrm{I})^{-1}\left(\frac{1}{u_n}\right)\mathrm{I} \tag{7.5}$$

where $\tilde{c}_{10}$ a constant coefficient. Using (7.1) and renaming $\tilde{c}_{10}$ to c_{10},

$$\begin{aligned}\mathcal{R} = {} & (c_1 u_{n-1} + c_2 u_n + c_3 u_{n+1})\mathrm{D}^{-1} + (c_4 u_{n-1} + c_5 u_n + c_6 u_{n+1})\mathrm{I} \\ & + (c_7 u_{n-1} + c_8 u_n + c_9 u_{n+1})\mathrm{D} + c_{10} u_n (u_{n+1} - u_{n-1})(\mathrm{D}-\mathrm{I})^{-1}\left(\frac{1}{u_n}\right)\mathrm{I}\end{aligned} \tag{7.6}$$

is the candidate recursion operator for (2.3).

Step 3 (Determine the coefficients). Starting from (7.6), we use (2.32) with $F = u_n(u_{n+1} - u_{n-1})$ to compute $\mathcal{R}'[F]$. The partial results will not be shown due to length. Using (2.21), we compute

$$F' = -u_n \mathrm{D}^{-1} + (u_{n+1} - u_{n-1})\mathrm{I} + u_n \mathrm{D}. \tag{7.7}$$

Then we compute $\mathcal{R} \circ F'$ and $F' \circ \mathcal{R}$. After substituting the pieces into (2.30) we simplify the resulting expression using rules such as

$$(\mathrm{D}-\mathrm{I})^{-1}\mathrm{D} = \mathrm{D}(\mathrm{D}-\mathrm{I})^{-1} = \mathrm{I} + (\mathrm{D}-\mathrm{I})^{-1}, \tag{7.8}$$

$$(\mathrm{D}-\mathrm{I})^{-1}\mathrm{D}^{-1} = \mathrm{D}^{-1}(\mathrm{D}-\mathrm{I})^{-1} = -\mathrm{D}^{-1} + (\mathrm{D}-\mathrm{I})^{-1}. \tag{7.9}$$

We further simplify by recursively using formulas like

$$\mathrm{D}U(u_n)(\mathrm{D}-\mathrm{I})^{-1}V(u_n)\mathrm{I} = U(u_{n+1})V(u_n)\mathrm{I} + U(u_{n+1})(\mathrm{D}-\mathrm{I})^{-1}V(u_n)\mathrm{I}, \tag{7.10}$$

$$(\mathrm{D}-\mathrm{I})^{-1}U(u_n)V(u_n)\mathrm{D} = U(u_{n-1})V(u_{n-1})\mathrm{I} + (\mathrm{D}-\mathrm{I})^{-1}U(u_{n-1})V(u_{n-1})\mathrm{I}. \tag{7.11}$$

Finally, we equate like terms to obtain a linear system for the c_i. Substituting the solution

$$c_1 = c_3 = c_4 = c_7 = c_9 = 0, \quad c_2 = c_5 = c_6 = c_8 = c_{10} = 1 \tag{7.12}$$

into (7.6) we obtain the final result

$$\mathcal{R} = u_n \mathrm{D}^{-1} + (u_n + u_{n+1})\mathrm{I} + u_n \mathrm{D} + u_n (u_{n+1} - u_{n-1})(\mathrm{D}-\mathrm{I})^{-1}\frac{1}{u_n}\mathrm{I}. \tag{7.13}$$

A straightforward computation confirms that $\mathcal{R}G^{(1)} = G^{(2)}$ with $G^{(1)}$ in (2.25) and $G^{(2)}$ in (2.26).

8. Computation of matrix recursion operators

We construct the recursion operator (2.35) for (2.4). Now all the terms in (2.30) are 2×2 matrix operators.

Step 1 (Determine the rank of the recursion operator). The difference in rank of symmetries is again used to compute the rank of the elements of the recursion operator. Using (2.7), (2.27) and (2.28),

$$\operatorname{rank} \mathbf{G}^{(1)} = \begin{pmatrix} 2 \\ 3 \end{pmatrix}, \quad \operatorname{rank} \mathbf{G}^{(2)} = \begin{pmatrix} 3 \\ 4 \end{pmatrix}. \tag{8.1}$$

Assuming that $\mathcal{R}\mathbf{G}^{(1)} = \mathbf{G}^{(2)}$, we use the formula

$$\operatorname{rank} \mathcal{R}_{ij} = \operatorname{rank} G_i^{(k+1)} - \operatorname{rank} G_j^{(k)}, \tag{8.2}$$

to compute a rank matrix associated to the operator

$$\operatorname{rank} \mathcal{R} = \begin{pmatrix} 1 & 0 \\ 2 & 1 \end{pmatrix}. \tag{8.3}$$

Step 2 (Determine the form of the recursion operator). As in the scalar case, we build a candidate $\mathcal{R}_0$:

$$\mathcal{R}_0 = \begin{pmatrix} (\mathcal{R}_0)_{11} & (\mathcal{R}_0)_{12} \\ (\mathcal{R}_0)_{21} & (\mathcal{R}_0)_{22} \end{pmatrix}, \tag{8.4}$$

with

$$\begin{aligned}
(\mathcal{R}_0)_{11} &= (c_1 u_n + c_2 u_{n+1})\mathrm{I}, \\
(\mathcal{R}_0)_{12} &= c_3 \mathrm{D}^{-1} + c_4 \mathrm{I}, \\
(\mathcal{R}_0)_{21} &= (c_5 u_n^2 + c_6 u_n u_{n+1} + c_7 u_{n+1}^2 + c_8 v_{n-1} + c_9 v_n)\mathrm{I} \\
&\qquad + (c_{10} u_n^2 + c_{11} u_n u_{n+1} + c_{12} u_{n+1}^2 + c_{13} v_{n-1} + c_{14} v_n)\mathrm{D}, \\
(\mathcal{R}_0)_{22} &= (c_{15} u_n + c_{16} u_{n+1})\mathrm{I}.
\end{aligned}$$

Analogous to the scalar case, the elements of matrix $\mathcal{R}_1$ are linear combinations with constant coefficients $\tilde{c}_{jk}$) of all suitable products of symmetries and covariants sandwiching $(\mathrm{D} - \mathrm{I})^{-1}$. Hence,

$$\sum_j \sum_k \tilde{c}_{jk} \mathbf{G}^{(j)} (\mathrm{D} - \mathrm{I})^{-1} \otimes \rho_n^{(k)\prime}, \tag{8.5}$$

where $\otimes$ denotes the matrix outer product, defined as

$$\begin{pmatrix} G_1^{(j)} \\ G_2^{(j)} \end{pmatrix} (\mathrm{D} - \mathrm{I})^{-1} \otimes \begin{pmatrix} \rho_{n,1}^{(k)\prime} & \rho_{n,2}^{(k)\prime} \end{pmatrix} = \begin{pmatrix} G_1^{(j)} (\mathrm{D} - \mathrm{I})^{-1} \rho_{n,1}^{(k)\prime} & G_1^{(j)} (\mathrm{D} - \mathrm{I})^{-1} \rho_{n,2}^{(k)\prime} \\ G_2^{(j)} (\mathrm{D} - \mathrm{I})^{-1} \rho_{n,1}^{(k)\prime} & G_2^{(j)} (\mathrm{D} - \mathrm{I})^{-1} \rho_{n,2}^{(k)\prime} \end{pmatrix}.$$

Only the pair $(\mathbf{G}^{(1)}, \rho_n^{(0)\prime})$ can be used, otherwise the ranks in (8.3) would be exceeded. Using (2.24) and (2.16) we compute

$$\rho_n^{(0)\prime} = \begin{pmatrix} 0 & \frac{1}{v_n}\mathrm{I} \end{pmatrix}, \tag{8.6}$$

Therefore, using (8.5) and renaming $\tilde{c}_{10}$ to c_{17},

$$\mathcal{R}_1 = \begin{pmatrix} 0 & c_{17}(v_{n-1} - v_n)(\mathrm{D} - \mathrm{I})^{-1} \frac{1}{v_n}\mathrm{I} \\ 0 & c_{17} v_n (u_n - u_{n+1})(\mathrm{D} - \mathrm{I})^{-1} \frac{1}{v_n}\mathrm{I} \end{pmatrix}. \tag{8.7}$$

Adding (8.4) and (8.7) we obtain

$$\mathcal{R} = \begin{pmatrix} \mathcal{R}_{11} & \mathcal{R}_{12} \\ \mathcal{R}_{21} & \mathcal{R}_{22} \end{pmatrix}, \tag{8.8}$$

with

$$\begin{aligned}
\mathcal{R}_{11} &= (c_1 u_n + c_2 u_{n+1})\mathrm{I} \\
\mathcal{R}_{12} &= c_3 \mathrm{D}^{-1} + c_4 \mathrm{I} + c_{17}(v_{n-1} - v_n)(\mathrm{D} - \mathrm{I})^{-1}\frac{1}{v_n}\mathrm{I}, \\
\mathcal{R}_{21} &= (c_5 u_n^2 + c_6 u_n u_{n+1} + c_7 u_{n+1}^2 + c_8 v_{n-1} + c_9 v_n)\mathrm{I} \\
&\quad + (c_{10} u_n^2 + c_{11} u_n u_{n+1} + c_{12} u_{n+1}^2 + c_{13} v_{n-1} + c_{14} v_n)\mathrm{D}, \\
\mathcal{R}_{22} &= (c_{15} u_n + c_{16} u_{n+1})\mathrm{I} + c_{17} v_n (u_n - u_{n+1})(\mathrm{D} - \mathrm{I})^{-1}\frac{1}{v_n}\mathrm{I}.
\end{aligned}$$

Step 3 (Determine the coefficients). All the terms in (2.30) need to be computed. The strategy is similar to the scalar case, yet the computations are much more cumbersome. Omitting the details, the result is: $c_2 = c_5 = c_6 = c_7 = c_8 = c_{10} = c_{11} = c_{12} = c_{13} = c_{15} = 0$, $c_1 = c_3 = c_4 = c_9 = c_{14} = c_{16} = -1$, and $c_{17} = 1$. Substitution the constants into (8.8) gives

$$\mathcal{R} = \begin{pmatrix} -u_n \mathrm{I} & -\mathrm{D}^{-1} - \mathrm{I} + (v_{n-1} - v_n)(\mathrm{D} - \mathrm{I})^{-1}\frac{1}{v_n}\mathrm{I} \\ -v_n \mathrm{I} - v_n \mathrm{D} & -u_{n+1}\mathrm{I} + v_n(u_n - u_{n+1})(\mathrm{D} - \mathrm{I})^{-1}\frac{1}{v_n}\mathrm{I} \end{pmatrix}. \tag{8.9}$$

It is straightforward to verify that $\mathcal{R}G^{(1)} = G^{(2)}$ with $G^{(1)}$ in (2.27) and $G^{(2)}$ in (2.28).

9. More examples

Example 9.1. The modified Volterra lattice [**3, 13**],

$$\dot{u}_n = u_n^2 (u_{n+1} - u_{n-1}), \tag{9.1}$$

has two non-polynomial densities $\rho_n^{(0)} = 1/u_n$ and $\rho_n^{(1)} = \ln(u_n)$, and infinitely many polynomial densities. The first two symmetries,

$$G^{(1)} = u_n^2 (u_{n+1} - u_{n-1}), \tag{9.2}$$

$$G^{(2)} = u_n^2 u_{n+1}^2 (u_n + u_{n+2}) - u_{n-1}^2 u_n^2 (u_{n-2} + u_n), \tag{9.3}$$

are linked by the recursion operator

$$\mathcal{R} = u_n^2 \mathrm{D}^{-1} + 2u_n u_{n+1}\mathrm{I} + u_n^2 \mathrm{D} + 2u_n^2 (u_{n+1} - u_{n-1})(\mathrm{D} - \mathrm{I})^{-1}\frac{1}{u_n}\mathrm{I}. \tag{9.4}$$

Example 9.2. The AL lattice (2.2) has infinitely many densities [**8**] and symmetries [**9**]. The recursion operator is of the form (8.8) with

$$\begin{aligned}
\mathcal{R}_{11} &= P_n \mathrm{D}^{-1} - u_n \Delta^{-1} v_{n+1}\mathrm{I} - u_{n-1} P_n \Delta^{-1}\frac{v_n}{P_n}\mathrm{I}, \\
\mathcal{R}_{12} &= -u_n u_{n-1}\mathrm{I} - u_n \Delta^{-1} u_{n-1}\mathrm{I} - u_{n-1} P_n \Delta^{-1}\frac{u_n}{P_n}\mathrm{I}, \\
\mathcal{R}_{21} &= v_n v_{n+1}\mathrm{I} + v_n \Delta^{-1} v_{n+1}\mathrm{I} + v_{n+1} P_n \Delta^{-1}\frac{v_n}{P_n}\mathrm{I}, \\
\mathcal{R}_{22} &= (u_n v_{n+1} + u_{n-1} v_n)\mathrm{I} + P_n \mathrm{D} + v_n \Delta^{-1} v_{n-1}\mathrm{I} + v_{n+1} P_n \Delta^{-1}\frac{u_n}{P_n}\mathrm{I},
\end{aligned}$$

where $P_n = 1 + u_n v_n$ and $\Delta = \mathrm{D} - \mathrm{I}$.

This recursion operator has an inverse which is quite exceptional. For further information see e.g. [**25**].

References

1. M. J. Ablowitz and P. A. Clarkson, *Solitons, nonlinear evolution equations and inverse scattering*, London Math. Soc. Lecture Note Ser. **149**, Cambridge University Press, Cambridge, U.K., 1991.
2. M. J. Ablowitz and J. F. Ladik, *Nonlinear differential-difference equations*, J. Math. Phys. **16** (1975), 598–603.
3. V. E. Adler, A. B. Shabat, and R. I. Yamilov, *Symmetry approach to the integrability problem*, Theoret. and Math. Phys. **125** (2000), 1603–1661.
4. D. Baldwin, W. Hereman, and J. Sayers, *Symbolic algorithms for the Painlevé test, special solutions and recursion operators for nonlinear PDEs*, this volume.
5. A. H. Bilge, *On the equivalence of linearization and formal symmetries as integrability tests for evolution equations*, J. Phys. A **26** (1993), 7511–7519.
6. Ü. Göktaş and W. Hereman, *Symbolic computation of conserved densities for systems of nonlinear evolution equations*, J. Symbolic Comput. **24** (1997), 591–621.
7. ———, **Mathematica package** *InvariantsSymmetries.m* is available since 1997 from **http://www.mathsource.com/cgi-bin/msitem?0208-932**.
8. ———, *Computation of conserved densities for nonlinear lattices*, Phys. D **123** (1998), 425–436.
9. ———, *Algorithmic computation of higher-order symmetries for nonlinear evolution and lattice equations*, Adv. Comput. Math. **11** (1999), 55–80.
10. Ü. Göktaş, W. Hereman, and G. Erdmann, *Computation of conserved densities for systems of nonlinear differential-difference equations*, Phys. Lett. A **236** (1997), 30–38.
11. W. Hereman and Ü. Göktaş, *Integrability tests for nonlinear evolution equations*, Computer Algebra Systems: A Practical Guide (M. Wester, ed.), Wiley, New York, 1999, pp. 211–232.
12. W. Hereman, Ü. Göktaş, M. D. Colagrosso, and A. J. Miller, *Algorithmic integrability tests for nonlinear differential and lattice equations*, Comput. Phys. Comm. **115** (1998), 428–446.
13. M. S. Hickman and W. A. Hereman, *Computation of densities and fluxes of nonlinear differential-difference equations*, Proc. Roy. Soc. Lond. Ser. A **459** (2003), 2705–2729.
14. P. E. Hydon and E. L. Mansfield, *A variational complex for difference equations*, Found. Comput. Math. **4** (2004), 187–217.
15. M. Kac and P. van Moerbeke, *On an explicitly soluble system of nonlinear differential equations related to certain Toda lattices*, Adv. Math. **16** (1975), 160–169.
16. E. L. Mansfield and P. E. Hydon, *On a variational complex for difference equations*, The Geometrical Study of Differential Equations (Washington, DC, 2000) (J. A. Leslie and T. P. Robart, eds.), Proc. NSF-CBMS Conf., Comtemp. Math., vol. 285, Amer. Math. Soc., Providence, R.I., 2001, pp. 121–129.
17. E. L. Mansfield and G. R. W. Quispel, *Towards a variational complex for the finite element method*, this volume.
18. A. G. Meshkov, *Computer package for investigation of the complete integrability*, Proc. Inst. Maths. NAS Ukraine **30** (2000), 35–46.
19. W. Oevel, H. Zhang, and B. Fuchssteiner, *Mastersymmetries and multi-Hamiltonian formulations for some integrable lattice systems*, Progr. Theoret. Phys. **81** (1989), 294–308.
20. P. J. Olver, *Applications of lie groups to differential equations*, 2nd. ed., Grad. Texts in Math. **107**, Springer, New York, 1993.
21. J. A. Sanders and J. P. Wang, *Integrable systems and their recursion operators*, Nonlinear Anal. **47** (2001), 5213–5240.
22. ———, *On recursion operators*, Phys. D **149** (2001), 1–10.
23. M. Toda, *Theory of nonlinear lattices*, Springer, Berlin, 1981.
24. J. P. Wang, *Symmetries and conservation laws of evolution equations*, Ph.D. Thesis, Thomas Stieltjes Institute for Mathematics, Amsterdam, 1998.
25. H. Zhang, G. Tu, W. Oevel, and B. Fuchssteiner, *Symmetries, conserved quantities, and hierarchies for some lattice systems with soliton structure*, J. Math. Phys. **32** (1991), 1908–1918.

Department of Mathematical and Computer Sciences, Colorado School of Mines, Golden, Colorado 80401–1887, USA.
E-mail address: `whereman@mines.edu`

Department of Mathematics, Faculty of Sciences, Free University, De Boelelaan 1081a, 1081 HV Amsterdam, The Netherlands
E-mail address: `jansa@cs.vu.nl`

Physics Department 59-33, California Institute of Technology, Pasadena, California 91125
E-mail address: `jack@its.caltech.edu`

Institute of Mathematics, Statistics & Actuarial Science, University of Kent, Canterbury, CT2 7NF, UK.
E-mail address: `J.Wang@kent.ac.uk`

Centre de Recherches Mathématiques
CRM Proceedings and Lecture Notes
Volume **39**, 2005

On the Numerical Analysis of Rapid Oscillation

A. Iserles

ABSTRACT. This paper surveys recent advances in the allied challenges of discretizing highly oscillatory ordinary differential equations and computing numerical quadrature of highly oscillatory integrals, and attempts to sketch the mathematical foundations of a general approach to these issues. Having described the Magnus, Cayley and Neumann methods for highly oscillatory ordinary differential equations and analysed the Filon quadrature of highly oscillatory integrals, its overall conclusion is that, once numerical methods are properly crafted, their accuracy increases in the presence of high oscillation.

1. Introduction

It is an article of faith in computational mathematics and scientific computing that highly oscillatory problems, ubiquitous in applications, are among the most intractable phenomena in numerical modelling. The contention of this paper is that this point of view is misleading, indeed woefully wrong. It is not just that properly crafted methods can discretize rapidly oscillating differential and integral equations perfectly well: the truth of the matter is that *such methods are likely to perform significantly better the higher the oscillation.* High oscillation is an ally of a numerical analyst!

In order to understand how to design efficient methods in the presence of high oscillation, we need first to understand why standard numerical methods perform so poorly in this setting. The reason is indeed quite elementary: the (local) error of a standard p-order time-stepping method can be expressed in the form $h^{p+1}\mathrm{D}(\boldsymbol{y},t)+O(h^{p+2})$, where D is either a $(p+1)$st derivative of the highly oscillating solution $\boldsymbol{y}$ or a $(p+1)$st-order elementary differential, composed of derivatives. However, each differentiation of a highly oscillatory function roughly multiplies locally the amplitude by the frequency. Therefore, the higher the oscillation and larger the frequency, the worse the local error. Moreover, the standard remedy to large error, increasing the order of the method, is a non-starter since, the larger the order, the greater the damage wrought by high oscillation.

The *Airy equation*

$$\frac{\mathrm{d}^2y}{\mathrm{d}t^2}+ty=0,\ t\geq 0,\quad y(0)=1,\ \frac{\mathrm{d}y(0)}{\mathrm{d}t}=0, \tag{1.1}$$

2000 *Mathematics Subject Classification.* Primary: 65L05, 41A55; Secondary: 65D30.
This is the final form of the paper.

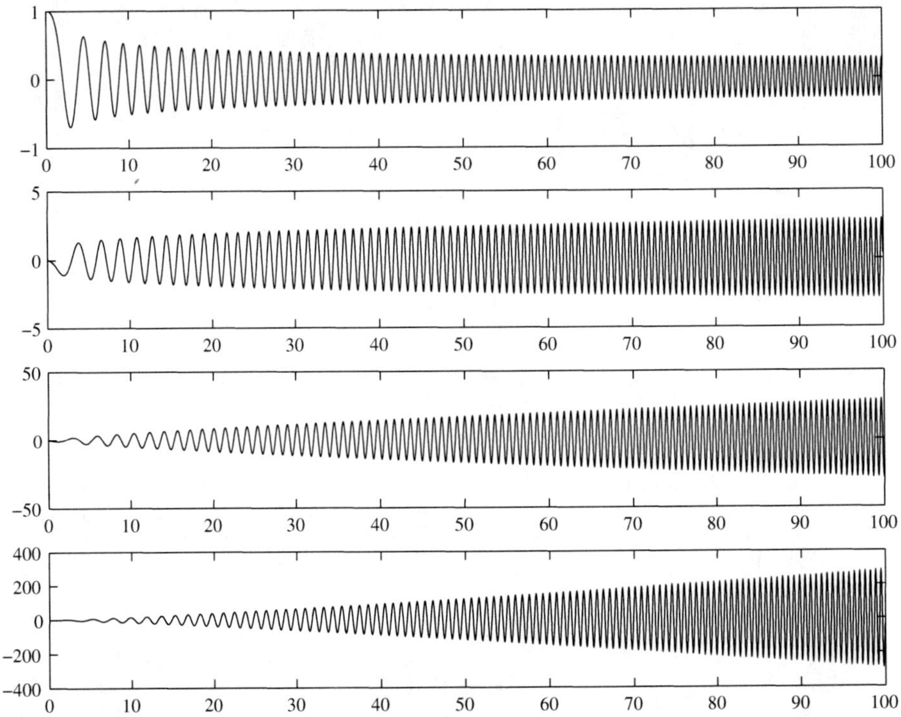

FIGURE 1. The solution of the Airy equation (1.1) and its first, second and third derivatives.

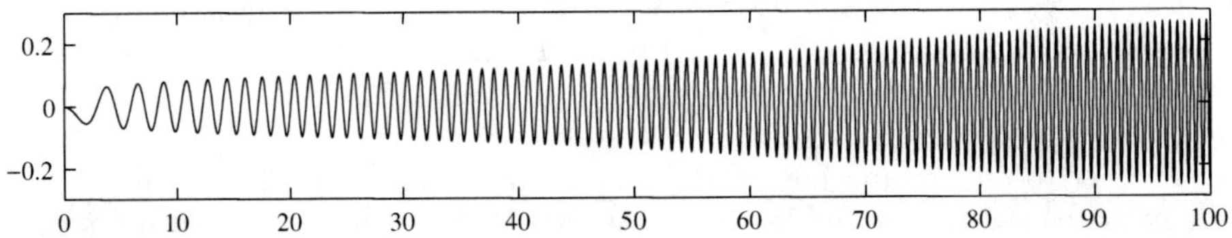

FIGURE 2. The error in solving (1.1)) with a three-stage explicit Runge–Kutta method and time-step $h = 1/20$.

is an excellent point of departure for our exposition. Figure 1 displays the exact solution y and its first three derivatives: it can be easily proved by the WKB method, that

$$y(t) \sim t^{-1/4} \sin\left(\frac{2}{3}t^{3/2}\right), \qquad \frac{\mathrm{d}y(t)}{\mathrm{d}t} \sim t^{1/4} \cos\left(\frac{2}{3}t^{3/2}\right),$$
$$\frac{\mathrm{d}^2 y(t)}{\mathrm{d}t^2} \sim -t^{3/4} \sin\left(\frac{2}{3}t^{3/2}\right), \qquad t \gg 1$$

and so on [**12**]. It is evident how fast the amplitude grows, once the function is differentiated. The consequences are for all to see in Figure 2, displaying the absolute error in the solution of (1.1) (converted into a first-order system) by the explicit third-order Runge–Kutta method with the tableau

$$\begin{array}{c|ccc} 0 & & & \\ \frac{1}{2} & \frac{1}{2} & & \\ 1 & -1 & 2 & \\ \hline & \frac{1}{6} & \frac{2}{3} & \frac{1}{6} \end{array}$$

[**9**]. The principal error term consists of elementary differentials, among them the fourth derivative of y, and it is the latter that dominates error accumulation. This phenomenon has been analysed qualitatively in [**12**] using WKB analysis. For example, for the Airy equation (1.1) and *any* classical numerical method (multistep or Runge–Kutta scheme, a truncated Taylor expansion...) of an *even* order p the global error is of the form $ct^{p/2+5/4}\chi(t)$, $t \gg 1$, where χ is a highly oscillating function in the interval $[-1,1]$ and $c \neq 0$ is method-dependent.

Needless to say, the numerical solution in Figure 2 is useless and the standard 'panic button' is of little use: decreasing the step-size further or using variable-step strategy rapidly leads to unacceptably small steps [**13**].

A key to the development of *effective* numerical methods for highly oscillatory problems is to stand the reasoning of the last paragraph on its head: *if differentiating a rapidly oscillating function is so bad, since the amplitude grows, surely integrating such a function must be good.* Thus, we need somehow to reformulate the underlying differential equation so that its numerical solution entails integration of known highly oscillatory functions.

We commence from a highly oscillatory ordinary differential (ODE) system

$$\frac{\mathrm{d}\boldsymbol{y}}{\mathrm{d}t} = \boldsymbol{f}(t,\boldsymbol{y}), \; t \geq t_0, \quad \boldsymbol{y}(t_0) = \boldsymbol{y}_0. \tag{1.2}$$

To time-step from t_n to $t_{n+1} = t_n + h$, it is convenient to rewrite (1.2) in the form

$$\frac{\mathrm{d}\boldsymbol{y}}{\mathrm{d}t} = A(t)\boldsymbol{y} + \boldsymbol{g}(t,\boldsymbol{y}), \; t \geq t_n, \quad \boldsymbol{y}(t_n) = \boldsymbol{y}_n,$$

where $\boldsymbol{g}$ is, in some sense, small. A good choice is to let A be the Jacobian matrix or its approximation.

Although it is possible to construct counterexamples, in most cases of interest the matrix A is stable (no eigenvalues have positive real parts) and it has some large eigenvalues in $\mathrm{i}\mathbb{R}$. In other words, the solution of the 'frozen' system

$$\frac{\mathrm{d}\boldsymbol{y}}{\mathrm{d}t} = A(\tilde{t})\boldsymbol{y}, \; t \geq t_n, \quad \boldsymbol{y}(t_n) = \boldsymbol{y}_n,$$

where $\tilde{t} \geq t_n$, is highly oscillatory. To time-step (1.2) from t_n to $t_{n+1} = t_n + h$, we choose $\alpha \geq 0$, set $\tilde{A} = A(t_n + \alpha h)$, represent the solution in the form

$$\boldsymbol{y}(t_n + \tau) = \mathrm{e}^{\tau\tilde{A}}\boldsymbol{x}(\tau), \quad \tau \geq 0 \tag{1.3}$$

and let $\boldsymbol{y}_{n+1} = \mathrm{e}^{h\tilde{A}}\boldsymbol{x}_1$, where $\boldsymbol{x}_1$ is our approximation of $\boldsymbol{x}(h)$. For nonlinear equations it makes sense to choose $\alpha = 0$ but if the system (1.2) is linear and the Jacobian matrix is independent of $\boldsymbol{y}$, a superior strategy is to exploit time symmetry and set $\alpha = \frac{1}{2}$.

Following [**12**], we solve for the function $\boldsymbol{x}$ in (1.3), observing that it obeys the modified differential equation

$$\frac{\mathrm{d}\boldsymbol{x}}{\mathrm{d}\tau} = B(\tau)\boldsymbol{x} + \mathrm{e}^{-\tau\tilde{A}}\boldsymbol{g}(t_n+\tau, \mathrm{e}^{\tau\tilde{A}}\boldsymbol{x}),\ \tau \geq 0, \quad \boldsymbol{x}(0) = \boldsymbol{y}_n. \tag{1.4}$$

where

$$B(\tau) = \mathrm{e}^{-\tau\tilde{A}}[A(t_n+\tau) - \tilde{A}]\mathrm{e}^{\tau\tilde{A}}$$

At a first glance, the ODE system (1.4) is nothing but a means of scrambling (1.2) into a more complicated form, a familiar mathematical vice. Yet, it has one redeeming virtue in comparison with the original system: the matrix $B(\tau) = \mathrm{e}^{-\tau\tilde{A}}[A(t+\tau) - \tilde{A}]\mathrm{e}^{\tau\tilde{A}}$ is itself highly oscillatory as a function of τ. This can be exploited by numerical methods that are based upon the integration of the vector field. Such methods are described in Section 2.

The careful reader would have noticed that converting the task in hand to the integration of highly oscillating functions is all very well, but in last analysis such integration is likely to be replaced by quadrature. This is again a numerical problem often foiled by high oscillation. Indeed, the situation is considerably worse than most practitioners would imagine. Suppose that we approximate the integral

$$I_h[f] = \int_0^h f(x)\mathrm{e}^{\mathrm{i}\omega x}\,\mathrm{d}x = h\int_0^1 f(hx)\mathrm{e}^{\mathrm{i}h\omega x}\,\mathrm{d}x, \quad \omega > 0, \tag{1.5}$$

by a Gauss–Christoffel quadrature [**3,8**]. Thus, we choose distinct quadrature nodes $c_1, c_2, \ldots, c_\nu \in [0,1]$, replace the integrand by its polynomial interpolation at the above points and integrate exactly. The outcome is the quadrature formula

$$Q_h^{\mathrm{GC}}[f] = h\sum_{l=1}^{\nu} b_l f(c_l h)\mathrm{e}^{\mathrm{i}c_l h\omega}. \tag{1.6}$$

How good is (1.6)? Everybody is familiar with the standard analysis of quadrature methods: by design, the method is at least of order ν and this can be improved, provided that the collocation polynomial satisfies favorable orthogonality conditions. This, however, makes sense *only* for suitably small *natural frequency* $h\omega$, when (1.5) is not oscillatory at all. Otherwise, a trivial argument can be used to argue that (1.6) is complete nonsense. For any $f \in \mathrm{L}_1[0,1]$ we know from the Riemann–Lebesgue Lemma that, for fixed $h > 0$, $\lim_{h\omega\to\infty} I_h[f] = 0$. (Actually, we know more: if f is smooth then $I_h[f] \sim O(\omega^{-1})$, $h\omega \to \infty$—this is a consequence of the van der Corput Lemma and will be used in Section 3.) On the other hand, for large $h\omega$, quadrature (1.6) consists of sampling a virtually-random highly oscillating function $f(hx)\mathrm{e}^{\mathrm{i}h\omega x}$ at ν fixed points in $[0,1]$ and forming a linear combination. Seen from this perspective, it is evident that a Gauss–Christoffel quadrature carries the error of $O(h)$ for $h\omega \gg 1$: the procedure of using a random number generator instead, scaling its output by h, is equally useless and much cheaper. If you wish to rescue Gauss–Christoffel at all costs, the only remedy is to split (1.5) into sufficiently many sub-intervals, each with small natural frequency, and apply (1.6) separately in each sub-interval.

The situation is typical of naive numerical approach to high oscillation: familiar and trusted methods are useless and the remedy appears to be to purge high oscillation by zooming into sufficiently small subproblems. This is *the exact opposite* of the approach advocated in this paper, to use algorithms that thrive on rapid oscillation. Thus, in Section 3 we present an analysis, further elaborated in [**10,11**], of the *Filon method*, a superior alternative to (1.6) whose performance improves in the presence of high oscillation.

2. Highly oscillatory ODE solvers

Our point of departure is the modified ODE system (1.4) but, for ease of exposition, we commence by assuming $\boldsymbol{g} \equiv \mathbf{0}$, namely that (1.2) is a linear system. In Subsection 2.4 we will comment briefly how to extend the methods to a fully nonlinear setting.

2.1. The Magnus method. Given the matrix linear ODE system

$$\frac{\mathrm{d}Y}{\mathrm{d}t} = B(t)Y,\ t \geq t_0, \quad Y(t_0) = Y_0, \tag{2.1}$$

we represent its solution in the form $Y(t) = \mathrm{e}^{\Omega(t)}Y_0$ and treat Ω as a new unknown. It is possible to show that

$$\Omega' = B - \frac{1}{2}[\Omega, B] + \frac{1}{12}[\Omega, [\Omega, B]] - \frac{1}{720}\left[\Omega, \left[\Omega, [\Omega, [\Omega, B]]\right]\right] + \cdots, \quad t \geq 0, \tag{2.2}$$

where $\Omega(0) = O$. A crucial feature of (2.2) renders it of an interest in geometric numerical integration. Let $\mathcal{G}$ be a finite-dimensional Lie group and $\mathfrak{g}$ the corresponding Lie algebra. If $Y_0 \in \mathcal{G}$ and $B(t) \in \mathfrak{g}$ for $t \geq t_0$ then $Y(t)$ evolves in $\mathcal{G}$ and it is often important to retain this structural feature of (2.1) under discretization. However, $\mathcal{G}$ is, in general, a nonlinear construct and the scope for retention of its structure by a direct application of numerical solvers to (2.1) is limited. On the other hand, $\mathfrak{g}$ is a *linear* space and, as long as we solve (2.2) by employing just linear-space operations and commutators, we are assured of respecting its structure. Hence the interest in using (2.2) as a platform for the development of *Lie-group methods*, an endeavour described at length in [**16**].

A powerful technique to solve (2.2) has been introduced by Wilhelm Magnus [**21**] and analysed in detail from the numerical standpoint by Iserles and Nørsett [**17**]. Thus, the function Ω can be expanded in the form

$$\begin{aligned}\Omega(t) = &\int_0^t B(x)\,\mathrm{d}x - \frac{1}{2}\int_0^t\int_0^{x_1}[B(x_2), B(x_1)]\,\mathrm{d}x_2\,\mathrm{d}x_1 \\ &+ \frac{1}{4}\int_0^t\int_0^{x_1}\int_0^{x_2}[[B(x_3), B(x_2)], B(x_1)]\,\mathrm{d}x_3\,\mathrm{d}x_2\,\mathrm{d}x_1 \\ &+ \frac{1}{2}\int_0^t\int_0^{x_1}\int_0^{x_1}[B(x_3), [B(x_2), B(x_1)]]\,\mathrm{d}x_3\,\mathrm{d}x_2\,\mathrm{d}x_1 + \cdots.\end{aligned} \tag{2.3}$$

Truncating (2.3) results in a high-quality numerical method. We observe, following the discussion in Section 1, that the algorithm consists of (multivariate) integration, hence is highly suitable for rapidly oscillating ODE systems, once we apply it to the linear system

$$\frac{\mathrm{d}x}{\mathrm{d}\tau} = B(\tau)\boldsymbol{x},\ \tau \geq 0, \quad \boldsymbol{x}(0) = \boldsymbol{y}_n, \quad \text{where} \quad B(t) = \mathrm{e}^{-\tau\tilde{A}}[A(t_n + \tau) - \tilde{A}]\mathrm{e}^{\tau\tilde{A}}.$$

In particular, we obtain a seventh-order method for the highly oscillatory ODE (1.2) (with $\boldsymbol{g} \equiv \mathbf{0}$) by letting $\boldsymbol{y}_{n+1} = \mathrm{e}^{h\tilde{A}}\mathrm{e}^{\tilde{\Omega}(h)}\boldsymbol{y}_n$, where

$$\tilde{\Omega}(h) = \int_0^h B(x)\,\mathrm{d}x - \frac{1}{2}\int_0^h\int_0^{x_1}[B(x_2), B(x_1)]\,\mathrm{d}x_2\,\mathrm{d}x_1. \tag{2.4}$$

2.2. The Cayley method. A major shortcoming of the Magnus method is that it requires the calculation of a matrix exponential, a task that is prohibitively expensive when the number of variables is large. In our context, we need to compute the exponential and its inverse once per step, to transform the variables in line with (1.3), while using the Magnus expansion means further computation of another exponential, that of $\widetilde{\Omega}(h)$. Although a matrix exponential can be suitably approximated and two exponentials can be conveniently amalgamated by using the BCH formula [**16**], arguably a superior alternative is to dispose of the exponential altogether. In place of the transformation (1.3) we may thus use the *Cayley transformation*

$$\boldsymbol{y}(t_n+\tau) = \left(I - \tfrac{1}{2}\tau\tilde{A}\right)^{-1}\left(I + \tfrac{1}{2}\tau\tilde{A}\right)\boldsymbol{x}(\tau), \quad \tau \geq 0.$$

This replaces (1.4) with the ODE system

$$\frac{\mathrm{d}x}{\mathrm{d}\tau} = B(\tau)\boldsymbol{x} + R^{-1}(\tau\tilde{A})\boldsymbol{g}\big(t_n+\tau, R(\tau\tilde{A})\boldsymbol{x}\big), \ \tau \geq 0, \quad \boldsymbol{x}(0) = \boldsymbol{y}_0,$$

where

$$B(\tau) = \left(I + \tfrac{1}{2}\tau\tilde{A}\right)^{-1}\left(I - \tfrac{1}{2}\tau\tilde{A}\right)\left[A(t_n+\tau) - \tilde{A} - \tfrac{1}{4}\tau^2 A(t_n+\tau)\tilde{A}^2\right]\left(I - \tfrac{1}{2}\tilde{A}\right)^{-2},$$

while $R(z) = \left(1+\frac{1}{2}z\right)/\left(1-\frac{1}{2}z\right)$.

The above procedure might seem advantageous, since forming $B(\tau)$ requires the solution of two linear algebraic systems, a task which is easier than the calculation of the exponential and allows greater latitude in the exploitation of possible sparsity and structure in the matrix $\tilde{A}$. Unfortunately, the cubic term in the square brackets means that, unlike the matrix that we have encountered in the previous subsection, B need not be small. This has been confirmed by unpublished numerical experiments and motivates us to abandon this approach,

The exponential transformation (1.3) is here to stay, at least for the time being. However, we can follow the reasoning in [**4**] and avoid the computation of the second exponential, associated with the Magnus method. The matrix $\tilde{A}$ has often a great deal of structure that renders it amenable to an affordable calculation (or approximation) of its exponential: one example will feature in the next subsection and several publications have been devoted to effective approximation of the exponential in this setting [**16, 18, 22**]. On the other hand, the matrix $\widetilde{\Omega}(h)$ of the last subsection is likely to have neither serendipitous structure nor sparsity, hence the calculation of its exponential is usually a more formidable task. The following technique dispenses with this calculation altogether.

Taking a leaf from [**15**], we solve (1.4) by a Cayley transformation: letting $\boldsymbol{x}(\tau) = R\big(\Omega(\tau)\big)\boldsymbol{y}_n$, we observe that the matrix function Ω obeys the ODE

$$\Omega' = A - \tfrac{1}{2}[\Omega, A] - \tfrac{1}{4}\Omega A\Omega, \ t \geq 0, \quad \Omega(0) = O. \tag{2.5}$$

(We have assumed for simplicity, and consistently with the framework of Subsection 2.1, that $\boldsymbol{g} \equiv \boldsymbol{0}$.) The solution of (2.5) can be expanded similarly to that of the equation (2.2),

$$\begin{aligned}\tilde{\Omega}(h) = \int_0^h B(x)\,\mathrm{d}x - \frac{1}{2}\int_0^h\int_0^{x_1}[B(x_2),B(x_1)]\,\mathrm{d}x_2\,\mathrm{d}x_1 \\ + \frac{1}{4}\int_0^h\int_0^{x_1}\int_0^{x_2}[[B(x_3),B(x_2)],B(x_1)]\,\mathrm{d}x_3\,\mathrm{d}x_2\,\mathrm{d}x_1 \\ - \frac{1}{4}\int_0^h\int_0^{x_1} B(x_2)\,\mathrm{d}x_2 B(x_1)\int_0^{x_1} B(x_2)\,\mathrm{d}x_2\,\mathrm{d}x_1 + \cdots.\end{aligned}$$

Observe that the first two terms above coincide with the truncated Magnus expansion (2.4). Since for all intents and purposes this is all we need, we conclude that there is little *practical* difference between Magnus and Cayley: the only real distinction is that, in place of calculating $\exp\widetilde{\Omega}(h)$, we compute $R(\widetilde{\Omega}(h))$.

Note further that the Cayley transformation is a Lie-group method only for some Lie groups [**15**]. Yet, this is of little relevance in the context of highly oscillatory problems. The important feature of the method is that it reduces time-stepping to the computation of integrals of highly oscillatory functions.

2.3. The Neumann method. The Magnus method is highly suitable for small systems, while the scope of the Cayley method is wider. However, none can be advocated for the very large systems originating in a semi-discretization of evolutionary partial differential equations. This subsection introduces a technique, based on the Neumann expansion, as applied to the linear *Schrödinger equation*

$$\mathrm{i}\frac{\mathrm{d}u}{\mathrm{d}t} = \nabla u + \lambda(t,\boldsymbol{x})u, \; t \geq 0, \quad \boldsymbol{x} \in \Theta,$$

with appropriate initial and boundary conditions. We assume here that $\lambda(t,\boldsymbol{x}) \gg 1$ for $t \gg 1$, rendering the solution highly oscillatory. For simplicity, we assume just two spatial dimensions, but the method can be easily generalized.

Discretizing the Laplacian with the standard *five-point stencil*, we obtain the linear ODE system

$$\mathrm{i}\frac{\mathrm{d}\boldsymbol{u}}{\mathrm{d}t} = \left[\frac{1}{(\Delta x)^2}S + \Lambda(t)\right]\boldsymbol{u}, \quad t \geq 0,$$

where Λ is a diagonal matrix. More importantly, the 'discrete Laplacian' S is a partitioned Toeplitz, symmetric, tridiagonal (TST) matrix, made out of smaller TST matrices. Its eigenvector matrix Q is known and, more pertinently, a product of the form $Q\boldsymbol{v}$ can be formed rapidly using the *fast Fourier transform* (FFT) [**14**].

To time-step from t_n to $t_{n+1} = t_n + \Delta t$, we first rewrite the ODE in the form

$$\frac{\mathrm{d}\boldsymbol{u}}{\mathrm{d}t} = [A(t) + \tilde{\Lambda}(t)]\boldsymbol{u}, \quad t \geq 0,$$

where

$$A(t) = -\frac{\mathrm{i}}{(\Delta x)^2}S - \mathrm{i}\hat{\lambda}(t)I, \quad \tilde{\Lambda}(t) = \mathrm{i}[\hat{\lambda}(t) - \Lambda(t)]$$

and $\hat{\lambda}(t) = \int_\Theta \lambda(t,\boldsymbol{x})\,\mathrm{d}\boldsymbol{x}/\mathrm{vol}\,\Theta$. Note that A shares the eigenvectors of S, while the diagonal matrix $\tilde{\Lambda}$ is small. Letting $\tilde{A} = A\big(t_n + \frac{1}{2}\Delta t\big)$, we transform variables

$$\boldsymbol{u}(t_n + \tau) = \mathrm{e}^{\tau\tilde{A}}\boldsymbol{v}(\tau), \quad \tau \geq 0,$$

whereby

$$\frac{\mathrm{d}\boldsymbol{v}}{\mathrm{d}\tau} = B(\tau)\boldsymbol{v}, \ \tau \geq 0, \quad \boldsymbol{v}(0) = \boldsymbol{u}^n, \tag{2.6}$$

and

$$B(\tau) = \mathrm{e}^{-\tau\tilde{A}}[A(t_n+\tau) - \tilde{A} - \tilde{\Lambda}(t_n+\tau)]\mathrm{e}^{\tau\tilde{A}}, \quad \tau \geq 0.$$

The computation of $\exp(\tau\tilde{A})$ can be accomplished with FFT, hence is affordable even for large systems. However, the matrix B is dense, with far less structure than, say, $\tilde{A}$, and both Magnus and Cayley transformations are prohibitively expensive. We propose instead to use the *Neumann expansion* [**2**, p. 516], occasionally known in an ODE context as the *Peano expansion* [**7**, p. 152]:

$$\boldsymbol{v}(\Delta t) = \sum_{m=0}^{\infty} \mathcal{N}_m(\Delta t)\boldsymbol{u}^n, \tag{2.7}$$

where $\mathcal{N}_0 \equiv I$ and

$$\begin{aligned}\mathcal{N}_m(t) &= \int_0^t B(x)\mathcal{N}_{m-1}(x)\,\mathrm{d}x \\ &= \int_0^t \int_0^{x_1} \cdots \int_0^{x_{m-1}} B(x_m)B(x_{m-1})\cdots B(x_1)\,\mathrm{d}x_m\,\mathrm{d}x_{m-1}\cdots\mathrm{d}x_1, \quad m \in \mathbb{N}.\end{aligned}$$

The expansion (2.7) possesses a number of useful features. Firstly, it consists of repeated integration of highly oscillatory functions, hence is suitable for rapidly oscillating systems. Secondly, it does not require the computation of exponentials. This, together with the low cost of forming the matrix B in (2.6), means that the expense of time-stepping is fairly moderate. Thirdly, the standard obstacle to the usefulness of Neumann series in, for example, the numerical solution of integral equations, but also for ODEs, is the convergence requirement $\|\boldsymbol{K}\| < 1$, where

$$\boldsymbol{K}[F] = \int_0^t B(x)F(x)\,\mathrm{d}x$$

[**1**]. This, however, is not a problem in the present setting *precisely because we are integrating highly oscillatory functions.*

2.4. Nonlinear equations. Let $\mathrm{e}^{\Omega(\tau)}$ be the fundamental solution of (2.1). The solution of (1.4) can be written using *nonlinear variation of constants*

$$\boldsymbol{x}(\tau) = \mathrm{e}^{\Omega(\tau)}\boldsymbol{y}_n + \int_0^\tau \mathrm{e}^{\Omega(\tau-\xi)}\mathrm{e}^{\xi\tilde{A}}\boldsymbol{g}(t_n+\xi, \mathrm{e}^{\xi\tilde{A}}\boldsymbol{x}(\xi))\,\mathrm{d}\xi.$$

This is an implicit equation, but it can be solved in few *waveform relaxation* iterations,

$$\begin{aligned}\boldsymbol{x}^{[0]}(\tau) &\equiv \boldsymbol{y}_n, \\ \boldsymbol{x}^{[l+1]}(\tau) &= \mathrm{e}^{\Omega(\tau)}\boldsymbol{y}_n + \int_0^\tau \mathrm{e}^{\Omega(\tau-\xi)}\mathrm{e}^{\xi\tilde{A}}\boldsymbol{g}(t_n+\xi, \mathrm{e}^{\xi\tilde{A}}\boldsymbol{x}^{[l]}(\xi)\,\mathrm{d}\xi, \quad l \in \mathbb{Z}_+.\end{aligned}$$

Typically, two or three iterations suffice and we observe that each requires integration with highly oscillating integrands.

This brief section represents work in progress, which is close in spirit to recent methods for stiff ordinary differential equations, pioneered by Kassam and Trefethen [**19**].

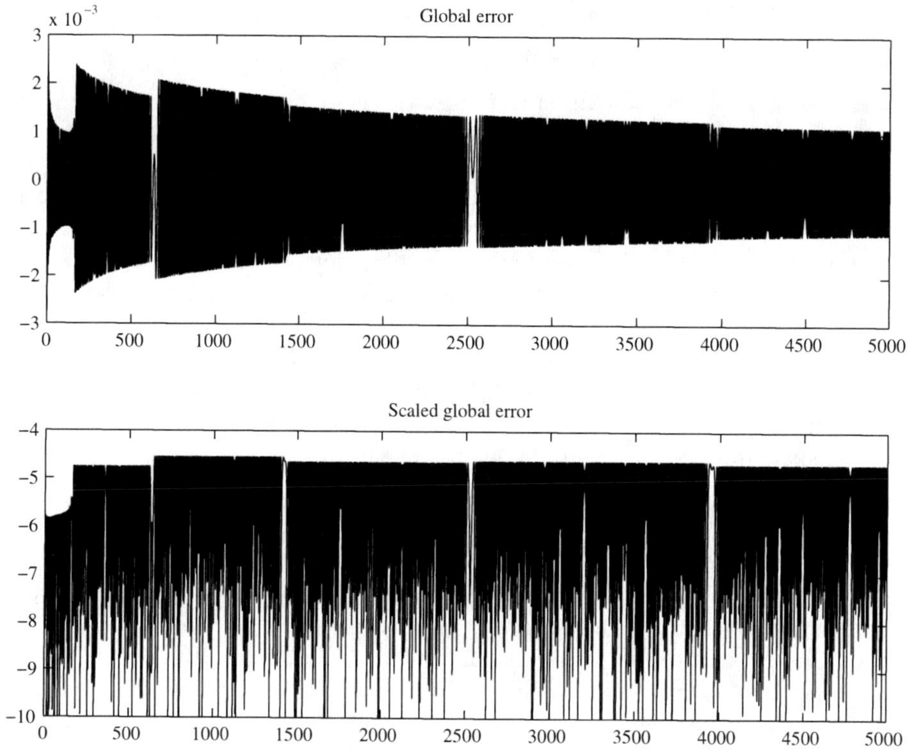

FIGURE 3. The error in solving (1.1) with the single-term Magnus method and $h = \frac{1}{4}$. The top plot displays the global error $|y_n - y(t_n)|$ and the bottom the scaled error $\log|y_n - y(t_n)| + \frac{1}{4}\log t$.

2.5. An example: the Airy equation. Converting the linear oscillator

$$\frac{\mathrm{d}^2 y}{\mathrm{d}t^2} + g(t)y = 0,\ t \geq 0, \quad y(0) = y_0, \quad \frac{\mathrm{d}y(0)}{\mathrm{d}t} = y_0',$$

where we assume that $g(t) \gg 1$ for $t \gg 1$, into a two-variable system and subjecting it to the transformation (1.4) results in

$$B(\tau) = [g(t_n + \tau) - \omega^2]\begin{bmatrix} (2\omega)^{-1}\sin 2\omega\tau & \omega^{-2}\sin^2\omega\tau \\ -\cos^2\omega\tau & -(2\omega)^{-1}\sin 2\omega\tau \end{bmatrix},$$

where $\omega = \sqrt{g(t_n + \frac{1}{2}h)} \gg 1$.

We evaluate just the lowest-order methods, applied to the Airy oscillator (1.1). Thus, Magnus and Cayley, both with just a single integral, yield

$$\boldsymbol{y}_{n+1} = \mathrm{e}^{h\tilde{A}}\mathrm{e}^{\widetilde{\Omega}(h)}\boldsymbol{y}_n \quad \text{and} \quad \boldsymbol{y}_{n+1} = \mathrm{e}^{h\tilde{A}}R(\widetilde{\Omega}(h))\boldsymbol{y}_n$$

respectively, where

$$\widetilde{\Omega}(h) = (\sin h\omega_n - h\omega_n\cos h\omega_n)\begin{bmatrix} \frac{\cos h\omega_n}{8\omega_n^3} & \frac{\sin h\omega_n}{4\omega_n^4} \\ \frac{\sin h\omega_n}{4\omega_n^2} & -\frac{\cos h\omega_n}{8\omega_n^3} \end{bmatrix}, \qquad \omega_n = \sqrt{t_n + \frac{1}{2}h}.$$

Moreover, the Neumann expansion with a single term reduces to

$$\boldsymbol{y}_{n+1} = \mathrm{e}^{h\tilde{A}}[I + \widetilde{\Omega}(h)]\boldsymbol{y}_n.$$

In the present case, one can think of the Cayley and Neumann expansions as approximating Magnus through the replacement of e^z by its [1/1] and [1/0] Padé approximations respectively.

Figure 3 displays the global error in solving the Airy equation (1.1) with the above Magnus method and the step size $h = \frac{1}{4}$. The contrast with Figure 2 could not have been greater and it represents a tangible demonstration of the advantage of the approach of this paper. It has been proved in [**12**] that the global error when (1.1) is solved with Magnus accumulates like $ct^{-1/4}\chi(t)$, where χ oscillates in $[-1, 1]$. This is confirmed in the bottom plot in Figure 3.

Both Cayley and Neumann produce results which are virtually indistinguishable from Figure 3. Specifically, the Cayley and Magnus solutions differ by just $\approx 10^{-10}$, whether Neumann departs from Magnus by at most $\approx 10^{-5}$.

3. Quadrature of highly oscillatory integrals

Discretizing highly oscillatory ODEs by converting the problem into the integration of highly oscillatory functions is a powerful approach which can be easily spoilt by bad quadrature. Occasionally, like in Subsection 2.5, we can evaluate integrals in an explicit form, but this is an exception and, in general, we need to use quadrature. Having already observed in Section 1 that a Gauss–Christoffel quadrature (1.6) is unsuitable for our needs, we look elsewhere for numerical salvation. Fortunately, quality quadrature methods for highly oscillatory integrals are available in the literature, e.g. the Zamfirescu method [**25**], the Levin method [**20**] and, in particular, the Filon method [**5, 6**]. With the sole exception of Levin's method, the analysis of these algorithms was lacking until recently [**10, 11**]. We simply have had no idea how good such methods can be.

3.1. The Filon method. In the sequel we focus on the Filon method, since it is the most general and natural of quadrature methods for highly oscillatory integrals and also since it is the most comprehensively understood. It is, indeed, remarkable that a method introduced (by Louis Napoleon George Filon, [**5**]) in 1928 and then either forgotten or often misunderstood turns out to be so effective.

The idea is really simple: to approximate the integral

$$I[f] = \int_0^1 f(x)K(\omega, x)\mathrm{d}x, \tag{3.1}$$

where K oscillates rapidly for $\omega \gg 1$, we choose quadrature nodes $c_1 < c_2 < \cdots < c_\nu$ in $[0, 1]$ and interpolate $f(x)$ (rather than interpolating $f(x)K(\omega, x)$ in Gauss–Christoffel) there by a polynomial of degree $\nu - 1$. Provided that we can evaluate the moments

$$\mu_m(\omega; K) = \int_0^1 x^m K(\omega, x)\,\mathrm{d}x, \quad m \in \mathbb{Z}_+,$$

explicitly for $0 \le m \le \nu - 1$, substitution of the interpolating polynomial in place of f in (3.1) results in the *Filon quadrature*

$$Q^{\mathrm{F}}[f] = \sum_{l=1}^{\nu} b_l(\omega) f(c_l). \tag{3.2}$$

We note in passing that moments can be evaluated explicitly, e.g. with standard symbolic algebra software, for a wide range of functions K of practical interest, e.g. $\mathrm{e}^{\mathrm{i}\omega(x-d)^m}$ for $d \in [0,1]$ and $m \in \{1,2,3\}$, $\mathrm{e}^{\mathrm{i}\omega|x-d|}/|x-d|^\alpha$ for $d \in [0,1]$ and $\alpha < 1$, the Bessel oscillator $\mathrm{J}_\nu(\omega x)$ and the Airy oscillator $\mathrm{Ai}(-\omega x)$.

Numerical experiments with (3.2) are remarkable, while elementary and intuitive arguments (sadly missing from some early treatments of Filon's method, that should be better left in their obscurity) imply that the method is likely to be very effective. In particular, by design the method integrates *exactly* all polynomials f of degree $\le \nu - 1$ and, unlike $f(x)K(\omega,x)$, a smooth function f can be interpolated very well by such polynomials. Having said this, there is a need for more formal analysis, both to understand the method better and to develop suitable strategies for good choice of quadrature nodes.

3.2. Regular oscillators. We commence our analysis from the regular oscillator $K(\omega,x) = \mathrm{e}^{\mathrm{i}\omega x}$, i.e.

$$I[f] = \int_0^1 f(x)\mathrm{e}^{\mathrm{i}\omega x}\,\mathrm{d}x. \tag{3.3}$$

This is equivalent to setting $h = 1$ in (1.5): it is easy to incorporate an additional small parameter $h > 0$ into the theory but we opt for (3.3) for simplicity's sake.

It is interesting to consider the quadrature error $E[f] = Q^{\mathrm{F}}[f] - I[f]$ in three regimes: $0 < \omega \ll 1$, $\omega = O(1)$ and $\omega \gg 1$. The first corresponds to a non-oscillatory regime and it is of interest to check how much (if at all) Filon is worse than Gauss–Christoffel in this setting. The second regime is consistent with a persistent numerical 'folk wisdom', namely that for optimal results a function should be sampled a fixed number of times per period. Finally, the regime $\omega \gg 1$ addresses our main concern, highly oscillatory problems.

A comprehensive analysis in [**10**] determines $E[f]$ in all three regimes for arbitrary smooth functions f.

(1) Suppose that the Gauss–Christoffel method for (3.3) with $\omega = 0$, using the same nodes, is of order $p \in \{\nu, \nu+1, \ldots, 2\nu\}$: this is equivalent to

$$\int_0^1 x^m \gamma(x)\,\mathrm{d}x = 0, \qquad m = 0,1,\ldots,p-\nu-1,$$

where $\gamma(t) = \prod_{l=1}^{\nu}(t - c_l)$. Then so is the Filon quadrature (3.2) for $0 < \omega \ll 1$.

(2) If $\omega = O(1)$ then (3.2) recovers only polynomials of degree $\nu - 1$, in other words is of order ν. Thus, the received 'folk wisdom' is perversely wrong: this is the *worst* regime.

(3) If $\omega \gg 1$ then, for a general choice of quadrature nodes, the error is $E[f] = O(\omega^{-1})$ but, if $c_1 = 0$ and $c_\nu = 1$ then $E[f] = O(\omega^{-2})$. The method is still of order ν but, remarkably, for high frequencies it is not polynomial recovery but the oscillation itself that kills the error. The higher the oscillation, the better Filon's quadrature and this most welcome phenomenon can be greatly enhanced by choosing the endpoints as quadrature nodes.

3.3. Irregular oscillators and the geometric model. Let g be a smooth, real function and set $K(\omega,x)=\mathrm{e}^{\mathrm{i}\omega g(x)}$: in place of (3.3) we thus integrate

$$I[f]=\int_0^1 f(x)\mathrm{e}^{\mathrm{i}\omega g(x)}\,\mathrm{d}x. \tag{3.4}$$

In the sequel we focus on the case $\omega\gg 1$, since it is both the most relevant to our discussion and the most comprehensively understood. Our exposition follows [**11**].

In order to examine the behaviour of $E[f]$ for $\omega\gg1$, it is highly relevant to mention the main step used in [**10**] to elicit the error asymptotics for $g(x)=x$, namely studying the asymptotic behaviour of the moments,

$$\mu_m(\omega;\mathrm{e}^{\mathrm{i}\omega x})\sim\frac{1^m\mathrm{e}^{\mathrm{i}\omega}-0^m}{\mathrm{i}\omega}+O(\omega^{-2}),\qquad m\in\mathbb{Z}_+,\ \omega\gg1. \tag{3.5}$$

In a manner made precise in [**10**], choosing the nodes $c_1=0$ and $c_\nu=1$ eliminates the influence of the 0^m and 1^m terms in (3.5) on the error, rendering $E[f]=O(\omega^{-2})$. With far greater generality, it has been established in [**11**] that if the moments of K obey the *geometric model*

$$\mu_m(\omega;K)\sim\sum_{j=1}^r v_j(\omega)\omega^{-\alpha_j}d_j^m+O(\omega^{-\beta}),\qquad m\in\mathbb{Z}_+,\ \omega\gg1, \tag{3.6}$$

where $d_1,d_2,\ldots,d_r\in[0,1]$, $v_1,v_2,\ldots,v_r\in\mathrm{L}_\infty[0,\infty)$ and

$$0<\alpha_1,\alpha_2,\ldots,\alpha_r<\beta$$

then, using *all* the points $d_1,d_2,\ldots,d_r$ as quadrature nodes yields $E[f]\sim O(\omega^{-\beta})$. In a manner that can be made precise, choosing each d_j eliminates the influence of $v_j(\omega)\omega^{-\alpha_j}$ on the error.

We have already mentioned an example of K that is consistent with the geometric model (3.6), namely $K(\omega,x)=\mathrm{e}^{\mathrm{i}\omega x}$. In that case $r=2$, $d_1=0$, $d_2=1$, $v_1(x)\equiv-1$, $v_2(x)=\mathrm{e}^{\mathrm{i}x}$, $\alpha_1=\alpha_2=1$ and $\beta=2$. Ref. [**11**] establishes very broad conditions for the consistency of $K(\omega,x)=\mathrm{e}^{\mathrm{i}\omega g(x)}$ with the geometric model.

The simplest case is when $g'\neq0$ in $[0,1]$ and either g'' does not change sign there, whereby we let $s=1$, or there exists $s\geq3$ so that $g^{(s)}\neq0$ in $[0,1]$. (Such condition is always satisfied when g is a polynomial.) Then (3.6) is satisfied with $r=2$, $d_1=0$, $d_2=1$, $\alpha_1=\alpha_2=1/s$ and $\beta=2/s$. The proof uses the van der Corput Lemma from harmonic analysis [**24**].

The case when g' vanishes at distinct points in $(0,1)$ is more complicated. The van der Corput lemma is combined with the method of stationary phase from perturbation theory, using bump functions. Firstly, suppose that g' vanishes at $\tilde d\in(0,1)$ and nowhere else in $[0,1]$. In other words, there exists an integer $\delta\geq1$ such that

$$g'(\tilde d)=g''(\tilde d)=\cdots=g^{(\delta-1)}(\tilde d)=0,\quad g^{(\delta)}(\tilde d)\neq0.$$

The method of stationary phase can be used to prove that there exist functionals a_i such that

$$I[f]\sim\omega^{1/\delta}\sum_{i=1}^\infty a_i[f]\omega^{-i/\delta},\qquad\omega\gg1,$$

for every function f with a sufficiently small compact support in a neighbourhood of $\tilde d$ [**24**].

Suppose now that g' vanishes at the $r-2$ distinct points $d_j \in (0,1)$ and that the lowest-order nonzero derivative at d_j is $g^{(\delta_j)}$. We use bump functions to 'stitch' compactly-supported functions together, adding contributions from the endpoints $d_1 = 0$ and $d_r = 1$. It then follows that g obeys the quadratic model with $\alpha_1 = \alpha_r = 1$ and $\alpha_j = 1/\delta_j$ for $j = 2, \ldots, r-1$. Also the value of β can be determined and it depends on the parity of the integers δ_j [**11**].

The analysis may break down when g' vanishes at an endpoint, e.g. in the case $K(\omega, x) = \mathrm{e}^{\mathrm{i}\omega x^2}$. Otherwise it provides the optimal strategy for the choice of quadrature points in the Filon method: we need $\nu \geq r$ and

$$d_1, d_2, \ldots, d_r \in \{c_1, c_2, \ldots, c_\nu\}.$$

3.4. Multivariate integration. Higher-order methods from Section 2 require multivariate integration, a notoriously difficult computational task. Fortunately, the special form of Magnus and Cayley integrals renders them amenable to a very effective quadrature [**16**] and a future publication will extend this to the Neumann expansion.

The first stage is to reduce the expense in function evaluations. Thus, suppose that $c_1, c_2, \ldots, c_\nu \in [0,1]$ are given quadrature nodes and that we have evaluated $B(c_l h)$, $l = 1, 2, \ldots, \nu$. It is elementary that the first integral in (2.3) can be integrated with a Gauss–Christoffel quadrature of order p, say. In other words, there exist weights $b_l^{[1]}$, $l = 1, 2, \ldots, \nu$ so that

$$h \sum_{l=1}^{\nu} b_l^{[1]} B(c_l h) = \int_0^h B(x)\,\mathrm{d}x + O(h^{p+1}).$$

It was proved in [**17**] that there also exist weights $b_{i,j}^{[2]}$, $i, j = 1, 2, \ldots, \nu$, such that

$$h^2 \sum_{i=1}^{\nu} \sum_{j=1}^{\nu} b_{i,j}^{[2]} [B(c_i h), B(c_j h)] = \int_0^h \int_0^{x_1} [B(x_2), B(x_1)]\,\mathrm{d}x_2\,\mathrm{d}x_1 + O(h^p).$$

Similarly for *all* Magnus or Cayley integrals: with just ν evaluations of the matrix B we can evaluate the entire truncated expansion to order p.

The drawback of this approach is that, although the number of function evaluations is very small, this is not the case with the volume of linear algebra, since we need to evaluate a very large number of commutators. The number of commutators, however, can be reduced a very great deal by techniques pioneered by Munthe-Kaas and Owren and by Blanes, Casas and Ros [**16**].

All this is true for 'standard' Magnus or Cayley, where Lagrange points represent the obvious choice of quadrature nodes. It has been proved in [**10**] that this approach remains equally valid for the Filon quadrature and highly oscillating integrands.

4. Conclusion

The present review of recent (and some new) results in the computation of highly oscillatory problems represents a snapshot of work in progress, rather than a mature and comprehensive theory. Yet, assembling all the above results does more than just presenting a number of disjoint algorithms, useful as they might be. General principles emerge.

Firstly, highly oscillatory differential equations should be solved by algorithms that integrate highly oscillating functions. There might be diverse ways of designing such algorithms: we have described three in Section 2 but more and better might well exist.

Secondly, quadrature of highly oscillatory functions is possible, indeed easy, but its optimal implementation calls for an understanding of the asymptotic behaviour of moments.

Thirdly, and most importantly, high oscillation is not an enemy of computation. Any discretization method that becomes less accurate (or more expensive) in the presence of rapid oscillation is, arguably, a *bad* method. And good methods do exist, although much is yet to be done in this area. The rule of a thumb is that high oscillation should assist precision and lower the cost of computation. The main lesson of this paper is that this is an attainable goal.

References

1. N. D. Aparicio, S. J. A. Malham, and M. Oliver, *Precomputation methods and the Evans function*, Heriot–Watt Univ., Edinburgh, 2003.
2. K. E. Atkinson, *The numerical solution of integral equations of the second kind*, Cambridge Univ. Press, Cambridge, 1997.
3. P. J. Davis and P. Rabinowitz, *Methods of numerical integration*, 2nd, Academic Press, Orlando, Fl., 1980.
4. F. Diele and S. Ragni, *The global error of Magnus methods based on the Cayley map for some oscillatory problems*, Tech. Report, University of Bari, 2002.
5. L. N. G. Filon, *On a quadrature formula for trigonometric integrals*, Proc. Roy. Soc. Edinburgh Sect. A **49** (1928), 38–47.
6. E. A. Flinn, *A modification of Filon's method of numerical integration*, J. Assoc. Comput. Mach. **7** (1960), 181–184.
7. F. R. Gantmacher, *Applications of the theory of matrices*, Wiley Interscience, New York, 1959.
8. W. Gautschi, *Construction of Gauss–Christoffel quadrature formulas*, Math. Comp. **22** (1968), 251–270.
9. E. Hairer, S. P. Nørsett, and G. Wanner, *Solving ordinary differential equations.* I. *Nonstiff problems*, 2nd ed., Springer Verlag, Berlin, 1993.
10. A. Iserles, *On the numerical quadrature of highly-oscillating integrals.* I. *Fourier transforms*, IMA J. Numer. Anal. **24**, 2004, 365–391.
11. ———, *On the numerical quadrature of highly-oscillating integrals.* II. *Irregular oscillators*, IAM J. Numer. Anal., 2005 (to appear).
12. ———, *On the global error of discretization methods for highly-oscillatory ordinary differential equations*, BIT **42** (2002), 561–599.
13. ———, *Think globally, act locally: Solving highly-oscillatory ordinary differential equations*, Appl. Num. Anal. **43** (2002), 145–160.
14. ———, *A first course in the numerical analysis of differential equations*, Cambridge Univ. Press, Cambridge, 1996.
15. ———, *On Cayley-transform methods for the discretization of Lie-group equations*, Found. Comput. Maths **1**, (2001), 129–160.
16. A. Iserles, H. Z. Munthe–Kaas, S. P. Nørsett, and A. Zanna, *Lie-group methods*, Acta Numer. **9** (2000), 215–365.
17. A. Iserles and S. P. Nørsett, *On the solution of linear differential equations in Lie groups*, Philos. Trans. Royal Soc. London Ser. A **357** (1999), 983–1020.
18. A. Iserles and A. Zanna, *Efficient computation of the matrix exponential by generalized polar decompositions*, DAMTP, University of Cambridge, 2002, NA2002/09.
19. A.-K. Kassam and L. N. Trefethen, *Fourth-order time stepping for stiff* PDE*s*, SIAM J. Sci. Comput., 2004 (to appear).
20. D. Levin, *Fast integration of rapidly oscillatory functions*, J. Comput. Appl. Math. **67** (1996), 95–101.

21. W. Magnus, *On the exponential solution of differential equations for a linear operator*, Comm. Pure Appl. Math. **7** (1954), 649–673.
22. H. Munthe-Kaas, R. G. W. Quispel, and A. Zanna, *Generalized polar decompositions on Lie groups with involutive automorphisms*, Found. Comp. Math. **1** (2001), 297–324.
23. F. W. J. Olver, *Asymptotics and special functions*, Academic Press, New York, 1974.
24. E. Stein, *Harmonic analysis: real-variable methods, orthogonality, and oscillatory integrals*, Princeton Univ. Press, Princeton, N.J., 1993.
25. I. Zamfirescu, *An extension of Gauss's method for the calculation of improper integrals*, Acad. R. P. Romîne Stud. Cerc. Mat. **14** (1963), 615–631 (Romanian).

Department of Applied Mathematics and Theoretical Physics, Centre for Mathematical Sciences, Wilberforce Rd, Cambridge CB3 0WA, United Kingdom.

E-mail address: ai@damtp.cam.ac.uk

Centre de Recherches Mathématiques
CRM Proceedings and Lecture Notes
Volume **39**, 2005

On Conservation Properties of Semidiscrete Canonical Hamiltonian Equations

R. Kozlov

ABSTRACT. There are many evolution partial differential equations which can be cast into Hamiltonian form. Conservation laws of these equations are related to one-parameter Hamiltonian symmetries admitted by the PDEs [**11**]. The same result holds for semidiscrete Hamiltonian equations [**7**]. In this paper we briefly review semidiscrete canonical Hamiltonian equations, considered in [**8**] in greater details. Using symmetries, we find conservation laws for the semidiscretized nonlinear wave equation.

1. Introduction

Many partial differential equations (PDEs) of nondissipative continuum mechanics can be presented in Hamiltonian form (see [**10**] and references therein). It is well known that conservation laws of Hamiltonian PDEs are related to one-parameter Hamiltonian symmetries [**11**]. The analog of this result holds in the semidiscrete case with no change in the statement although the framework must be modified [**7**].

In this paper we consider canonical Hamiltonian equations

$$\mathbf{v}_t = \frac{\delta\mathcal{H}}{\delta\mathbf{w}}, \quad \mathbf{w}_t = -\frac{\delta\mathcal{H}}{\delta\mathbf{v}}, \tag{1.1}$$

which form a special type of the Hamiltonian equations. For instance, the nonlinear wave equation

$$v_{tt} = \Delta v - V'(v) \tag{1.2}$$

can be rewritten in the form (1.1). We will introduce semidiscrete analogs of the canonical Hamiltonian PDEs and show how one can use the Hamiltonian form of Noether's theorem to find conservation laws of these equations.

To find conservation laws of semidiscrete equations with the help of Noether's theorem, we need to know symmetries in the evolutionary form. Different methods which can be used to find symmetries of discrete and semidiscrete equations are discussed, for example, in [**9**]. The most successful application was shown for linear difference equations [**4, 5, 9**], where for the considered equations symmetry algebras

2000 *Mathematics Subject Classification.* Primary: 37K05; Secondary: 70S10.
This is the final form of the paper.

isomorphic to those of the underlying continuous equations were found. However, it is not known how to find all symmetries admitted by nonlinear semidiscrete equations.

One of the possibilities, which we exploit in this paper, is to make use of the admitted Lie point symmetries. Many Lie point symmetries of semidiscrete equations can be easily found as Lie point symmetries of the underlying continuous equations preserved under the space discretization. In both continuous and semidiscrete cases these symmetries are given by the same vector fields. Using factorization, one can obtain corresponding evolutionary operators. If the latter are Hamiltonian symmetries, they let us find conservation laws. Obviously, for our purpose we will not be interested in all symmetries, but only in Hamiltonian ones.

The layout of the paper is as follows: In Section 2 we briefly introduce Hamiltonian equations and specify canonical Hamiltonian equations. Symmetries and the Hamiltonian form of Noether's theorem are discussed in Section 3. In Section 4 we examine a semidicretization of the equation (1.2) and find its conservation laws. In final Section 5 we make concluding remarks. In particular, we mention the connection between Euler–Lagrange equations and canonical Hamiltonian equations.

2. Hamiltonian equations

For simplicity we will consider the case of one space coordinate x. We assume that the solutions are sufficiently smooth, all variational derivatives tend to zero as the solution tends to zero and the solution and a number of its space derivatives tend to zero as $|x| \to \infty$. We suppose that the solution decays fast enough so that all integrals and sums make sense.

2.1. Hamiltonian partial differential equations. Many systems of evolution equations

$$\mathbf{u}_t = K(x, \mathbf{u}^{(m)}),$$

where $\mathbf{u}$ stands for N dependent variables $\mathbf{u} = (u^1, u^2, \dots, u^N)^T$ and $\mathbf{u}^{(m)} = (\mathbf{u}, \mathbf{u}_1, \mathbf{u}_2, \dots, \mathbf{u}_m)$ represents $\mathbf{u}$ and a finite set of derivatives of $\mathbf{u}$ with respect to space coordinate x, can be cast into the Hamiltonian form

$$\text{(2.1)} \quad \mathbf{u}_t = \mathcal{D}\left(\frac{\delta \mathcal{H}}{\delta \mathbf{u}}\right), \quad \mathcal{H}[\mathbf{u}] = \int H(x, \mathbf{u}^{(n)})\, dx, \quad \frac{\delta H}{\delta \mathbf{u}} = \left(\frac{\delta H}{\delta u^1}, \frac{\delta H}{\delta u^2}, \dots, \frac{\delta H}{\delta u^N}\right)^T$$

with the help of the Hamiltonian functional $\mathcal{H}[\mathbf{u}]$, variational operator $\delta \cdot / \delta \mathbf{u}$ and the linear operator $\mathcal{D}$ [**11**]. We denote as $\mathcal{F}$ the space of functionals

$$\int P(t, x, \mathbf{u}^{(k)})\, dx, \quad k \in \mathbb{N}.$$

The operator $\mathcal{D}$ must be Hamiltonian, i.e., it forms the Poisson bracket

$$\text{(2.2)} \qquad \{\mathcal{P}, \mathcal{L}\} = \int \left(\frac{\delta \mathcal{P}}{\delta \mathbf{u}}\right)^T \mathcal{D}\left(\frac{\delta \mathcal{L}}{\delta \mathbf{u}}\right) dx$$

satisfying the conditions of *skew-symmetry*

$$\text{(2.3)} \qquad \{\mathcal{P}, \mathcal{L}\} = -\{\mathcal{L}, \mathcal{P}\}$$

and the *Jacobi identity*

$$\text{(2.4)} \qquad \{\{\mathcal{P}, \mathcal{L}\}, \mathcal{R}\} + \{\{\mathcal{R}, \mathcal{P}\}, \mathcal{L}\} + \{\{\mathcal{L}, \mathcal{R}\}, \mathcal{P}\} = 0$$

for all functionals $\mathcal{P}, \mathcal{L}, \mathcal{R} \in \mathcal{F}$.

The variational derivatives of a functional can be found by the action of the Euler operators on the integrand

$$\frac{\delta \mathcal{H}}{\delta u^i} = E^i(H), \quad i = 1, \ldots, N, \tag{2.5}$$
$$E^i = \frac{\partial \cdot}{\partial u^i} - D_x\left(\frac{\partial \cdot}{\partial u_1^i}\right) + D_x^2\left(\frac{\partial \cdot}{\partial u_2^i}\right) + \cdots + (-D_x)^n\left(\frac{\partial \cdot}{\partial u_n^i}\right) + \cdots,$$

where D_x is the total space derivative operator.

2.2. Canonical Hamiltonian equations. Canonical Hamiltonian equations form a subset of the equations (2.1) characterized by an even-dimensional space of dependent variables $N = 2n$, $\mathbf{u} = (v^1, \ldots, v^n, w^1, \ldots, w^n)^T$, and the canonical Hamiltonian operator

$$J = \begin{pmatrix} 0_n & I_n \\ -I_n & 0_n \end{pmatrix}, \tag{2.6}$$

where I_n is the $n \times n$ identity matrix and 0_n is the $n \times n$ zero matrix. Thus these Hamiltonian equations have the form (1.1). It is easy to see that the Poisson bracket generated by operator J

$$\{\mathcal{P}, \mathcal{L}\} = \int \sum_{i=1}^{n} \left(\frac{\delta \mathcal{P}}{\delta v^i}\frac{\delta \mathcal{L}}{\delta w^i} - \frac{\delta \mathcal{P}}{\delta w^i}\frac{\delta \mathcal{L}}{\delta v^i}\right) dx$$

satisfies skew-symmetry (2.3) and Jacobi identity (2.4).

2.3. Semidiscrete Hamiltonian equations. Given a Hamiltonian PDE, we attempt to discretize both the Poisson bracket and the Hamiltonian functional so that we preserve Hamiltonian structure.

To consider semidiscrete equations we introduce a two-dimensional mesh which is uniform in space and continuous in time. Let us denote the mesh points as $\{x_i(t)\}$, $i \in \mathbb{Z}$, $t \geq 0$ and define mesh Ω by two conditions:

$$\Omega : x_{i+1}(t) - x_i(t) = x_i(t) - x_{i-1}(t), \quad x_i(t+\tau) = x_i(t), \qquad i \in \mathbb{Z},\ t, \tau \geq 0. \tag{2.7}$$

The first equation requires the space mesh to be uniform for any fixed time. The second equation requires that only vertical mesh lines in the time-space plane are considered.

Now we can introduce discrete space derivatives $\underset{h}{u}{}_1^i = \underset{+h}{D} u^i$, $\underset{h}{u}{}_2^i = \underset{-h}{D} \underset{h}{u}{}_1^i$, $\ldots$, $\underset{h}{u}{}_{2k+1}^i = \underset{-h}{D} \underset{h}{u}{}_{2k}^i$, $\underset{h}{u}{}_{2k+2}^i = \underset{-h}{D} \underset{h}{u}{}_{2k+1}^i, \ldots, i = 1, \ldots, l$, where $\underset{+h}{D}$ and $\underset{-h}{D}$ are the right and left discrete differentiation operators

$$\underset{+h}{D} = \frac{S_+ - 1}{h}; \quad \underset{-h}{D} = \frac{1 - S_-}{h}, \tag{2.8}$$

defined with the help of the right shift S_+ and left shift S_- operators

$$S_+ f(x) = f(x+h), \quad S_- f(x) = f(x-h). \tag{2.9}$$

We will consider the space of discrete derivatives $\underset{h}{\mathbf{u}}{}^{(m)} = (\mathbf{u}, \underset{h}{\mathbf{u}}{}_1, \underset{h}{\mathbf{u}}{}_2, \ldots, \underset{h}{\mathbf{u}}{}_m)$ and functionals of the form

$$\underset{h}{\mathcal{P}} = \sum_{\Omega} \underset{h}{P}(t, x, h, \underset{h}{\mathbf{u}}{}^{(m)})h, \tag{2.10}$$

where the summation is taken over all space points of mesh Ω for some fixed time. We denote the space of such functionals as $\underset{h}{\mathcal{F}}$.

We assume that the Hamiltonian operator $\mathcal{D}$ can be approximated by an operator $\underset{h}{\mathcal{D}}$ such that the discrete bracket

$$\{\underset{h}{\mathcal{P}}, \underset{h}{\mathcal{L}}\}_h = \sum_{\Omega} \left(\frac{\delta \underset{h}{\mathcal{P}}}{\delta \mathbf{u}} \right)^T \underset{h}{\mathcal{D}} \left(\frac{\delta \underset{h}{\mathcal{L}}}{\delta \mathbf{u}} \right) h \tag{2.11}$$

defines a Poisson bracket for functionals from $\underset{h}{\mathcal{F}}$, i.e. the bracket $\{\cdot,\cdot\}_h$ is skew-symmetric and satisfies the Jacobi identity. Then we can choose some approximation $\underset{h}{\mathcal{H}}$ of the Hamiltonian $\mathcal{H}$ to obtain the set of semidiscrete Hamiltonian equations

$$\dot{\mathbf{u}}_j = \underset{h}{\mathcal{D}} \left(\frac{\delta \underset{h}{\mathcal{H}}}{\delta \mathbf{u}_j} \right), \qquad j \in \mathbb{Z} \tag{2.12}$$

which approximate equation (2.1) on mesh (2.7).

For the space discretization of the canonical Hamiltonian PDEs we can keep the canonical operator J since it generates a discrete Poisson bracket, namely

$$\{\underset{h}{\mathcal{P}}, \underset{h}{\mathcal{L}}\}_h = \sum_{\Omega} \sum_{i=1}^{n} \left(\frac{\delta \underset{h}{\mathcal{P}}}{\delta v^i} \frac{\delta \underset{h}{\mathcal{L}}}{\delta w^i} - \frac{\delta \underset{h}{\mathcal{P}}}{\delta w^i} \frac{\delta \underset{h}{\mathcal{L}}}{\delta v^i} \right) h,$$

and take a discretization $\underset{h}{\mathcal{H}}$ of the Hamiltonian functional $\mathcal{H}$. This procedure provides us semidiscrete canonical Hamiltonian equations

$$\dot{\mathbf{v}}_j = \frac{\delta \underset{h}{\mathcal{H}}}{\delta \mathbf{w}_j}, \qquad \dot{\mathbf{v}}_j = \frac{\delta \underset{h}{\mathcal{H}}}{\delta \mathbf{v}_j}, \qquad j \in \mathbb{Z}, \tag{2.13}$$

where we use vector notations $\mathbf{v} = (v^1, \ldots, v^n)^T$, $\mathbf{w} = (w^1, \ldots, w^n)^T$.

3. Symmetries and conservation laws

3.1. Invariance of semidiscrete equations. Let $\underset{h}{Z}$ be the space of sequences of variables $(t, x, h, \mathbf{u}, \underset{h}{\mathbf{u}_1}, \underset{h}{\mathbf{u}_2}, \ldots)$ and $\underset{h}{\mathcal{A}}$ be the space of analytic functions of a finite number of variables z from $\underset{h}{Z}$.

Invariance of the semidiscrete equations

$$\dot{\mathbf{u}} = \mathbf{F}(z), \qquad F_i \in \underset{h}{\mathcal{A}} \tag{3.1}$$

defined in the points of some two-dimensional mesh Ω was considered in [**7**]. Symmetries of equations (3.1) are transformations generated by vector fields of the form

$$X = \xi^t(z)\frac{\partial}{\partial t} + \xi^x(z)\frac{\partial}{\partial x} + \eta^i(z)\frac{\partial}{\partial u^i} + \cdots, \qquad \xi^t, \xi^x, \eta^i \in \underset{h}{\mathcal{A}}, \tag{3.2}$$

which leave the equations and the mesh invariant. The infinitesimal criterion of invariance can be presented by the three conditions

$$X(\dot{\mathbf{u}} - \mathbf{F}(z)) = 0; \tag{3.3}$$

$$\underset{-h}{D}\underset{+h}{D}(\xi^x) = 0; \quad D_t(\xi^x) = 0; \tag{3.4}$$

which are to be satisfied on the solutions of (3.1). Condition (3.3) requires the invariance of Eqs. (3.1) while conditions (3.4) that of mesh Ω. The operator X must be prolonged on all variables appearing in the equation (3.1)

$$X = \xi^t \frac{\partial}{\partial t} + \xi^x \frac{\partial}{\partial x} + \eta^i \frac{\partial}{\partial u^i} + \phi^i \frac{\partial}{\partial \dot{u}^i} + \underset{h}{\zeta_1^i} \frac{\partial}{\partial \underset{h}{u_1^i}} + \underset{h}{\zeta_2^i} \frac{\partial}{\partial \underset{h}{u_2^i}} + \cdots + \underset{+h}{D}(\xi^x) \frac{\partial}{\partial h}. \tag{3.5}$$

On the uniform in space grid the coefficients of the prolonged operator are defined by the prolongation formulas:

$$\begin{gathered} \phi^i = D_t(\eta^i) - \dot{u}^i D_t(\xi^t) - u_1^i D_t(\xi^x), \\ \underset{h}{\zeta_1^i} = \underset{+h}{D}(\eta^i) - S_+(\dot{u}^i) \underset{+h}{D}(\xi^t) - \underset{h}{u_1^i} \underset{+h}{D}(\xi^x), \\ \underset{h}{\zeta_2^i} = \underset{-h}{D} \underset{+h}{D}(\eta^i) - 2\underset{h}{u_2^i} \underset{+h}{D}(\xi^x) - \frac{1}{h} S_+(\dot{u}^i) \underset{+h}{D}(\xi^t) + \frac{1}{h} S_-(\dot{u}^i) \underset{-h}{D}(\xi^t), \ldots . \end{gathered} \tag{3.6}$$

Note that $u_1^i = D_x(u^i)$ is the "continuous" derivative. It is supposed to be in some discrete representation, for example, $\underset{h}{\widetilde{D}_0}(u)$, which will be introduced below.

Operators of the form (3.2) are called Lie–Bäcklund (or *generalized*) symmetries. It is a difficult task to find Lie–Bäcklund operators admitted by discrete equations. However, Lie point symmetries

$$X = \xi^t(t, x, \mathbf{u}) \frac{\partial}{\partial t} + \xi^x(t, x, \mathbf{u}) \frac{\partial}{\partial x} + \eta^i(t, x, \mathbf{u}) \frac{\partial}{\partial u^i} + \cdots , \tag{3.7}$$

where coefficients ξ^t, ξ^x and η depend only on dependent and independent variables, are easier to detect since such symmetries are given by the same vector fields in the continuous and discrete cases [**2**]. Practically, one can check whether Lie point symmetries admitted by the underlying PDEs are admitted by the semidiscrete equations or not. Although this procedure does not guarantee that we find all symmetries, it lets us avoid solving discrete determining equations.

3.2. Factorization of operators. Following [**2, 3**], let us consider a special operation of left multiplication of a Lie–Bäcklund operator by an analytic function $\tilde{\xi}(z) \in \underset{h}{\mathcal{A}}$:

$$\tilde{\xi} * X = \tilde{\xi}\xi^t \frac{\partial}{\partial t} + \tilde{\xi}\xi^x \frac{\partial}{\partial x} + \tilde{\xi}\eta^i \frac{\partial}{\partial u^i} + \cdots + \underset{+h}{D}(\tilde{\xi}\xi^x) \frac{\partial}{\partial h}. \tag{3.8}$$

The first coordinates in operator $\tilde{\xi} * X$ are multiplied by $\tilde{\xi}(z)$ while the remaining coordinates are computed according to the prolongation formulas (3.6).

The operator

$$\xi^t(z) * D_t + \xi^x(z) * \underset{h}{D_x}, \tag{3.9}$$

where D_t is the total time derivative operator and $\underset{h}{D_x}$ is a discrete presentation of the total space derivative operator D_x, plays the role of an ideal of the Lie algebra of operators (3.2) [**7**]. There are several possibilities to choose operator $\underset{h}{D_x}$. One can take a representation based on the right or left discrete derivative [**2**, **3**]

$$
(3.10)\qquad
\begin{aligned}
D^{+} &= \frac{\partial}{\partial x} + \underset{+h}{\widetilde{D}}(u)\frac{\partial}{\partial u} + \cdots, \qquad \underset{+h}{\widetilde{D}} = \sum_{n=1}^{\infty} \frac{(-h)^{n-1}}{n} \underset{+h}{D}^{n},\\
D^{-} &= \frac{\partial}{\partial x} + \underset{-h}{\widetilde{D}}(u)\frac{\partial}{\partial u} + \cdots, \qquad \underset{-h}{\widetilde{D}} = \sum_{n=1}^{\infty} \frac{h^{n-1}}{n} \underset{-h}{D}^{n},
\end{aligned}
$$

or the discrete representation based on the central difference derivative [**7**]

$$
(3.11)\quad D^{0} = \frac{\partial}{\partial x} + \underset{h}{\widetilde{D}_0}(u)\frac{\partial}{\partial u} + \cdots, \qquad \underset{h}{\widetilde{D}_0} = \sum_{k=0}^{\infty} \alpha_{2k+1} h^{2k} \underset{h}{D_0}^{2k+1}, \qquad \underset{h}{D_0} = \frac{S_+ - S_-}{2h}
$$

with coefficients

$$
\alpha_{2k+1} = (-1)^k \frac{1}{2}\frac{3}{4}\cdots\frac{2k-1}{2k}\frac{1}{2k+1} = \frac{(-1)^k}{2k+1}\frac{(2k-1)!!}{2^k k!},
$$
$$
(2k-1)!! = 1\cdot 3\cdot 5\cdots(2k-1).
$$

Let us mention that operator $\underset{h}{\widetilde{D}_0}$ can be presented in terms of powers of the shift operator S_+ :

$$
(3.12)\qquad \underset{h}{\widetilde{D}_0} = \frac{1}{2h}\sum_{p=0}^{\infty} c_p (S_{2p+1} - S_{-2p-1}), \qquad S_k = (S_+)^k
$$

where the coefficients are given by the series

$$
(3.13)\qquad
\begin{aligned}
c_p &= (-1)^p \sum_{k=p}^{\infty} (-1)^k \frac{\alpha_{2k+1}}{2^{2k}} \frac{(2k+1)!}{(k-p)!(k+1+p)!}\\
&= (-1)^p \sum_{k=p}^{\infty} \frac{1}{k+1+p}\frac{1}{2^{2k}}\frac{((2k-1)!!)^2}{(k-p)!(k+p)!}
\end{aligned}
$$

Lemma 3.1. *The series* (3.13) *defining coefficients* c_p *for* $p \in \mathbb{N}$ *are convergent.*

One can easily check that the operators $\underset{+h}{\widetilde{D}}$ and $\underset{-h}{\widetilde{D}}$ can not be presented in the form (3.12) with convergent coefficients.

Lemma 3.2. *The operator* $\underset{h}{\widetilde{D}_0}$ *is skew-adjoint, i.e.*

$$
\sum_{i=-\infty}^{\infty} v_i \underset{h}{\widetilde{D}_0} u_i h = - \sum_{i=-\infty}^{\infty} u_i \underset{h}{\widetilde{D}_0} v_i h
$$

for $\{u_i\}$, $\{v_i\}$, $i \in \mathbb{Z}$ *such that* $u_i, v_i \to 0$ *as* $i \to \infty$.

In what follows we will use D^0. With the help of the ideal $\xi^t * D_t + \xi^x * D^0$ we can find the evolutionary operator corresponding to the operator (3.2)

$$
(3.14)\quad \overline{X} = X - \xi^t * D_t + \xi^x * D^0 = \bar{\eta}^i \frac{\partial}{\partial u^i} + \cdots, \qquad \bar{\eta}^i = \eta^i - \xi^t \dot{u}^i - \xi^x \underset{h}{\widetilde{D}_0}(u^i)
$$

On the solutions of equations (2.12) we can exclude time derivatives $\dot{u}^i$, $i = 1, \ldots, N$ from the coefficients of the evolutionary operator.

It is important to note that although the factorization let us find evolutionary operators corresponding to given Lie point operators, generally, the obtained evolutionary operators does not have to be admitted by the semidiscrete equations which admit the original nonevolutionary operators. In the continuous case $\xi^t * D_t = \xi^t \cdot D_t$ and $\xi^x * D_x = \xi^x \cdot D_x$, i.e., the operation $*$ is equivalent to a left multiplication of the prolonged operator [**2**]. The operators D_t and D_x are admitted by all differential equations and, consequently, the operators $\xi^t * D_t$ and $\xi^x * D_x$ are also admitted by all equations. It follows that an evolutionary operator is admitted if it corresponds to an admitted nonevolutionary operator [**11**].

This result does not hold in the semidiscrete case. In the general case operator $\xi^t * D_t = \xi^t \cdot D_t$ is admitted, but operator $\xi^x * D^0$ is not. Thus, having obtained an evolutionary operator, one has to check that this operator or used in the factorization operator $\xi^x * D^0$ is admitted by the considered semidiscrete equations.

We can provide only a very restricted class of operators $\xi^x * D^0$ which are admitted by an arbitrary discrete equation.

Lemma 3.3. *For ξ^x such that $\underset{+h}{D}(\xi^x) = 0$ we have*

$$\xi^x * D^0 = \xi^x \cdot D^0$$

and, consequently, this operator is admitted by all difference equations.

3.3. Conservation laws. For a system of Hamiltonian equations (2.13) considered on the grid (2.7) we have the following types of the conservation laws:

3.3.1. *Conservation of symplecticity* [**8**]. Due to their canonical form semidiscrete equations possess the conservation of symplecticity

$$\frac{d}{dt}\underset{h}{\omega} = 0, \quad \underset{h}{\omega} = \sum_{j=-\infty}^{\infty} d\mathbf{v}_j \wedge d\mathbf{w}_j h = \sum_{j=-\infty}^{\infty} \sum_{i=1}^{n} dv_j^i \wedge dw_j^i h, \tag{3.15}$$

where $d\mathbf{v}_j = (dv_j^1, \dots, dv_j^n)^T$ and $d\mathbf{w}_j = (dw_j^1, \dots, dw_j^n)^T$ are solutions of the variational equations

$$\begin{aligned} d\dot{v}_j^i &= \sum_{k=1}^{n} \sum_{l} \frac{\partial}{\partial w_{j+l}^k} \left(\frac{\delta \underset{h}{\mathcal{H}}}{\delta w_j^i} \right) dw_{j+l}^k + \sum_{k=1}^{n} \sum_{l} \frac{\partial}{\partial v_{j+l}^k} \left(\frac{\delta \underset{h}{\mathcal{H}}}{\delta w_j^i} \right) dv_{j+l}^k, \\ d\dot{w}_j^i &= -\sum_{k=1}^{n} \sum_{l} \frac{\partial}{\partial w_{j+l}^k} \left(\frac{\delta \underset{h}{\mathcal{H}}}{\delta v_j^i} \right) dw_{j+l}^k - \sum_{k=1}^{n} \sum_{l} \frac{\partial}{\partial v_{j+l}^k} \left(\frac{\delta \underset{h}{\mathcal{H}}}{\delta v_j^i} \right) dv_{j+l}^k. \end{aligned}$$

We suppose that $d\mathbf{v}_j, d\mathbf{w}_j \to 0$ as $j \to \infty$. The 2-form (3.15) is a generalization of the symplectic structure for the Hamiltonian ODEs [**12**] to the infinite set of semidiscrete equations.

Underlying equations (1.1) possess the conservation of symplecticity

$$\frac{d}{dt}\omega = 0, \quad \omega = \int d\mathbf{v} \wedge d\mathbf{w}\, dx = \int \sum_{i=1}^{n} dv^i \wedge dw^i\, dx \tag{3.16}$$

that is the continuous limit of (3.15). In this case $d\mathbf{v}$ and $d\mathbf{w}$ are solutions of the variational equations for (1.1)

$$dv^i_t = \widehat{d}\left(\frac{\delta\mathcal{H}}{\delta w^i}\right) = \sum_{k=1}^{n}\sum_{l}\frac{\partial}{\partial w^k_l}\left(\frac{\delta\mathcal{H}}{\delta w^i}\right)dw^k_l + \sum_{k=1}^{n}\sum_{l}\frac{\partial}{\partial v^k_l}\left(\frac{\delta\mathcal{H}}{\delta w^i}\right)dv^k_l;$$

$$dw^i_t = -\widehat{d}\left(\frac{\delta\mathcal{H}}{\delta v^i}\right) = -\sum_{k=1}^{n}\sum_{l}\frac{\partial}{\partial w^k_l}\left(\frac{\delta\mathcal{H}}{\delta v^i}\right)dw^k_l - \sum_{k=1}^{n}\sum_{l}\frac{\partial}{\partial v^k_l}\left(\frac{\delta\mathcal{H}}{\delta v^i}\right)dv^k_l;$$

where the operator $\widehat{d}$ denotes the vertical differential of the vertical form it acts on. In our case we have vertical 0-forms, i.e. functions. We assume that $d\mathbf{v}$, $d\mathbf{w}$ and their space derivatives which appear in the variational equations decay at $x \to \infty$.

3.3.2. *Conservation of distinguished functionals.*

Definition. For a given Hamiltonian operator $\underset{h}{\mathcal{D}}$ a distinguished functional is a functional $\underset{h}{\mathcal{C}}(x, h, \underset{h}{\mathbf{u}}^{(n)})$ such that

$$\underset{h}{\mathcal{D}}\left(\frac{\delta \underset{h}{\mathcal{C}}}{\delta \mathbf{u}}\right) = 0 \tag{3.17}$$

It follows that a functional is distinguished if and only if its Poisson bracket with every other functional is trivial:

$$\{\underset{h}{\mathcal{C}}, \underset{h}{\mathcal{H}}\}_h = 0 \quad \text{for any} \quad \underset{h}{\mathcal{H}} \in \underset{h}{\mathcal{F}}. \tag{3.18}$$

Distinguished functionals are conserved by semidiscrete equations originating from any Hamiltonian functional. For the canonical bracket there are no nontrivial distinguished functionals so that we do not get conservation laws of this type.

3.3.3. *Hamiltonian form of Noether's theorem.*

Definition. The Hamiltonian vector field associated with a functional $\underset{h}{\mathcal{P}}$ is the unique smooth vector field $X_{\mathcal{P}}$ satisfying

$$X_{\mathcal{P}}(\underset{h}{\mathcal{F}}) = \{\underset{h}{\mathcal{F}}, \underset{h}{\mathcal{P}}\}_h \tag{3.19}$$

In the coordinate form it can be presented as the operator

$$X_{\mathcal{P}} = \underset{h}{\mathcal{D}}\left(\frac{\delta \underset{h}{\mathcal{P}}}{\delta u^i}\right)\frac{\partial}{\partial u^i} \tag{3.20}$$

Some symmetries (3.2) are given as Hamiltonian vector fields or are equivalent to Hamiltonian vector fields under factorization (3.14). Such vector fields let us use the following theorem [**7**].

Theorem 3.4. *For a Hamiltonian system of semidiscrete evolution equations* (2.12) *a Hamiltonian vector field $X_{\mathcal{P}}$ determines a generalized symmetry of the system if and only if there is an equivalent functional $\underset{h}{\widetilde{\mathcal{P}}} = \underset{h}{\mathcal{P}} - \underset{h}{\mathcal{C}}$, differing from $\underset{h}{\mathcal{P}}$ only by a time-dependent distinguished functional $\underset{h}{\mathcal{C}}(t, x, h, u^{(n)})$, such that $\underset{h}{\widetilde{\mathcal{P}}}$ determines a conservation law.*

For the operator J generating the canonical bracket a Hamiltonian vector field has the form

(3.21) $$X_P = \frac{\delta \underset{h}{\mathcal{P}}}{\delta w^i}\frac{\partial}{\partial v^i} - \frac{\delta \underset{h}{\mathcal{P}}}{\delta v^i}\frac{\partial}{\partial w^i},$$

where $\underset{h}{\mathcal{P}}$ is the generating functional. The canonical bracket has only trivial time-dependent functionals, i.e. functions $f(t)$. Thus, for a Hamiltonian symmetry (3.21) there corresponds a conservation law $\underset{h}{\widetilde{\mathcal{P}}} = \underset{h}{\mathcal{P}} - f(t)$, where the function $f(t)$ needs to be found with the help of the considered equations. In a particular case when the Hamiltonian functional and the considered Hamiltonian symmetry are time independent we obtain a linear function $f(t) = at + b$, $a, b = \text{const}$.

4. The nonlinear wave equation

In this section we consider the nonlinear wave equation

(4.1) $$v_{tt} = v_{xx} - V'(v),$$

where $V(v)$ is some smooth function. For simplicity, we consider the case of scalar v. With the help of a new variable $w = v_t$, the equation (4.1) can be rewritten as the system

(4.2) $$\begin{cases} v_t = w; \\ w_t = v_{xx} - V'(v). \end{cases}$$

This is a canonical Hamiltonian system generated by the Hamiltonian functional

(4.3) $$\mathcal{H} = \int \left(\frac{w^2}{2} + \frac{v_x^2}{2} + V(v) \right) dx.$$

Let us take the following approximation of the Hamiltonian functional

(4.4) $$\mathcal{H}_h = \sum_{\Omega} H[v, w]h, \quad H = \frac{w^2}{2} + \frac{\underset{h}{v_1^2}}{2} + V(v).$$

It provides us the system of the semidiscrete equations

(4.5) $$\begin{cases} \dot{v} = w; \\ \dot{w} = \underset{h}{v_2} - V'(v). \end{cases}$$

For arbitrary $V(v)$ the admitted transformation group for (4.5) is two-dimensional. Its Lie algebra is spanned by the operators

(4.6) $$X_1 = \frac{\partial}{\partial t}, \quad X_2 = \frac{\partial}{\partial x}.$$

Lorentz transformation

$$X_3 = x\frac{\partial}{\partial t} + t\frac{\partial}{\partial x},$$

admitted by underlying system (4.2) with arbitrary $V(v)$ [**1**], is lost under discretization since it brakes the mesh invariance. Thus in the case of arbitrary $V(v)$ Noether's theorem gives two conservation laws:

(1) The time translation X_1 leads to the conservation of the Hamiltonian functional $\underset{h}{\mathcal{H}}$. Indeed, the factorization of operator X_1 gives the evolutionary vector field

$$\overline{X}_1 = w\frac{\partial}{\partial v} + \left(\underset{h}{v_2} - V'(v)\right)\frac{\partial}{\partial w},$$

which is generated by the Hamiltonian functional. Thus we obtain conservation of Hamiltonian functional (4.4) that stands for the conservation of energy.

(2) The space translation X_2 corresponds to the evolutionary operator

$$\overline{X}_2 = \underset{h}{\widetilde{D}_0}(v)\frac{\partial}{\partial v} + \underset{h}{\widetilde{D}_0}(w)\frac{\partial}{\partial w}. \tag{4.7}$$

Checking condition (3.21) for the coefficients of operator $\overline{X}_2$, we find the generating functional

$$\underset{h}{\mathcal{P}_2} = \sum_{\Omega} w\underset{h}{\widetilde{D}_0}(v)h = -\sum_{\Omega} v\underset{h}{\widetilde{D}_0}(w)h. \tag{4.8}$$

This functional is a conservation law of the semidiscrete equations (4.5). In the continuous limit it corresponds to the functional

$$\mathcal{P}_2 = \int wv_x\,dx = -\int vw_x\,dx, \tag{4.9}$$

which is a conservation law of the Eqs. (4.2). In physical applications this conservation law is referred as linear momentum.

In order to obtain this conservation law we need to consider "nonlocal" functionals, i.e. functionals which are defined as double sums over the mesh points, since operator $\underset{h}{\widetilde{D}_0}$ is a sum over infinitely many mesh points.

For the quadratic potential $V(v) = Cv^2/2$ there is an infinite series of conservation laws for the wave system (4.5). In this case the semidiscrete system admits the infinite set of Hamiltonian operators

$$Y_k = \underset{h}{D_0}(\underset{h}{v_{2k}})\frac{\partial}{\partial v} + \underset{h}{D_0}(\underset{h}{w_{2k}})\frac{\partial}{\partial w}, \qquad k = 0, 1, \dots,$$

which provides us the infinite set of the conserved functionals

$$\underset{h}{R_k} = \sum_{\Omega} w\underset{h}{D_0}(\underset{h}{v_{2k}})h = -\sum_{\Omega} v\underset{h}{D_0}(\underset{h}{w_{2k}})h$$

In the continuous limit these conservation laws correspond to the functionals

$$R_k = \int vw_{(2k+1)}\,dx,$$

which are conserved quantities for the system (4.2).

Since for the quadratic potential we get the system of linear equations their solutions possess a superposition principle. It is reflected in the invariance with respect to the symmetry

$$Z = \alpha(t,x)\frac{\partial}{\partial v} + \alpha_t(t,x)\frac{\partial}{\partial w},$$

where the function $\alpha(t,x)$ is an arbitrary solution of the equation

$$\alpha_{tt}(t,x) - \frac{\alpha(t,x+h) - 2\alpha(t,x) + \alpha(t,x-h)}{h^2} - C\alpha(t,x). \tag{4.10}$$

The operator Z is Hamiltonian. It corresponds to the functional

$$\underset{h}{T} = \sum_{\Omega} \big(\alpha(t,x)w - \alpha_t(t,x)v\big)h,$$

which is a conservation law of the semidiscrete linear system if function $\alpha(t,x)$ satisfies the equation (4.10). In the continuous limit this functional goes into the functional

$$T = \int \big(\alpha(t,x)w - \alpha_t(t,x)v\big)\,dx,$$

which is a conservation law of (4.2) if the function $\alpha(t,x)$ satisfies the equation

$$\alpha_{tt}(t,x) = \alpha_{xx}(t,x) - C\alpha(t,x). \tag{4.11}$$

5. Conclusions

In the paper we considered semidiscrete canonical Hamiltonian equations and showed how to find conservation laws of such equations using Noether's theorem. Our interest to canonical Hamiltonian equations is also motivated by their connection with Euler–Lagrange equations [**10**].

Many PDEs can be presented as Euler–Lagrange equations for appropriate Lagrangian functionals [**6, 11**] and can be rewritten as canonical Hamiltonian equations (1.1). Let us illustrate this on the example of the nonlinear wave equation (4.1) which is the Euler–Lagrange equation

$$\frac{\delta L}{\delta v} = 0, \qquad \frac{\delta}{\delta v} = \frac{\partial}{\partial v} - D_t \frac{\partial}{\partial v_t} - D_x \frac{\partial}{\partial v_x} \tag{5.1}$$

for the Lagrangian functional

$$\mathcal{L} = \iint L(v, v_t, v_x)\,dx\,dt, \qquad L = \frac{v_t^2}{2} - \frac{v_x^2}{2} - V(v). \tag{5.2}$$

With the help of the Legendre transformation

$$w = \frac{\partial L}{\partial v_t} = v_t, \qquad \mathcal{H} = \int \left(\frac{\partial L}{\partial v_t} v_t - L\right) dx$$

we obtain the Hamiltonian functional (4.3). It generates equations (4.2), which are equivalent to (4.1).

Similar connection between Euler–Lagrange equations and canonical Hamiltonian equations can be established in the semidiscrete case. Let us consider the semidiscretization of functional (5.2)

$$\underset{h}{\mathcal{L}} = \int \sum_{\Omega} \underset{h}{L}(v, \dot{v}, \underset{h}{v_1})h\,dt, \qquad \underset{h}{L} = \frac{\dot{v}^2}{2} - \frac{\underset{h}{v_1}^2}{2} - V(v).$$

Its Euler–Lagrange equation

$$\frac{\delta \underset{h}{\mathcal{L}}}{\delta v} = 0, \qquad \frac{\delta}{\delta v} = \frac{\partial}{\partial v} - D_t \frac{\partial}{\partial \dot{v}} - \underset{-h}{D} \frac{\partial}{\partial \underset{h}{v_1}} \tag{5.3}$$

has the form

$$\ddot{v} = \underset{h}{v_2} - V'(v). \tag{5.4}$$

We can introduce the discrete Hamiltonian functional

$$w = \frac{\partial \underset{h}{L}}{\partial \dot{v}} = \dot{v}, \quad \underset{h}{\mathcal{H}} = \sum_{\Omega} \left(\frac{\partial L}{\partial \dot{v}} \dot{v} - \underset{h}{L} \right) h,$$

which in our case is the functional (4.4)). It generates the semidiscrete evolution equations (4.5), which are equivalent to (5.4).

Thus, looking for conservation laws of the semidiscrete Euler–Lagrange equations, one can consider the equivalent canonical Hamiltonian systems and find conservation laws in the Hamiltonian framework. They have the form

$$\int \mathcal{P}\, dx \quad \text{and} \quad \sum_{\Omega} \underset{h}{\mathcal{P}} h \tag{5.5}$$

in the continuous and semidiscrete cases correspondingly. If necessary, the conservation laws can be rewritten in term of the original variables. For example, the conservation laws of the semidiscrete wave system (4.5) with an arbitrary potential $V(v)$ found in Section 4 can be rewritten in terms of variables used in the Lagrangian approach

$$\underset{h}{\mathcal{P}_1} = \sum_{\Omega} \left(\frac{\dot{v}^2}{2} + \frac{\underset{h}{v_1^2}}{2} + V(v) \right) h, \quad \underset{h}{\mathcal{P}_2} = \sum_{\Omega} \dot{v} \underset{h}{\widetilde{D}_0}(v) h.$$

For a quadratic potential we also have the infinite series of the conservation laws

$$\underset{h}{R_k} = \sum_{\Omega} \dot{v} \underset{h}{D_0}(\underset{h}{v_{2k}}) h = - \sum_{\Omega} v \underset{h}{D_0}(\underset{h}{\dot{v}_{2k}}) h, \quad k = 0, 1, 2, \ldots$$

and the conservation laws

$$\underset{h}{T} = \sum_{\Omega} \big(\alpha(t, x) \dot{v} - \alpha_t(t, x) v \big) h,$$

where the function $\alpha(t, x)$ must satisfy the equation (4.10).

One of the advantages of Lagrangian approach over Hamiltonian one is a possibility to find local conservation. In the Hamiltonian framework we can find only global conservation laws of the form (5.5). This advantage of the Lagrangian approach is difficult to represent for the semidiscrete equations because as we have seen some conservation laws have densities which involve discrete presentations of the continuous derivatives and, consequently, are not local.

For simplicity we considered only the case of one-dimensional space. The extension to the multi-dimensional space is straightforward.

References

1. W. F. Ames, R. L. Anderson, V. A. Dorodnitsyn, E. V. Ferapontov, R. K. Gazizov, N. H. Ibragimov, and S. R. Svirshchevskii, CRC *handbook of Lie group analysis of differential equations. 1. Symmetries, exact solutions and conservation laws* (N. Ibragimov, ed.), CRC Press, Boca Raton, FL, 1994.
2. V. A. Dorodnitsyn, *Transformation groups in difference spaces*, Current Problems in Mathematics. Newest Results, vol. 34, Itogi Nauki i Tekhniki, Akad. Nauk SSSR, Vsesoyuz. Inst. Nauchn. i Tekhn. Inform., Moscow, 1989, pp. 149–191 (Russian); English transl., J. Soviet Math. **55** (1991), no. 1, 1490–1517.
3. V. A. Dorodnitsyn, *Newton's group and commutative properties of Lie–Bäcklund operators in finite difference space*, Preprint of Keldysh Inst. Appl. Math., no. 175, Moscow, 1988 (Russian).

4. R. Floreanini, J. Negro, L. M. Nieto, and L. Vinet, *Symmetries of the heat equation on the lattice*, Lett. Math. Phys. **36** (1996), no. 4, 351–355.
5. R. Floreanini and L. Vinet, Lie symmetries of finite-difference equations, J. Math. Phys. **36** (1995), no 12, 7024–7042.
6. N. H. Ibragimov, *Transformation groups applied to mathematical physics*, Math. Appl. (Soviet Ser.), vol. 3, D. Reidel Publi. Co., 1995.
7. R. Kozlov, *Conservation laws of semidiscrete Hamiltonian equations*, J. Math. Phys. **42** (2001), no. 4, 1708–1727.
8. ———, *Conservation laws of semidiscrete canonical Hamiltonian equations*, J. Phys. A **34** (2001), no. 17, 3651–3669.
9. D. Levi, L. Vinet, and P. Winternitz, *Lie group formalism for difference equations*, J. Phys. A **30** (1997), 633–649.
10. J. E. Marsden and T. S. Ratiu, *Introduction to mechanics and symmetry. A basic exposition of classimathcal mechanimathcal systems*, Texts Appl. Math., vol. 17, Springer-Verlag, New York, 1994.
11. P. J. Olver, *Applications of Lie groups to differential equations*, Springer, New York, 1986.
12. J. M. Sanz–Serna and M. P. Calvo, *Numerical Hamiltonian problems*, Appl. Math. Math. Comput., no. 7. Chapman & Hall, London, 1994.

Department of Informatics, University of Oslo, 0371, Oslo, Norway.
E-mail address: kozlov@ifi.uio.no

Centre de Recherches Mathématiques
CRM Proceedings and Lecture Notes
Volume **39**, 2005

Discrete Lie Symmetries for Difference Equations

D. Levi and M. A. Rodríguez

Abstract. We apply two of the methods previously introduced to find discrete symmetries of differential equations to the case of difference and differential-difference equations.

1. Introduction

Continuous Lie symmetries have always played a very important role in the study of differential equations [**15**]. They have been introduced by Sophus Lie as a tool to unify all solution techniques for ordinary differential equations. In particular they are useful to

- get special solutions by *symmetry reduction*;
- classify equations according to their symmetry;
- prove equivalence of equations under point transformations.

They have been extended with success to the case of differential—difference and difference—difference equations[**10–12, 16**]. As is evident from the results presented in Ref. [**10, 11**] all the steps necessary to obtain the Lie symmetries are not significantly different in the case of equations on the lattice from the continuous case. We just need to prolong the action of the infinitesimal generators to all points of the lattice involved in the lattice equation rather than to all derivatives. This prolongation is obtained by just calculating the infinitesimal coefficients in the new points.

As was shown in [**13**] the invariance of a differential-difference equation with respect to a discrete shift of the lattice index $\widetilde{n} = n + N$ provides a reduction of the equation to a system of N ordinary differential equations. Moreover, discrete symmetries, i.e. symmetries associated to a discrete finite group, are very important in Quantum Mechanics. In this field one considers *parity, charge conjugation, rotations by* π, etc... and discrete symmetries providing selection rules.

2000 *Mathematics Subject Classification.* Primary: 39A, 58D19,76M60.

This work was done while D L was visiting the Departamento de Física Teórica of Universidad Complutense de Madrid (Spain). His visit was financed by the *Ministerio of Educación, Cultura y Deportes* of Spain (SAB01-0140). This work was also partially supported by Ministry of Science and Technology (Spain) under grant BFM2002-02646 and NATO under grant PST.CLG.978431.

This is the final form of the paper.

Discrete symmetries are usually easy to guess but difficult to find in a systematic way. Few techniques are available and have been applied to get discrete symmetries for differential equations. We will consider here discrete symmetries obtained using techniques proper of the continuous Lie symmetries. This is the reason why in the title of this paper we speak of "discrete Lie symmetries".

One technique is obtained by considering the normalizer of the continuous Lie point symmetries of the equation. This is a well known technique and can be found in many textbooks (see for instance [**2**]). The application to the case of differential equations has been considered in all details by Hydon [**4**]. The normalizer can be obtained only in the case when a nontrivial non commuting group of continuous Lie point symmetries exists.

A different technique has been introduced by Gaeta and Rodríguez [**1**] in 1996. They considered a modification of Lie technique which, as we will see in Section 2, depends on a starting ansatz. Thus it does not provide a complete result but it can be applied also when no continuous symmetry is present.

There are at least two other methods to construct discrete symmetries. One of them is based on a formulation of differential equations through differential forms [**7**] and the other [**14, 17, 18**] solves the determining equations for the group transformations directly. Reid, Weih and Wittkopf [**18**] obtained the discrete symmetries by solving the determining equations for the group transformations directly. In more recent works [**14, 17**] they simplified the work taking advantage of the infinitesimal results; however, the method requires extremely heavy calculations.

In Section 2 we present two methods introduced to obtain discrete symmetries. In Section 3 we present two examples of the derivation of discrete Lie symmetries for discrete equations.

2. The calculus of discrete symmetries

We will now discuss briefly two of the methods presented in the Introduction for determining the discrete symmetries of differential equations. One of them was proposed by Hydon [**4, 5**] and it is essentially the classical method of constructing the normalizer of a group. The other, due to Gaeta and Rodríguez [**1**], is a modification of Lie's method, defining a discrete symmetry as a discretization of the parameter of a continuous symmetry.

2.1. Automorphisms of the symmetry algebra. Let us consider a differential equation for a dependent variable u and N independent variables $\mathbf{x} = (x_1, \ldots, x_N)$, with a Lie group of symmetries, G and its corresponding Lie algebra $\mathcal{G}$. We construct all the automorphisms of this Lie algebra. Some of them will, obviously, correspond to continuous symmetries. Others will be essentially new and will define discrete symmetries of the equation. And, finally, others will be ignored as they are not symmetries of the equation.

Let $\mathcal{G}$ be a Lie algebra of finite dimension n and $\mathcal{B} = \{X_1, \ldots, X_n\}$ a basis of $\mathcal{G}$. The commutation relations in this basis can be written as:

$$[X_i, X_j] = c_{ij}^k X_k, \tag{2.1}$$

where c_{ij}^k are the structure constants of $\mathcal{G}$ and here and below summation over repeated indices is implied.

The defining equation for an automorphism $\phi\colon \mathcal{G} \to \mathcal{G}$ is

$$\phi[X, Y] = [\phi(X), \phi(Y)], \tag{2.2}$$

Let Φ be the matrix representation of ϕ in the basis $\mathcal{B}$ ($\det \Phi \neq 0$). Then

$$\phi(X_i) = \Phi^j{}_i X_j \tag{2.3}$$

and Eq. (2.2) on the base $\mathcal{B}$ reads:

$$c_{ij}^k \Phi^l{}_k = \Phi^m{}_i \Phi^r{}_j c_{mr}^l. \tag{2.4}$$

If we define the matrices of the adjoint representation

$$C(i)^k{}_j = c_{ij}^k, \tag{2.5}$$

Eq. (2.4) can also be written as:

$$\Phi C(i) = \Phi^l{}_i C(l) \Phi. \tag{2.6}$$

As automorphisms related through conjugation by an element of the symmetry group are equivalent, the matrix Φ can be simplified by conjugations with respect to any symmetry transformation.

Let us consider those elements $g_i \in G$ which are generated by an element X_i of the basis $\mathcal{B}$ of the corresponding Lie algebra $\mathcal{G}$,

$$g_i = \mathrm{e}^{\lambda X_i}. \tag{2.7}$$

Conjugation provides the transformation

$$\phi \to \mathrm{e}^{-\lambda X_i} \phi \mathrm{e}^{\lambda X_i}. \tag{2.8}$$

We can consider the transformation of an element of the basis $\mathcal{B}$ and we have:

$$\mathrm{e}^{-\lambda X_i} X_j \mathrm{e}^{\lambda X_i} = A(i, \lambda)^k{}_j X_k, \tag{2.9}$$

where

$$A(i, \lambda) = \mathrm{e}^{\lambda C(i)}. \tag{2.10}$$

We can apply the transformations (2.10)):

$$\Phi \to \mathrm{e}^{\lambda C(i)} \Phi \tag{2.11}$$

to simplify the matrix Φ.

We can now construct a representation of the automorphism in the space of variables. Let X be a vector field given by

$$X = \sum_{a=1}^{N} \xi^a(\mathbf{x}, u) \partial_{x_a} + \varphi(\mathbf{x}, u) \partial_u \tag{2.12}$$

which is a symmetry of the equation under study. By a symmetry transformation:

$$(\mathbf{x}, u) \to \big(\widehat{\mathbf{x}}(x, u), \widehat{u}(x, u)\big) \tag{2.13}$$

the vector field X is transformed into a new vector field

$$\widehat{X} = \sum_{a=1}^{N} \widehat{\xi}^a(\widehat{\mathbf{x}}, \widehat{u}) \partial_{\widehat{x}_a} + \widehat{\varphi}(\widehat{\mathbf{x}}, \widehat{u}) \partial_{\widehat{u}}. \tag{2.14}$$

As the transformation is a symmetry of the equation, $\widehat{X}$ must have the form (2.12).

If we consider a basis of the symmetry algebra, $\{X_i\}$, $i = 1, \ldots, n$, the transformed vector fields are:

$$\widehat{X}_i = \xi_i^a(\widehat{\mathbf{x}}, \widehat{u}) \partial_{\widehat{x}_a} + \varphi_i(\widehat{\mathbf{x}}, \widehat{u}) \partial_{\widehat{u}} \tag{2.15}$$

and, as the transformation is an automorphism of the algebra,

(2.16) $$\widehat{X}_i = \Phi^j{}_i X_j.$$

Applying (2.16) to the new variables, we get:

(2.17) $$\xi^b_j(\mathbf{x},u)\frac{\partial\widehat{x}_a}{\partial x_b} + \varphi_j(\mathbf{x},u)\frac{\partial\widehat{x}_a}{\partial u} = (\Phi^{-1})^i{}_j\xi^a_i(\widehat{\mathbf{x}},\widehat{u}),$$
$$\xi^b_j(\mathbf{x},u)\frac{\partial\widehat{u}}{\partial x_b} + \varphi_j(\mathbf{x},u)\frac{\partial\widehat{u}}{\partial u} = (\Phi^{-1})^i{}_j\varphi_i(\widehat{\mathbf{x}},\widehat{u}).$$

The solution of Eqs. (2.17) provides the automorphism as a transformation in our space of variables and functions. Then, we must check if the automorphism is a symmetry of the equation and if it does correspond to a continuous symmetry. Among the remaining automorphisms we can find the discrete symmetries.

2.2. Determining equation for a discrete symmetry. Let's review the method of the discretization of the parameter of a continuous transformation.

Let us consider a differential equation:

(2.18) $$\partial_J u = f(\mathbf{x}, u, \partial_{J'}u, \dots)$$

where $\partial_J u = \partial^{j_1}_{x_1}\cdots\partial^{j_N}_{x_N}u$, $J = (j_1,\dots,j_N)$ and the infinitesimal Lie point symmetry transformation

(2.19) $$\widehat{x}_i(\lambda) = x_i + \lambda\xi_i(\mathbf{x},u),$$
$$\widehat{u}(\widehat{\mathbf{x}},\lambda) = u(\mathbf{x}) + \lambda\phi(\mathbf{x},u),$$

so that the associated symmetry vector field is given by

(2.20) $$X = \sum_{a=1}^{N}\xi_a(\mathbf{x},u)\partial_{x_a} + \varphi(\mathbf{x},u)\partial_u.$$

If we introduce the extended equation

(2.21) $$\partial_J\widehat{u} = F(\widehat{\mathbf{x}},\widehat{u},\partial_{J'}\widehat{u},\dots;\lambda),$$

taking into account (2.19), F will satisfy the partial differential equation

(2.22) $$\frac{\partial F}{\partial\lambda} + \sum_{a=1}^{N}\xi_a\frac{\partial F}{\partial\widehat{x}_a} + \sum_{J'}\phi^{J'}\frac{\partial F}{\partial\widehat{u}_{J'}} = \phi^J\Big|_{\partial_J\widehat{u}=F}.$$

where $\phi^{J'}$ are the J'-prolongations of the vector field (2.20). In order to recover the original function, we impose the initial condition:

(2.23) $$F(\widehat{\mathbf{x}},\widehat{u},\partial_{J'}\widehat{u},\dots;0) = f(\mathbf{x},u,\partial_{J'}u,\dots)$$

and if we want to obtain the same function for a particular value λ_0 of the parameter and hence a discrete symmetry

(2.24) $$F(\widehat{\mathbf{x}},\widehat{u},\partial_{J'}\widehat{u},\dots;\lambda_0) = F(\widehat{\mathbf{x}},\widehat{u},\partial_{J'}\widehat{u},\dots;0) = f(\mathbf{x},u,\partial_{J'}u,\dots).$$

Let's notice that these discrete symmetries are obtained in terms of the Lie symmetries of the extended equation (2.21). Moreover, to apply this method, we do not need to know the continuous symmetries of the given equation (2.18). In fact, the determining equation provides a general solution which does not depend on the differential equation we are studying. The equation appears when we impose the boundary condition as in (2.23) and the discrete symmetries by requiring Eq. (2.24). However, in general we have to make some assumptions to get a solution of the equation, because we do not know the symmetries which appear in it. We will use this method to determine the discrete symmetries of the discrete Painlevé I equation [**8**].

3. Lie discrete symmetries of lattice equations

The construction of point symmetries of discrete equations is similar to the standard approach to point symmetries of continuous equations. The main difference lies in the form of the prolongation and in the way we solve the determining equations. Consequently, the procedure presented in the previous section for constructing discrete symmetries of continuous equations can be carried over in a straightforward manner to the discrete case, just by changing the form of the prolongation. In the following we will apply the methods discussed in Section 2 to two examples, the discrete Painlevé I equation [**8**] and the Toda equation [**19**]. The Painlevé equation has no continuous symmetry (for a generic choice of the parameters), and we will use the determining equation method to find its discrete symmetries. The Toda equation has a five dimensional continuous group of symmetries [**3, 12**] and we will construct the discrete symmetries by computing the automorphisms of the corresponding Lie algebra.

3.1. Discrete Painlevé equation. The discrete Painlevé I equation is given by

$$u_{n+1} + u_n + u_{n-1} = \frac{\alpha x_n + \beta}{u_n} + \gamma \tag{3.1}$$

where α, β and γ are arbitrary constants and $u_n = u(x_n)$. The equation which defines the lattice points can be written as:

$$x_{n+1} - x_n = h \tag{3.2}$$

so that $x_n = hn + x_0$. With the choice $u_n = -1 + h^2 w(x)$, $\alpha = -h^5$, $\beta = 3$ and $\gamma = 6$ Eq. (3.1) reduces, in the continuous limit as $h \to 0$, to

$$w_{xx} = 6w^2 + x. \tag{3.3}$$

which is the Painlevé I transcendent [**6**].

The infinitesimal symmetry generator is given by

$$\widehat{X}_n = \xi_n(x_n, u_n)\partial_{x_n} + \phi_n(x_n, u_n)\partial_{u_n}, \tag{3.4}$$

and its prolongation

$$\text{pr}\widehat{X}_n = \sum_{j=-1}^{1} \widehat{X}_{n+j}. \tag{3.5}$$

Depending on the values of the parameters (α, β and γ) we have various possibilities:

- If $\alpha \neq 0$ then $\xi_n(x_n, u_n) = 0$ and $\phi_n(x_n, u_n) = 0$ and thus no symmetry is present.
- If $\alpha = 0$ but $\beta \neq 0$ then $\xi_n(x_n, u_n) = K_0$ with K_0 an arbitrary constant and $\phi_n(x_n, u_n) = 0$, i.e., only space translations are possible.
- If $\alpha = 0$, $\beta = 0$ Eq. (3.1) is linear and we find for it a four dimensional symmetry group.

So, when $\alpha \neq 0$ the discrete Painlevé I equation (3.1) has no continuous symmetry and we cannot apply the method of the normalizer.

We define the extended λ dependent equation in the transformed space as

$$\widehat{u}_+ + \widehat{u}_- = F(\widehat{x}, \widehat{u}; \lambda), \tag{3.6}$$

where, for $\lambda = 0$

$$F(\widehat{x}, \widehat{u}; 0) = -u + \frac{\alpha x + \beta}{u} + \gamma. \tag{3.7}$$

For the sake of simplicity we have used the notation

$$x = x_n, \qquad x_+ = x_{n+1}, \qquad x_- = x_{n-1},$$
$$u = u(x_n), \quad u_+ = u(x_{n+1}), \quad u_- = u(x_{n-1}).$$

The generic Lie point transformation, written in terms of the infinitesimal symmetry generators, is:

$$\frac{d\widehat{x}}{d\lambda} = \xi(\widehat{x}, \widehat{u}), \qquad \frac{d\widehat{u}}{d\lambda} = \phi(\widehat{x}, \widehat{u}). \tag{3.8}$$

From Eq. (3.2) we have $\xi = K_0$. Differentiating Eq. (3.6) with respect to λ and taking into account Eq. (3.8) we get:

$$\frac{\partial F}{\partial \lambda} + K_0 \frac{\partial F}{\partial \widehat{x}} + \phi(\widehat{x}, \widehat{u}) \frac{\partial F}{\partial \widehat{u}} = \phi(\widehat{x}_+, \widehat{u}_+) + \phi(\widehat{x}_-, F - \widehat{u}_+). \tag{3.9}$$

By differentiating Eq. (3.9) with respect to $\widehat{u}_+$ and $\widehat{u}$ we get

$$\phi(\widehat{x}, \widehat{u}) = \phi^{(0)}(\widehat{x}) + \widehat{u}\phi^{(1)}(\widehat{x}). \tag{3.10}$$

Substituting Eq. (3.10) into Eq. (3.9) and requiring that the obtained equation be satisfied for any u_+, we get

$$\phi^{(1)}(\widehat{x}) = K_1 + K_2(-1)^{(\widehat{x} - \widehat{x}_0)/h}. \tag{3.11}$$

Equation (3.9) is now reduced to a linear partial differential equation of first order which can be solved on the characteristics:

$$\frac{\mathrm{d}\lambda}{1} = \frac{\mathrm{d}\widehat{x}}{K_0} = \frac{\mathrm{d}\widehat{u}}{\phi^{(0)} + \widehat{u}\,\phi^{(1)}} = \frac{\mathrm{d}F}{\phi_+^{(0)} + \phi_-^{(0)} + \phi_-^{(1)} F}, \tag{3.12}$$

where

$$\phi^{(i)} = \phi^{(i)}(x), \quad \phi_+^{(i)} = \phi^{(i)}(x_+), \quad \phi_-^{(i)} = \phi^{(i)}(x_-), \qquad i = 0, 1. \tag{3.13}$$

The first invariant is:

$$\widehat{x} = x + K_0\lambda. \tag{3.14}$$

The action of this transformation on the discrete Painlevé equation (3.1) corresponds just to a change in the parameter β. We could carry out the calculation with $K_0 \neq 0$, but for the sake of clarity of the presentation, we have set $K_0 = 0$. Consequently, $\phi^{(i)}$ will not depend on λ.

The next invariant is obtained by integrating Eq. (3.12) for $\widehat{u}$ as a function of λ. We have:

$$\widehat{u} = \mathrm{e}^{\lambda\phi^{(1)}} \left[u + \frac{\phi^{(0)}}{\phi^{(1)}} (1 - \mathrm{e}^{-\lambda\phi^{(1)}}) \right]. \tag{3.15}$$

The integration of Eq. (3.12) for F satisfying the boundary conditions (3.7) gives

$$F(\widehat{x},\widehat{u};\lambda) = \mathrm{e}^{\lambda\phi_-^{(1)}}\left[\frac{\phi^{(0)}}{\phi^{(1)}}(1-\mathrm{e}^{-\lambda\phi^{(1)}}) + \frac{\alpha\widehat{x}+\beta}{\widehat{u}\mathrm{e}^{-\phi^{(1)}\lambda} - \phi^{(0)}/\phi^{(1)}(1-\mathrm{e}^{-\lambda\phi^{(1)}})} + \gamma\right] - \widehat{u}\mathrm{e}^{\lambda[\phi_-^{(1)}-\phi^{(1)}]} - \frac{\phi_+^{(0)}+\phi_-^{(0)}}{\phi_-^{(1)}}(1-\mathrm{e}^{\lambda\phi_-^{(1)}}). \tag{3.16}$$

To get a discrete Lie symmetry we require that there exists a value of λ, say λ_0, such that

$$F(\widehat{x},\widehat{u};0) = F(\widehat{x},\widehat{u};\lambda_0). \tag{3.17}$$

If we want Eq. (3.17) to be satisfied, we need

$$(\phi_-^{(1)} - \phi^{(1)})\lambda_0 = 2\pi\mathrm{i}N, \quad N\in\mathbf{Z}, \tag{3.18}$$

$$\phi^{(0)}(1-\mathrm{e}^{-\lambda_0\phi^{(1)}}) = 0. \tag{3.19}$$

From Eq. (3.18) we obtain $K_2 = 0$ and then, $\phi^{(1)} = K_1$. Equation (3.19) is solved by requiring one or the other of the following two conditions:

$$\lambda_0 K_1 = 2\pi\mathrm{i}N, \quad N\in\mathbf{Z}, \tag{3.20}$$

$$\phi^{(0)} = 0. \tag{3.21}$$

In the case of condition (3.20), Eq. (3.17) is satisfied. However, this provides no discrete symmetry.

In the case of the second condition (3.21), if $\gamma \neq 0$, Eq. (3.17) implies Eq. (3.20), i.e., no discrete symmetry is present. If $\gamma = 0$, then we have $K_1\lambda_0 = \mathrm{i}\pi N$, providing a discrete symmetry $\widehat{u} = \pm u$. Note, however, that in this case, the continuum limit of this difference equation is not Painlevé I.

3.2. Discrete symmetries of the Toda equation. Let us consider as our second example the Toda equation:

$$u_{tt} = \mathrm{e}^{u_+-u} - \mathrm{e}^{u-u_-} \tag{3.22}$$

where $x_\pm = x \pm h$, $u_\pm = u(x\pm h, t)$ and h is the lattice step (see Eq. (3.2)).

As it is well known [**3, 9, 12**], the following operators form a basis of the symmetry algebra of the Toda equation:

$$X_1 = \partial_u, \quad X_2 = \partial_x, \quad X_3 = \partial_t, \quad X_4 = t\partial_u, \quad X_5 = t\partial_t - \frac{2x}{h}\partial_u. \tag{3.23}$$

The nonzero commutation relations are:

$$[X_2, X_5] = -\frac{2}{h}X_1, \quad [X_3, X_4] = X_1, \quad [X_3, X_5] = X_3, \quad [X_4, X_5] = -X_4 \tag{3.24}$$

and the matrices $C(i)$ of the adjoint representation are:

$$C(1)=0,\quad C(2)=\begin{pmatrix}\cdot&\cdot&\cdot&\cdot&-\frac{2}{h}\\ \cdot&\cdot&\cdot&\cdot&\cdot\\ \cdot&\cdot&\cdot&\cdot&\cdot\\ \cdot&\cdot&\cdot&\cdot&\cdot\\ \cdot&\cdot&\cdot&\cdot&\cdot\end{pmatrix},\qquad C(3)=\begin{pmatrix}\cdot&\cdot&\cdot&1&\cdot\\ \cdot&\cdot&\cdot&\cdot&\cdot\\ \cdot&\cdot&\cdot&\cdot&1\\ \cdot&\cdot&\cdot&\cdot&\cdot\\ \cdot&\cdot&\cdot&\cdot&\cdot\end{pmatrix},$$

$$C(4)=\begin{pmatrix}\cdot&\cdot&-1&\cdot&\cdot\\ \cdot&\cdot&\cdot&\cdot&\cdot\\ \cdot&\cdot&\cdot&\cdot&\cdot\\ \cdot&\cdot&\cdot&\cdot&-1\\ \cdot&\cdot&\cdot&\cdot&\cdot\end{pmatrix},\quad C(5)=\begin{pmatrix}\cdot&\frac{2}{h}&\cdot&\cdot&\cdot\\ \cdot&\cdot&\cdot&\cdot&\cdot\\ \cdot&\cdot&-1&\cdot&\cdot\\ \cdot&\cdot&\cdot&1&\cdot\\ \cdot&\cdot&\cdot&\cdot&\cdot\end{pmatrix}. \tag{3.25}$$

Applying Eq. (2.6) and defining ${\Phi^i}_j = b_{ij}$ we get :

$$b_{21}=b_{23}=b_{24}=b_{31}=b_{41}=b_{51}=b_{53}=b_{54}=0. \tag{3.26}$$

The most significant remaining equations are:

$$b_{34}(b_{55}+1)=b_{44}(b_{55}-1)=b_{33}(b_{55}-1)=b_{43}(b_{55}+1)=0, \tag{3.27}$$

$$b_{11}=b_{33}b_{44}-b_{34}b_{43}. \tag{3.28}$$

The determinant of Φ must be different from zero (ϕ is an automorphism) and consequently, taking into account Eq. (3.26), we must have $b_{11}\neq 0$. Using Eq. (3.27) we conclude that $b_{55}=\pm 1$. We will distinguish two cases:

(a) $b_{55}=1$. In this case, $b_{34}=b_{43}=0$ and $b_{33}\neq 0$, $b_{44}\neq 0$. The other elements of the matrix Φ must satisfy the equations:

$$b_{32}=b_{42}=b_{52}=0,\quad b_{13}=b_{33}b_{45},\quad b_{14}=b_{35}b_{44},\quad b_{11}=b_{22}=b_{33}b_{44}. \tag{3.29}$$

The matrix $\Phi_1=\Phi$ is:

$$\Phi_1=\begin{pmatrix}b_{33}b_{44}&b_{12}&b_{33}b_{45}&b_{35}b_{44}&b_{15}\\ 0&b_{33}b_{44}&0&0&b_{25}\\ 0&0&b_{33}&0&b_{35}\\ 0&0&0&b_{44}&b_{45}\\ 0&0&0&0&1\end{pmatrix}. \tag{3.30}$$

(b) $b_{55}=-1$. Now, $b_{33}=b_{44}=0$ and $b_{34}\neq 0$, $b_{43}\neq 0$. The other elements satisfy the equations:

$$\begin{aligned} &b_{32}=b_{42}=b_{52}=0, && b_{13}=-b_{35}b_{43},\\ &b_{14}=-b_{34}b_{45}, && b_{11}=-b_{22}=-b_{34}b_{43}. \end{aligned} \tag{3.31}$$

The matrix $\Phi_2=\Phi$ is:

$$\Phi_2=\begin{pmatrix}-b_{34}b_{43}&b_{12}&-b_{35}b_{43}&-b_{34}b_{45}&b_{15}\\ 0&b_{34}b_{43}&0&0&b_{25}\\ 0&0&0&b_{34}&b_{35}\\ 0&0&b_{43}&0&b_{45}\\ 0&0&0&0&-1\end{pmatrix}. \tag{3.32}$$

To simplify the matrix Φ we will conjugate the automorphism using the continuous transformations in the adjoint representation.

The exponentials of the matrices $C(i)$ are easy to find. Using $C(3)$ we can put $b_{35}=0$, with $C(4)$, $b_{45}=0$, $C(5)$ gives $b_{12}=0$, and, finally using $C(2)$, $b_{15}=0$. Then, the simplified Φ_1 reads:

$$\Phi_1 = \begin{pmatrix} b_{33}b_{44} & 0 & 0 & 0 & 0 \\ 0 & b_{33}b_{44} & 0 & 0 & b_{25} \\ 0 & 0 & b_{33} & 0 & 0 \\ 0 & 0 & 0 & b_{44} & 0 \\ 0 & 0 & 0 & 0 & 1 \end{pmatrix}, \quad b_{33}, b_{44} \neq 0. \tag{3.33}$$

The same procedure can be used with Φ_2 and the result is:

$$\Phi_2 = \begin{pmatrix} -b_{34}b_{43} & 0 & 0 & 0 & 0 \\ 0 & b_{34}b_{43} & 0 & 0 & b_{25} \\ 0 & 0 & 0 & b_{34} & 0 \\ 0 & 0 & b_{43} & 0 & 0 \\ 0 & 0 & 0 & 0 & -1 \end{pmatrix}, \quad b_{34}, b_{43} \neq 0. \tag{3.34}$$

Let's realize the automorphisms in the space of the variables and functions of the Toda equation. We have to solve the following system of equations:

$$\begin{aligned} \tau_j(t,x,u)\frac{\partial \hat{t}}{\partial t} + \xi_j(t,x,u)\frac{\partial \hat{t}}{\partial x} + \varphi_j(t,x,u)\frac{\partial \hat{t}}{\partial u} &= (\Phi^{-1})^i{}_j \tau_i(\hat{t},\widehat{x},\widehat{u}), \\ \tau_j(t,x,u)\frac{\partial \widehat{x}}{\partial t} + \xi_j(t,x,u)\frac{\partial \widehat{x}}{\partial x} + \varphi_j(t,x,u)\frac{\partial \widehat{x}}{\partial u} &= (\Phi^{-1})^i{}_j \xi_i(\hat{t},\widehat{x},\widehat{u}), \\ \tau_j(t,x,u)\frac{\partial \widehat{u}}{\partial t} + \xi_j(t,x,u)\frac{\partial \widehat{u}}{\partial x} + \varphi_j(t,x,u)\frac{\partial \widehat{u}}{\partial u} &= (\Phi^{-1})^i{}_j \varphi_i(\hat{t},\widehat{x},\widehat{u}), \end{aligned}$$

where $j=1,\ldots,5$. In Case (a), from (3.35) for $X_1=\partial_u$ ($\mu=1/b_{33}$, $\nu=1/b_{44}$, $\sigma=-b_{25}/b_{33}b_{44}$) we get:

$$\frac{\partial \hat{t}}{\partial u} = 0, \quad \frac{\partial \widehat{x}}{\partial u} = 0, \quad \frac{\partial \widehat{u}}{\partial u} = \mu\nu. \tag{3.35}$$

For $X_2=\partial_x$

$$\frac{\partial \hat{t}}{\partial x} = 0, \quad \frac{\partial \widehat{x}}{\partial x} = \mu\nu, \quad \frac{\partial \widehat{u}}{\partial x} = 0. \tag{3.36}$$

For $X_3=\partial_t$

$$\frac{\partial \hat{t}}{\partial t} = \mu, \quad \frac{\partial \widehat{x}}{\partial t} = 0, \quad \frac{\partial \widehat{u}}{\partial t} = 0. \tag{3.37}$$

For $X_4=t\partial_u$

$$t\frac{\partial \hat{t}}{\partial u} = 0, \quad t\frac{\partial \widehat{x}}{\partial u} = 0, \quad t\frac{\partial \widehat{u}}{\partial u} = \nu\hat{t}. \tag{3.38}$$

For $X_5=t\partial_t-2x/h\partial_u$

$$t\frac{\partial \hat{t}}{\partial t} - \frac{2x}{h}\frac{\partial \hat{t}}{\partial u} = \hat{t}, \quad t\frac{\partial \widehat{x}}{\partial t} - \frac{2x}{h}\frac{\partial \widehat{x}}{\partial u} = \sigma, \quad t\frac{\partial \widehat{u}}{\partial t} - \frac{2x}{h}\frac{\partial \widehat{u}}{\partial u} = -\frac{2\widehat{x}}{h}. \tag{3.39}$$

The solution of Eqs. (3.35), (3.36), (3.37) is;

$$\hat{t} = \mu t + \alpha, \quad \widehat{x} = \mu\nu x + \beta, \quad \widehat{u} = \mu\nu u + \gamma. \tag{3.40}$$

Substituting in Eq. (3.38) we get $\alpha = 0$ and from Eq. (3.39) we get $\sigma = 0$ and $\beta = 0$. The translation in u is a continuous symmetry, so the possible discrete symmetries are:

$$\hat{t} = c_1 t, \quad \widehat{x} = c_2 x, \quad \widehat{u} = c_2 u, \quad c_1, c_2 \neq 0. \tag{3.41}$$

Applying Eq. (3.41) to the Toda equation (3.22) we get that $c_2 = \pm 1$, $c_1 = \pm 1$ are the only admissible solutions. When $c_2 = 1$, $c_1 = -1$, we obtain a discrete symmetry given by the transformation:

$$\hat{t} = -t, \quad \widehat{x} = x, \quad \widehat{u} = u. \tag{3.42}$$

When $c_2 = -1, c_1 = 1$, we get the discrete symmetry:

$$\hat{t} = t, \quad \widehat{x} = -x, \quad \widehat{u} = -u. \tag{3.43}$$

It is easy to check that the automorphism of Case (b) cannot be realized in the representation under consideration.

4. Conclusions

In this article we have shown that the two methods, the automorphisms of the symmetry algebra [**5**] and the determining equation for discrete symmetries [**1**], provide discrete symmetries, also in the case of discrete equations. It is worthwhile to notice that in the case of the Toda lattice, we have obtained a non obvious discrete symmetry (3.43) using these techniques.

Work is in progress to apply the other techniques mentioned at the beginning of Section 2 and for the construction of lattices and discrete equations with prescribed discrete symmetries, a problem of interest in Chemistry and Quantum Mechanics.

References

1. G. Gaeta and M. A. Rodríguez, *Discrete symmetries of differential equations*, J. Phys. A **29** (1996), 859–880.
2. I. M. Gel'fand, R. A. Minlos, and Z. Ya Shapiro, *Representations of the rotation and Lorentz groups and their applications*, Pergamon Press, London, 1963.
3. R. Hernández Heredero, D. Levi, M. A. Rodríguez, and P. Winternitz, *Lie algebra contractions and symmetries of the Toda hierarchy*, J. Phys. A **33** (2000), 5025–5040.
4. P. E. Hydon, *Discrete point symmetries of ordinary differential equations*, Proc. Roy. Soc. London Ser. A **454** (1998) 1961–1972.
5. ———, *How to construct the discrete symmetries of partial differential equations*, European J. Appl. Math. **11** (2000), 515–527.
6. E. L. Ince, *Ordinary differential equations*, Dover, New York, 1956.
7. A. V. Kistovich and Yu D. Chaschechkin, *Regular method for searching of differential equations discrete symmetries*, Regul. Chaotic Dyn. **6** (2001), 327–336.
8. D. Levi , O. Ragnisco, and M. A. Rodríguez, *On non-isospectral flows, Painlevé equations and symmetries of differential and difference equations*, Teoret. Mat. Fizika **93** (1992), 473–480.
9. D. Levi and M. A. Rodríguez, *Symmetry group of partial differential equations and of differential difference equations: The Toda lattice versus the Korteweg-de Vries equation*, J. Phys. A **25** (1992), L975–L979.
10. D. Levi, S. Tremblay, and P. Winternitz, *Lie point symmetries of difference equations and lattices*, J. Phys. A **33** (2001), 521–554.
11. ———, *Lie symmetries of multidimensional difference equations*, J. Phys. A **34** (2001), 9507–9524.
12. D. Levi and P. Winternitz, *Continuous symmetries of discrete equations*, Phys. Lett. A **152** (1991), 335–338.
13. ———, *Symmetries and conditional symmetries of differential difference equations*, J. Math. Phys. **34** (1993), 3713–3730.

14. I. G. Lisle and G. J. Reid, *Geometry and structure of Lie pseudogroups from infinitesimal defining systems*, J. Symbolic Comput. **26** (1998), 355–379.
15. P. J. Olver, *Applications of Lie groups to differential equations*, Springer, New York, 1986.
16. G. R. W. Quispel, H. W. Capel, and R. Sahadevan, *Continuous symmetries of differential-difference equations: The Kac-van Moerbeke equation and Painlevé reduction*, Phys. Lett. A **170** (1992), 379–383.
17. G. J. Reid, A. Boulton, and I. G. Lisle, *Characterizing Lie equations by their infinitesimal symmetries*, 1999 (unpublished)
18. G. J. Reid, D. T. Weih, and A. D. Wittkopf, *A point symmetry group of a differential equation which cannot be found using infinitesimal methods*, Modern Group Analysis: Advances Analytical and Computational Methods in Mathematical Physics (N. H. Ibraghimov, M. Torrisi, and A. Valenti, eds.) Kluwer, Dordrecht, 1993, pp. 93–99.
19. M. Toda, *Theory of nonlinear lattices* Springer, Berlin, 1981.

Dipartimento Ingegneria Elettronica, Università Roma Tre and INFN-Sezione di Roma Tre, Via della Vasca Navale 84, 00146–Roma, Italy.
E-mail address: levi@fis.uniroma3.it

Departamento de Física Teórica II, Facultad de Físicas, Universidad Complutense, 28040—Madrid, Spain.
E-mail address: rodrigue@fis.ucm.es

Centre de Recherches Mathématiques
CRM Proceedings and Lecture Notes
Volume **39**, 2005

Trivializations, Factorizations, and Geometric Integration for Pseudo-Rigid Bodies

D. Lewis

1. Introduction

The interaction between geometric mechanics and numerical analysis is a rich source of insights and challenges for both the geometer and the analyst. The highly developed structures of conservative systems with symmetry provide both a powerful set of tools for the design of numerical methods for the integration of differential equations and an imposing array of criteria to be used in assessing the performance of those methods. Here we illustrate the interplay between some key components of the geometric theory of conservative systems on Lie groups and the design of numerical schemes for such systems by focusing on a single system with a high ratio of symmetry to degrees of freedom and using a combination of conservation laws, bundle trivializations, and matrix factorizations to develop several versions of the dynamics that are particularly well-suited for efficient numerical implementation.

We study the dynamics of a generalized incompressible pseudo-rigid body, a canonical Hamiltonian system on the cotangent bundle of the Lie group $\mathrm{SL}(n)$ of $n \times n$ matrices with unit determinant. This system is the n dimensional, incompressible analog of a pseudo-rigid body, or affinely-rigid body, which is a highly simplified model of a three dimensional elastic body, in which the body undergoes only affine deformations. A pseudo-rigid body can be regarded as a three-director Cosserat model, analogous to two-director Cosserat models of elastic rods and one-director Cosserat models of elastic shells. See [**4, 17, 28, 29**], and references therein for detailed treatments of pseudo-rigid bodies. The pseudo-rigid model is equivariant with respect to the cotangent lift of the action of the rotation group $\mathrm{SO}(n)$ on $\mathrm{SL}(n)$, reflecting the material frame invariance of the continuum models it mimics.

The most important example of an incompressible pseudo-rigid body is the Dirichlet model of an incompressible fluid drop acted on by self-gravitation. This Hamiltonian system has an illustrious history and has spurred the development of several of the crucial constructs and techniques used in the analysis of mechanical

2000 *Mathematics Subject Classification.* Primary: 70H33, 65L80; Secondary: 34A26, 70G65.

Supported in part by NSF Grant DMS 01-04292 and by the UCSC Academic Senate Committee on Research.

This is the final form of the paper.

systems with symmetry. Special motions of this system were studied by Newton, Maclaurin, and Jacobi (see [**2**] and references therein); the general version was formulated by Dirichlet [**7**], who proved that it is an invariant subsystem of the Euler equations for a free boundary incompressible fluid moving under self-gravitational forces. Dedekind and Riemann [**5, 25**] completed the analysis of the steady motions of the system initiated by Newton, including a detailed nonlinear stability and bifurcation analysis utilizing energy methods that were generalized only in the latter part of the twentieth century. Cartan, Poincare, Chandrasekhar, and others have studied various aspects of the stability of the Riemann ellipsoids, as the steady motions of the Dirichlet model are known. (See [**2, 8**] and references therein.) Note that the Dirichlet model is equivariant with respect to the cotangent lifts of both left *and* right multiplication by rotation matrices; the right action, which we will not take into account here, corresponds to the particle relabelling symmetry of a fluid.

The key to the historical significance of the Dirichlet system is its rich geometric structure. In fact, although quantum versions of pseudo-rigid bodies have lead to a resurgence of interest in these models, the system is significant primarily because the features that facilitated the detailed analyses of the nineteenth century are common to many other successful models of mechanical systems. Many of the techniques employed in the analysis of the Dirichlet system were successfully generalized and eventually evolved into the cornerstones of modern Lie group theory and geometric mechanics. The Euler representations of rigid body and incompressible inviscid fluid motions, corresponding to left (body) trivialization of the (co)tangent bundle of the rotation group and right (spatial) trivialization of the (co)tangent bundle of the (pseudo-)group of volume-preserving diffeomorphisms, i.e. the group of particle relabelling symmetries, have played a crucial role in mechanics. Dirichlet and Riemann provided explicit, detailed treatments of the trivialization and reduction procedures in the setting of affine motions and brilliantly utilized the underlying symmetries and associated conservation laws in their analysis of the reduced system. There is still much to be learned from the Dirichlet and pseudo-rigid models. These systems provide a natural 'test track' for both analytic and numerical methods.

To obtain a convenient form of Hamilton's equations on the cotangent bundle of a Lie group G, one typically makes use of either the left or right trivialization of the cotangent bundle, obtaining a noncanonical Hamiltonian system on the direct product $G \times \mathfrak{g}^*$ of the group with the dual of the Lie algebra. Each trivialization has appealing features related to the symmetries and conservation laws of the system. For example, the angular momentum associated to the equivariance of the system with respect to spatial rotations, i.e. the action of $\mathrm{SO}(n)$ on the cotangent bundle induced by *left* multiplication, is the skew-symmetric component of the *right* trivialized momentum. While the different trivializations lead to equivalent systems, in that trajectories of one system are mapped onto trajectories of the other by an appropriate change of variables, there can be significant differences in the approximate discrete trajectories generated using numerical methods. For example, the construction of numerical schemes conserving angular momentum is straightforward in the right trivialization—it suffices that the momentum be updated using a symmetric matrix—but considerably more subtle in the left trivialization, since the angular momentum couples the configuration and momentum variables in that representation.

We depart from the traditional geometric mechanics framework and propose an alternative to the classical and familiar left and right trivializations of the tangent and cotangent bundles of a Lie group. Using the QR factorization, we introduce the 'centered trivialization', in which (co)tangent vectors are mapped to the (co)tangent space at the identity, i.e. the Lie algebra or its dual, by left multiplication by the orthogonal factor and right multiplication by the triangular factor. This centered trivialization makes explicit the symmetries of the system. In addition, the QR factorization facilitates the implementation of efficient Lie group integrators for the pseudo-rigid body dynamics, exploiting the existence of a family of particularly convenient analogs of the exponential maps of the (special) orthogonal group and the group of special upper triangular matrices. By combining this centered trivialization with restrictions to level sets of the angular momentum, we can easily construct numerical schemes that respect both the incompressibility constraint and the conservation of angular momentum.

Restricting the dynamics to a level set of the angular momentum explicitly enforces a key property of the system and reduces the number of variables to be tracked in numerical simulations. However, the restricted dynamics are no longer Hamiltonian. If one wishes to employ symplectic or Poisson integrators, it is possible to recapture the symplectic nature of the dynamics by dropping to a quotient of the angular momentum level set by an appropriate group action. This quotient has a very natural, explicit representation in the case at hand, but is no longer a Lie group. Nonetheless, the induced dynamical system on this quotient manifold can still be numerically approximated using Lie group methods for homogeneous manifolds.

2. Pseudo-rigid bodies

We consider a canonical Hamiltonian system on the cotangent bundle $T^* \mathrm{SL}(n)$ of the group of unit determinant $n \times n$ matrices, with Hamiltonian

$$H(p_g) = \tfrac{1}{2}\langle p_g, p_g M_{\mathrm{ref}}^{-1}\rangle + W(g^T g). \tag{2.1}$$

Here $\langle\ ,\ \rangle$ denotes the usual matrix inner product $\langle A, B\rangle = \mathrm{trace}(A^T B)$,

$$M_{\mathrm{ref}} := \int_{\mathcal{B}_{\mathrm{ref}}} \rho_{\mathrm{ref}}(X) X X^T dX$$

is the mass matrix of the reference body $\mathcal{B}_{\mathrm{ref}}$, and W is the stored energy of the material expressed in terms of the Cauchy–Green tensor $C(g) := g^T g$.

The Hamiltonian (2.1) is constructed using the Legendre transformation from $T\,\mathrm{SL}(n)$ to $T^*\,\mathrm{SL}(n)$ determined by the total energy

$$E(v_g) = \int_{\mathcal{B}_{\mathrm{ref}}} \left(\tfrac{1}{2}\rho_{\mathrm{ref}}(X)\|v_g X\|^2 + \widetilde{W}(g^T g, X)\right) dX$$

of an affine deformation of a nonlinearly elastic body $\mathcal{B}$ with reference configuration $\mathcal{B}_{\mathrm{ref}}$ and stored energy function $\widetilde{W}\colon \mathrm{Sym}_1(n) \times \mathcal{B}_{\mathrm{ref}} \to \mathbb{R}$ determining the elastic properties of the material forming the body. Here $\mathrm{Sym}_1(n)$ denotes the manifold of symmetric $n \times n$ matrices with unit determinant. (Note that $\mathrm{Sym}_1(n)$ is *not* a group.) The total energy of the system is computed by integrating the total energy of an affinely deformable continuum over the body, just as the kinetic energy of a rigid body is obtained by integrating the kinetic energies of all of the particles constituting the rigid body The material frame invariance of the elastic material, i.e. the equivariance of the Hamiltonian equations determined by (2.1) under the cotangent lift of the action of $\mathrm{SO}(n)$ on $\mathrm{SL}(n)$ by left multiplication is reflected in the dependence of the stored energy function on g only through the left Cauchy–Green tensor $C(g) \in \mathrm{Sym}_1(n)$.

We identify the cotangent bundle $T^*\,\mathrm{SL}(n)$ with the direct product $\mathrm{SL}(n) \times \mathrm{sl}(n)$, where $\mathrm{sl}(n)$ denotes the Lie algebra of $\mathrm{SL}(n)$; note that $\mathrm{sl}(n)$ is the space of traceless $n \times n$ matrices. We utilize the right trivialization of the generalized pseudo-rigid body equations, motivated by the simplicity of the expression for the angular momentum in this trivialization. We first implement the trivialization

$$T^*_g\,\mathrm{SL}(n) \ni p_g \mapsto \big(g, (d_eR_g)^*p_g\big) = (g, p_g g^T) \in \mathrm{SL}(n) \times \mathrm{sl}(n)^*$$

and then, for convenience, identify the dual $\mathrm{sl}(n)^*$ with $\mathrm{sl}(n)$ using the usual matrix inner product. If we let $\mathbb{P}_0 : \mathbb{R}^{n\times n} \to \mathrm{sl}(n)$ denote projection onto $\mathrm{sl}(n)$, specifically $\mathbb{P}_0 A := A - \frac{1}{3}(\operatorname{trace} A)\mathbf{1}$, then the trivialization map $\tau\colon T^*\,\mathrm{SL}(n) \to \mathrm{SL}(n) \times \mathrm{sl}(n)$ takes the form $\tau(p_g) := \big(g, \mathbb{P}_0(p_g g^T)\big)$, with inverse $\tau^{-1}(g,p) = pg^{-T} \in T^*_g\,\mathrm{SL}(n)$. The induced Hamiltonian $h := H \circ \tau^{-1}$ on $\mathrm{SL}(n) \times \mathrm{sl}(n)$ satisfies

$$h(g,p) = \tfrac{1}{2}\|p\|_g^2 + W(g^Tg), \tag{2.2}$$

where

$$\|p\|_g^2 := \langle p, pM(g)^{-1}\rangle \quad \text{for} \quad M(g) := gM_{\mathrm{ref}}g^T.$$

For any $g \in \mathrm{SL}(n)$, the trivialized Legendre transformation $\mathbb{L}_g$ at g is the isomorphism of $\mathrm{sl}(n)$ given by

$$\mathbb{L}_g p := \mathbb{P}_0\big(pM(g)^{-1}\big).$$

Let ∇W denote the gradient of the stored energy $W\colon \mathrm{Sym}_1(n) \to \mathbb{R}$ with respect to the standard matrix inner product, i.e. $\langle \nabla W(s), \delta s\rangle = dW(s)(\delta s)$ for all $\delta s \in T_s\,\mathrm{Sym}_1(n)$. Let $\mathrm{sym}_0(n)$ denote the space of symmetric $n \times n$ matrices with trace equal to zero. If we define the map $\delta W/\delta g\colon \mathrm{SL}(n) \to \mathrm{sym}_0(n)$ by

$$\frac{\delta W}{\delta g}(g) = 2\mathbb{P}_0\big(g\nabla W(g^Tg)g^T\big),$$

then

$$\begin{aligned} d(W \circ C)(g)(\xi g) &= \big\langle \nabla W\big(C(g)\big), dC(g)(\xi g)\big\rangle \\ &= \langle \nabla W(g^Tg), g^T(\xi + \xi^T)g\rangle \\ &= \langle 2g\nabla W(g^Tg)g^T, \xi\rangle \\ &= \left\langle \frac{\delta W}{\delta g}(g), \xi \right\rangle, \end{aligned}$$

for all $g \in \mathrm{SL}(n)$ and $\xi \in \mathrm{sl}(n)$, since ∇W is symmetric and the identity matrix is orthogonal to the space $\mathrm{sl}(n)$.

Remark. If the elastic material is isotropic, i.e. the stored energy function W is invariant with respect to multiplication by rotation matrices on both the left *and* right, then W is a function w of the invariants $\mathrm{I}(C) = \operatorname{trace}(C)$ and $\mathrm{II}(C) = \frac{1}{2}\big(\operatorname{trace}(C)^2 - \operatorname{trace}(C^2)\big)$. In this case, $\nabla\mathrm{I}(C) = \mathbf{1}$ and $\nabla\mathrm{II}(C) = \mathrm{I}\mathbf{1} - C$ imply that

$$\nabla W(C) = (w_{\mathrm{I}} + \mathrm{I}w_{\mathrm{II}})\mathbf{1} - w_{\mathrm{II}}C$$

and hence

$$\frac{\delta W}{\delta g}(g) = (w_{\mathrm{I}} + \mathrm{I}w_{\mathrm{II}})G - w_{\mathrm{II}}G^2, \quad \text{where} \quad G(g) := gg^T.$$

Note that $\mathrm{I}(C) = \mathrm{I}(G)$ and $\mathrm{II}(C) = \mathrm{II}(G)$; hence in this situation W and its derivatives can be expressed entirely in terms of the right Cauchy–Green tensor G, rather than the left Cauchy–Green tensor C.

Theorem 1. *The right trivialized form of Hamilton's equations is*

$$\left.\begin{aligned} \dot{g}g^{-1} &= x_g(g,p) := \mathbb{L}_g p, \\ \dot{p} &= x_p(g,p) := \mathbb{P}_0\big(pM(g)^{-1}p^T\big) - \frac{\delta W}{\delta g}(g) \end{aligned}\right\} \tag{2.3}$$

PROOF. The symplectic structure ω_R on $\mathrm{SL}(n) \times \mathrm{sl}(n)$ induced by pulling back the canonical symplectic structure on $T^*\,\mathrm{SL}(n)$ by the right trivialization is

$$\omega_R(g,p)\big((\xi_1 g, \delta p_1), (\xi_2 g, \delta p_2)\big) = \langle \delta p_2, \xi_1\rangle - \langle \delta p_1, \xi_2\rangle - \langle p, [\xi_1, \xi_2]\rangle, \tag{2.4}$$

where $\xi_j, \delta p_j \in \mathrm{sl}(n)$, $j = 1, 2$, and $[\ ,\ ]$ denotes the standard matrix commutator.

The Hamiltonian vector field $X_h(g,p) = \big(x_g(g,p)g, x_p(g,p)\big)$ is determined by the relation

$$\begin{aligned} dh(g,p)(\xi g, \delta p) &= \omega_R(g,p)\big(X_h(g,p), (\xi g, \delta p)\big) \\ &= \langle \delta p, x_g\rangle - \langle x_p, \xi\rangle - \langle p, [x_g, \xi]\rangle \end{aligned}$$

for all $g \in \mathrm{SL}(n)$ and $p, \xi, \delta p \in \mathrm{sl}(n)$. (See, e.g. [**1**].) For convenience, let $v = pM(g)^{-1}$. The identities

$$dh(g,p)(\xi g, \delta p) = \left\langle \frac{\delta W}{\delta g}(g) - v^T p, \xi \right\rangle + \langle v, \delta p\rangle$$

and

$$\langle a, [b,c]\rangle = \mathrm{trace}\big(a^T(bc - cb)\big) = \mathrm{trace}\big((a^T b - ba^T)c\big) = \langle [b^T, a], c\rangle \tag{2.5}$$

for any square matrices a, b, c yield $x_g(g,p) = \mathbb{P}_0(v) = \mathbb{P}_g p$ and

$$\mathbb{P}_0\left(-\frac{\delta W}{\delta g}(g) + v^T p\right) = \mathbb{P}_0\big(x_p(g,p) + [x_g(g,p)^T, p]\big) = x_p(g,p) + \mathbb{P}_0([v^T, p]).$$

Solving this equation for x_p yields (2.3). □

Note that x_p is symmetric, and hence the skew-symmetric component of p is constant. This conservation law is, in fact, the incarnation of Noether's theorem for the $\mathrm{SO}(n)$ equivariance of the system. We can utilize this conservation law by replacing the full right trivialized momentum p with the sum of its skew-symmetric component, which can be regarded as a parameter, and its symmetric component, which typically varies with time. Specifically, fixing a momentum value $m \in \mathrm{so}(n) \approx \mathrm{so}(n)^*$ and introducing the new variable $s \in \mathrm{sym}_0(n)$, we set $p = m + s$. This yields the momentum level set system

$$(\dot{g}, \dot{s}) = \big(x_g(g, m+s)g, x_p(g, m+s)\big) \tag{2.6}$$

on $\mathrm{SL}(n) \times \mathrm{sym}_0(n)$.

The explicit implementation of the conservation of angular momentum in (2.6) eliminates $\dim \mathrm{SO}(n) = n(n-1)/2$ variables from the description system. *Any* numerical simulation of the pseudo-rigid body model utilizing the representation (2.6) is necessarily momentum-preserving. One possible disadvantage of the momentum level set representation is the lack of a natural Hamiltonian characterization of the equations (2.6). This is not necessarily a liability, but it does rule out the use of symplectic or Poisson integrators in numerical simulations. By further reducing the number of dynamic variables, declaring some configuration information to be inessential and dropping it from the system, we can recapture the symplectic nature of the original system.

2.1. QR factorization and the centered trivialization. The system (2.3) is equivariant with respect to the action $u \cdot (g, p) = (ug, upu^T)$ of $\mathrm{SO}(n)$ on $\mathrm{SL}(n) \times \mathrm{sl}(n)$ induced by the canonical action of $\mathrm{SO}(n)$ on $T^* \mathrm{SL}(n)$ and the trivialization map τ. We can make this equivariance more explicit, and exploit it in our numerical schemes, by mapping the system (2.3) or its momentum level set analog (2.6) onto a product group by means of the QR factorization of a matrix into orthogonal and upper triangular factors. As we shall see in §3, this factorization yields versions of the pseudo-rigid body equations that are particularly well-suited for numerical implementation using Lie group methods.

Let $\mathrm{ST}_+(n)$ denote the Lie group of upper triangular matrices with positive diagonal entries and unit determinant. The product group $\mathrm{SO}(n) \times \mathrm{ST}_+$ acts freely and transitively on $\mathrm{SL}(n)$ by $(q, r) \cdot g = qgr^{-1}$; i.e., every element of $\mathrm{SL}(n)$ has a unique representation as a product qr, with $q \in \mathrm{SO}(n)$ and $r \in \mathrm{ST}_+$. If $g \in \mathrm{SL}(n)$ represents the current configuration of a linearly deformable elastic or fluid body, then the QR decomposition of g has the natural mechanical interpretation that q determines the attitude of the body, while r determines its shape. The map $(q, r) \mapsto qr \in \mathrm{SL}(n)$ can be used to pull back a mapping on $\mathrm{SL}(n)$ to a mapping on $\mathrm{SO}(n) \times \mathrm{ST}_+$. The QR factorization of an invertible matrix g can be constructed as follows: let r be the transpose of the Cholesky factorization of the left Cauchy–Green tensor $g^T g$ associated to g, so that $r^T r = g^T g$, and let $q = gr^{-1}$. Note that, in practice, other schemes for computing the factorization may be more efficient or stable. See, [**10**] for an explicit algorithm for the Cholesky factorization and detailed discussions of numerical algorithms for the QR factorization.

Remark. Other matrix decompositions can be used effectively in the analysis of the pseudo-rigid body equations. For example, the polar decomposition of a matrix into rotational and symmetric components can also be used to explicitly capture the equivariance of the system. However, given that the manifold $\mathrm{Sym}_1(n)$ of symmetric matrices with unit determinant is not a Lie group, we do not pursue that possibility here. (This manifold is, however, amenable to more general geometric integration methods.) The n dimensional Dirichlet problem, which is equivariant with respect to the action of $\mathrm{SO}(n) \times \mathrm{SO}(n)$ induced by left and right multiplication on $\mathrm{SL}(n)$, has been treated analytically by using, either explicitly or implicitly, the singular value decomposition. However, this decomposition does not lead to a global product structure, since the action on $\mathrm{SL}(n)$ fails to be free at points with repeated eigenvalues. The resulting complexities are not insurmountable, but the treatment is significantly more subtle than in the case of a free action. (See, e.g. [**8, 15**] for modern treatment of this approach. Riemann's classic paper [**25**] tacitly utilizes the singular value decomposition, treating points with continuous isotropy as a special case.) We restrict our attention here to the left action of $\mathrm{SO}(n)$ on $\mathrm{SL}(n)$, thereby avoiding any complications due to isotropy and the associated nonuniqueness of the singular value decomposition.

Using the QR factorization we can construct a 'centered trivialization' of the tangent bundle $T\,\mathrm{SL}(n)$, as an alternative to the left and right trivializations. The combination of the decomposition of the configuration variable with this centered trivialization yields versions of the pseudo-rigid body equations that appear to be of some analytic interest and, as we shall argue, have some clear advantages over the traditional representations of the system in numerical simulations. If we left trivialize the tangent bundle $T\,\mathrm{SO}(n)$ of the rotation group and right trivialize the tangent bundle $T\mathrm{ST}_+$ of the incompressible upper triangular group, setting

$$\Omega = q^{-1}\dot{q} \quad \text{and} \quad \rho = \dot{r}r^{-1},$$

then differentiating $g = qr$ with respect to time and utilizing the evolution equation for g in (2.3) yields

$$\Omega + \rho = q^{-1}(\dot{q}r + q\dot{r})r^{-1} = q^{-1}\dot{g}r^{-1} = q^{-1}(\mathbb{L}_{qr}p)q.$$

We introduce the centered trivialization

$$T^*_{qr}\,\mathrm{SL}(n) \ni p_{qr} \mapsto \big(qr, \mathbb{P}_0(q^T p_{qr} r^T)\big)$$

and let $\phi = q^T p q \in \mathrm{sl}(n)$ denote the centrally trivialized momentum corresponding to the right trivialized momentum p. Using the equivariance of the mass matrix map M, namely $M(qr) = qM(r)q^T$, and of the projection $\mathbb{P}_0$, we obtain the identity

$$\mathbb{L}_{qr}p = \mathbb{P}_0\big(pqM(r)^{-1}q^T\big) = q\mathbb{P}_0\big(\phi M(r)^{-1}\big)q^T = q(\mathbb{L}_r\phi)q^T$$

for all q, r, and p; thus $\Omega + \rho = \mathbb{L}_r\phi$. Since any element of $\mathrm{sl}(n)$ can be uniquely decomposed as the sum of elements of $\mathrm{so}(n)$ and $\mathrm{st}_+(n)$, Ω and ρ are uniquely determined by this equation.

We now use the equivariance of the map $x_p\colon \mathrm{SL}(n) \times \mathrm{sl}(n) \to \mathrm{sym}_0(n)$ with respect to the action $u \cdot (g, p) = (ug, upu^T)$ on $\mathrm{SL}(n) \times \mathrm{sl}(n)$ and the action $u \cdot s = usu^T$ on $\mathrm{sym}_0(n)$ to rewrite the evolution equation for ϕ. For any $q \in \mathrm{SO}(n)$, $r \in \mathrm{ST}_+(n)$, and $\phi \in \mathrm{sl}(n)$, we have

$$x_p(qr, q\phi q^T) = qx_p(r, \phi)q^T.$$

Differentiation of ϕ with respect to time thus yields

$$\begin{aligned}\dot{\phi} &= \dot{q}^T pq + q^T\dot{p}q + q^T p\dot{q}\\ &= (q^{-1}\dot{q})^T\phi + q^T x_p(qr, q\phi q^T)q + \phi q^{-1}\dot{q}\\ &= x_p(r, \phi) + [\phi, \Omega].\end{aligned}$$

Thus the centrally trivialized pseudo-rigid body equations are

$$(\dot{q}, \dot{r}, \dot{\phi}) = \big(q\Omega, \rho, r, x_p(r, \phi) + [\phi, \Omega]\big), \quad \text{where} \quad \Omega + \rho = \mathbb{L}_r\phi. \tag{2.7}$$

Remark. It is possible to construct hybrid equations, in which the configuration equations utilize the centered trivialization and the momentum equations are expressed in terms of the right trivialization, e.g.

$$(\dot{q}, \dot{r}, \dot{s}) = \big(q\Omega, \rho r, x_p(qr, m + s)\big), \quad \text{where} \quad \Omega + \rho = q^T\mathbb{L}_r(m + s)q. \tag{2.8}$$

Such hybrid versions may be useful, e.g., in numerical settings where it is desirable to remain as close to the right trivialized equations (2.3) as is practically possible, while accurately and efficiently capturing the nonlinear configuration constraint $g = qr \in \mathrm{SL}(n)$.

The projections of $\mathrm{sl}(n)$ onto the subalgebra $\mathrm{so}(n)$ and subspace $\mathrm{sym}_0(n)$ commute with the actions of the rotation group $\mathrm{SO}(n)$ on these spaces. Hence we can easily obtain a centrally trivialized version of the momentum level set form (2.6) of the pseudo-rigid body equations. If we introduce the variable $\sigma := q^T sq$ and the function $\mu(q) := q^T mq$, then $\phi = \sigma + \mu(q)$ and differentiation of σ with respect to time yields the centrally trivialized momentum level set equations

$$(\dot{q}, \dot{r}, \dot{\sigma}) = \big(q\Omega, \rho r, x_p\big(r, \mu(q) + \sigma\big) + [\sigma, \Omega]\big) \tag{2.9}$$

where $\Omega + \rho = \mathbb{L}_r\big(\mu(q) + \sigma\big)$.

2.2. Symplectic reduction. The centrally trivialized versions (2.7) and (2.9) of the pseudo-rigid body equations reveal the limited role of the rotational component q of the configuration in the evolution of the system. The rotational factor q influences the evolution of the system (2.9) only through the centered angular momentum $\mu(q)$. Given the rotational component $q(t)$ of a trajectory of (2.9), we obtain the equation

$$\frac{d}{dt}\mu(q) = \dot{q}^T mq + q^T m\dot{q} = \Omega^T \mu(q) + \mu(q)\Omega = [\mu(q), \Omega]$$

for the evolution of $\mu(q)$. If we regard μ as a variable, rather than a function of q, taking values in the coadjoint orbit $\mathcal{O}_m := \{\mu(q) : q \in \mathrm{SO}(n)\}$, we obtain the quotient system

$$(\dot{r}, \dot{\mu}, \dot{\sigma}) = \big(\rho r, [\mu, \Omega], x_p(r, \mu + \sigma) + [\sigma, \Omega]\big), \tag{2.10}$$

where $\Omega + \rho = \mathbb{L}_r(\mu + \sigma)$, on $P_m := \mathrm{ST}_+(n) \times \mathcal{O}_m \times \mathrm{sym}_0(n)$.

The system (2.10) is *not* equivalent to the original system (2.3), since the orientation matrix q cannot be fully reconstructed from the relationship $\mu = q^T mq$. This system is, however, Hamiltonian; it is the symplectically reduced system on $P_m \approx J^{-1}(m)/SO(n)_m$, where $J : T^* \mathrm{SL}(n) \to \mathrm{so}(n)^* \approx \mathrm{so}(n)$ is the momentum map associated to the cotangent lift of left multiplication by $\mathrm{SO}(n)$ on $\mathrm{SL}(n)$ and $\mathrm{SO}(n)_m = \{q \in \mathrm{SO}(n) : qm = m\}$ is the isotropy subgroup of m, with algebra $\mathrm{o}(n)_m = \{\xi \in \mathrm{sl}(n) : [m, \xi] = 0\}$. P_m inherits a nondegenerate symplectic structure from the canonical structure on $T^* \mathrm{SL}(n)$, via a procedure known as Meyer–Marsden–Weinstein symplectic reduction. See, e.g., [**1**] and references therein for a detailed description of the reduction procedure.

Theorem 2. *The equations* (2.10) *are Hamiltonian with respect to the Hamiltonian*

$$h_m(r, \mu, \sigma) := \tfrac{1}{2}\|\mu + \sigma_r^2 + W(r^T r) \tag{2.11}$$

and symplectic structure

$$\begin{aligned} \omega_m(r, \mu, \sigma)\big((\rho 1 r, [\mu, \xi_1], \delta\sigma_1), (\rho 2 r, [\mu, \xi_2], \delta\sigma_2)\big) \\ = \omega_R(r, \sigma)\big((\rho 1 r, \delta\sigma_1), (\rho 2 r, \delta\sigma_2)\big) - \langle \mu, [\rho 1 + \xi_1, \rho 2 + \xi_2]\rangle. \end{aligned} \tag{2.12}$$

PROOF. Let $\pi_m \colon J^{-1}(m) \to P_m$ denote the $\mathrm{SO}(n)$-invariant map

$$\pi_m(qr, m + s) := q^{-1} \cdot (qr, m, s) = (r, q^T mq, q^T sq)$$

and let $\iota_m \colon J^{-1}(m) \to \mathrm{SL}(n) \times \mathrm{sl}(n)$ denote the inclusion map. Then the symplectic structure on P_m induced by the right trivialized symplectic structure ω_R on $\mathrm{SL}(n) \times \mathrm{sl}(n)$ is determined by the relation $\pi_m^* \omega_m = \iota_m^* \omega_R$. (See, e.g. [**1**].) The linearization of the projection π_m satisfies

$$T_{(qr, q(\mu+\sigma)q^T)}\pi_m\big(q(\rho + \xi)r, q(\delta\sigma - [\sigma, \xi])q^T\big) = (\rho r, [\mu, \xi], \delta\sigma)$$

for any $q \in \mathrm{SO}(n)$, $r \in \mathrm{ST}_+(n)$, $\mu \in \mathcal{O}_m$, σ, $\delta\sigma \in \mathrm{sym}_0(n)$, $\rho \in \mathrm{st}_+(n)$, and $\xi \in \mathrm{so}(n)$. For notational convenience, let $v_j = \rho_j + \xi_j$, $j = 1, 2$. Equivariance of the symplectic structure ω_R implies that

$$\begin{aligned}
&\omega_m(r, \mu, \sigma)\big((\rho 1 r, [\mu, \xi_1], \delta\sigma_1), (\rho 2 r, [\mu, \xi_2], \delta\sigma_2)\big)\\
&\quad = \omega_R(qr, m + q\sigma q^T)\big((qv_1 r, q(\delta\sigma_1 - [\sigma, \xi_1])q^T)\big), \big((qv_2 r, q(\delta\sigma_2 - [\sigma, \xi_2])q^T)\big)\\
&\quad = \omega_R(r, \mu + \sigma)\big((v_1 r, \delta\sigma_1 - [\sigma, \xi_1]), (v_2 r, \delta\sigma_2 - [\sigma, \xi_2])\big)\\
&\quad = \langle \delta\sigma_2 - [\sigma, \xi_2], v_1\rangle - \langle \delta\sigma_1 - [\sigma, \xi_1], v_2\rangle - \langle \mu + \sigma, [v_1, v_2]\rangle\\
&\quad = \langle \delta\sigma_2, \rho_1\rangle - \langle \delta\sigma_1, \rho_2\rangle - \langle \sigma, [\rho 1, \rho 2]\rangle - \langle \mu, [v_1, v_2]\rangle,
\end{aligned}$$

which equals (2.12). Note that we have used the orthogonality of symmetric and skew-symmetric matrices in simplifying the term $\langle \sigma, [v_1, v_2]\rangle$. The rotational factor q does not appear explicitly in the bracket and the infinitesimal rotation ξ_j appears only in the term $\langle \mu, [v_1, v_2]\rangle$, which does not depend on the $\mathrm{so}(n)_m$ components of the v_j's. Thus the symplectic structure (2.12) is well-defined on P_m.

If we set $z = (r, \mu, \sigma)$, $\delta z = (\rho r, [\mu, \xi], \delta s)$, and $u = (\mu + \sigma)M(r)^{-1}$, then the Hamiltonian (2.11) has linearization

$$dh_m(z)(\delta z) = \langle u, [\mu, \xi] + \delta\sigma\rangle + \left\langle \frac{\delta W}{\delta g}(r) - u^T(m + \sigma), \rho \right\rangle.$$

The Hamiltonian vector field $X_{(h,m)}$ associated to h_m is determined by the relation

$$dh_m(z)(\delta z) = \omega_m(z)\big(X_{(h,m)}(z), \delta z\big),$$

i.e.

$$\begin{aligned}
0 &= \langle x_r(z) - u, \delta\sigma\rangle,\\
0 &= \langle x_r(z) + x_\mu(z) - u, [\mu, \xi]\rangle,\\
0 &= \left\langle x_\sigma(z) + [x_r(z)^T, \sigma] + [(x_r + x_\mu)(z)^T, \mu] + \frac{\delta W}{\delta g}(r) - u^T(\mu + \sigma), \rho \right\rangle
\end{aligned}$$

for all $\rho \in \mathrm{st}_+(n)$, $\Omega \in \mathrm{so}(n)$, and $\delta\sigma \in \mathrm{sym}_0(n)$. (Here we have used the identity (2.5).) The first equation is satisfied if $x_r(z)$ equals the upper triangular component of u; substituting this into the second equation shows that $x_\mu(z)$ equals $\Omega := u - x_r(z)$ modulo the isotropy subalgebra $\mathrm{so}(n)_\mu$. Substituting these expressions into the third equation and regrouping terms yields

$$\begin{aligned}
x_\sigma(z) &= -\frac{\delta W}{\delta g}(r) + u^T(\mu + \sigma) - [u^T, \mu + \sigma] + [\Omega^T, \sigma]\\
&= -\frac{\delta W}{\delta g}(r) + (\mu + \sigma)u^T + [\sigma, \Omega].
\end{aligned}$$ □

3. Numerical integration of the trivialized equations

There are many possible strategies for the numerical integration of the various forms of the pseudo-rigid body equations. A traditional linear update of the form

$$U_{\Delta t}(g, p) = (g, p) + \Delta t \widetilde{X}(g, p, \Delta t) \tag{3.1}$$

for some map $\widetilde{X}: (\mathbb{R}^{n\times n})^2 \times [0, \varepsilon) \to (\mathbb{R}^{n\times n})^2$ can easily be implemented by applying standard general-purpose algorithms to the right trivialized equations (2.3). However, a linear update such as (3.1) typically will not capture the key features of the pseudo-rigid body dynamics: incompressibility, conservation of energy and angular momentum, and symplecticity.

One economical approach to implementation of the incompressibility constraint is the use of a 'step and project' method: Given any choice of infinitesimal update $\widetilde{X}$, a 'naive' step (3.1) can be followed by a nonlinear projection to yield an update lying in $\mathrm{SL}(n) \times \mathrm{sl}(n)$. Specifically, given a naive update (3.1), we can use the nonlinear projection $\mathbb{P}_{\mathrm{SL}} \colon \mathrm{GL}(n) \times \mathrm{sl}(n) \to \mathrm{SL}(n) \times \mathrm{sl}(n)$ given by

$$\mathbb{P}_{\mathrm{SL}}(g, p) := \left(\frac{1}{\det g} g, p \right)$$

to construct a restricted update map taking values in $\mathrm{SL}(n) \times \mathrm{sl}(n)$. Given a time step Δt, there is a neighborhood $\mathcal{V}_{\Delta t}$ of $(\mathbf{1}, \mathbf{0}) \in \mathrm{SL}(n) \times \mathrm{sl}(n)$ such that $U_{\Delta t}$ maps $\mathcal{V}_{\Delta t}$ into $\mathrm{GL}(n) \times \mathrm{sl}(n)$ and hence the update $\widetilde{U}_{\Delta t} := \mathbb{P}_{\mathrm{SL}} \circ U_{\Delta t}$ is defined on $\mathcal{V}_{\Delta t}$. The projection $\mathbb{P}_{\mathrm{SL}}$ is equivariant with respect to the left and right actions of $\mathrm{SO}(n)$ on $\mathrm{SL}(n) \times \mathrm{sl}(n)$; hence if $\widetilde{X}$ is equivariant with respect to either or both of those actions, $U_{\Delta t}$ is also equivariant.

The step-and-project method described above has serious limitations. The time step Δt must be taken to be sufficiently small to capture the expected trajectories in $\mathcal{V}_{\Delta t}$. If ill-conditioned matrices, i.e. matrices that are near to the set of singular matrices, are to be considered, this restriction may impose a prohibitively small time step. In addition to this limitation, there is the risk that the 'after the fact' projection may impose a definite energy bias, either pumping energy into the system or dissipating energy.

Another approach to satisfying the incompressibility constraint is the use of Lie group methods. (See, e.g., [**11–13, 16, 18, 19, 21**] and references therein.) These methods make use of two key structures of a Lie group: the triviality of the tangent bundle $TG \approx G \times \mathfrak{g}$, where $\mathfrak{g} = T_e G$ is the Lie algebra of the group G, and the existence of the exponential map $\exp \colon \mathfrak{g} \to G$. Lie group integration schemes can be regarded as approximating solutions of nonlinear ODEs of the form

$$\dot{g} = d_0 \, \mathrm{mult}_g \, X(g),$$

where $X \colon G \to \mathfrak{g}$ is a left (respectively right) trivialized vector field and $mult_g$ denotes left (respectively right) multiplication by g, using the solutions $g(t) = \mathrm{mult}_{g_0}(\exp(t\xi))$ of 'linear' ODEs $\dot{g} = d_0 \, \mathrm{mult}_g \, \xi$, where $\xi \in \mathfrak{g}$ is constant. For example, a Lie group version of Euler's method is

$$U_{\Delta t}(g) = \mathrm{mult}_g(\exp(\Delta t X(g)).$$

Higher order, more efficient Lie group methods can be constructed by replacing the vector field X with a more general infinitesimal update map and replacing the exponential map with an 'algorithmic exponential'.

For most Lie groups, a closed form expression for the exponential map is not available and approximation of the exponential to near-machine precision can be prohibitively costly. While much work has been done in recent years to increase the efficiency of high order approximations of the exponential map for special classes of matrix groups (see, e.g., [**22, 24, 30**] and references therein), such approximations are typically still quite expensive unless the dimension of the group is low. In many situations, it appears to be advantageous to replace the matrix exponential with some other map that possesses some of the key properties of the exponential for the group in question, but is significantly faster to evaluate. Given a Lie group

G with Lie algebra $\mathfrak{g}$, we say that a (smooth) map $\operatorname{Exp}\colon \mathfrak{g} \to G$ is an *algorithmic exponential* if $\operatorname{Exp}(0) = \mathbf{1}$ and $d_0 \operatorname{Exp}$ is the identity map on $\mathfrak{g}$. (See [**12, 18, 19**].) The exponential map of a Lie group is equivariant with respect to the adjoint action of G on $\mathfrak{g}$ and the inner action on G, i.e. $\exp \circ \operatorname{Ad}_g = I_g \circ \exp$ for all $g \in G$, where $I_g(h) = ghg^{-1}$ and $\operatorname{Ad}_g = d_0 I_g$. While we do not require that an algorithmic exponential be equivariant with respect to these actions, it is a very desirable property and one that is possessed by several important examples.

A well-known equivariant algorithmic exponential for the (special) orthogonal group is the Cayley transform $\operatorname{cay}\colon o(n) \to O(n)$

$$\operatorname{cay} \xi = (\mathbf{1} - \tfrac{1}{2}\xi)^{-1}(\mathbf{1} + \tfrac{1}{2}\xi) = \exp(\xi) + \mathcal{O}(\|\xi\|^3). \tag{3.2}$$

The use of the Cayley transform in geometric integration schemes, particularly finite element schemes, was popularized by J. C. Simo and collaborators. (See [**18, 19, 26, 27**] and references therein for some representative examples.) A possible algorithmic exponential for $\mathrm{ST}_+(n)$ is

$$\operatorname{Exp}_{\mathrm{ST}_+(n)}(\xi)_{ij} := \begin{cases} e^{\xi_{ii}} = \exp(\xi)_{ii} \\ \xi_{ij} + \frac{1}{2}\sum_{i \le k \le j} \xi_{ik}\xi_{kj} = \exp(\xi)_{ij} + \mathcal{O}(\|\xi\|^3) & i < j \\ 0 & i > j \end{cases} \tag{3.3}$$

Note that higher order approximations of the true exponential for $\mathrm{ST}_+(n)$ are readily obtained by including more terms from the exponential series expansion for the off-diagonal terms.

The other key component of a Lie group method is the infinitesimal update map, which gives the algebra element to be 'exponentiated' to obtain the finite update. An *infinitesimal update map* for a left or right trivialized vector field $X\colon G \to \mathfrak{g}$ is a map $\Xi\colon G \times [0, \varepsilon) \to \mathfrak{g}$ satisfying the consistency condition $\Xi(\,\cdot\,, 0) = X$. Combining an infinitesimal update map $\Xi\colon G \times [0, \varepsilon) \to \mathfrak{g}$ and an algorithmic exponential Exp on some neighborhood $\mathcal{V}$ of 0 in $\mathfrak{g}$ yields the Lie group update map

$$U_{\Delta t}(g) := \operatorname{mult}_g\big(\operatorname{Exp}(\Delta t \Xi(g, \Delta t)\big) \tag{3.4}$$

for all g such that $\Delta t \Xi(g, \Delta t) \in \mathcal{V}$. Infinitesimal update maps can be modeled on traditional numerical methods, e.g. Runge–Kutta or Newmark methods, with possible correction terms to compensate for the effects of the algorithmic update. (See, e.g. [**11, 21**] and references therein.) We will not emphasize the design of infinitesimal updates here. Rather, we focus on the general advantages and disadvantages of our avatars of the pseudo-rigid body system as regards availability of efficient algorithmic exponentials and ease of enforcement of conservation of angular momentum.

We can implement a Lie group method for the full right trivialized pseudo-rigid body equations (2.3) using the true matrix exponential for $\mathrm{SL}(n)$ (in practice, some very high-precision numerical approximation of the exponential) and some infinitesimal update Ξ:

$$U_{\Delta t}(g, p) := \big(\operatorname{mult}_g\big(\exp(\xi_g)\big), p + \xi_p\big), \quad \text{where} \quad (\xi_g, \xi_p) = \Delta t \Xi(g, p, \Delta t). \tag{3.5}$$

However, the expense of the high-precision evaluation of the exponential is substantial for large n. With the exception of the symplectically reduced system (2.12), the pseudo-rigid body equations derived in §2 are all direct products of Lie groups and all of the groups involved, with the significant exception of $\mathrm{SL}(n)$ itself, possess convenient (algorithmic) exponentials: the Cayley transform (3.2) for $\mathrm{SO}(n)$, (3.3) for

$\mathrm{ST}_+(n)$, and the identity map for the vector spaces, which are additive groups. We can implement Lie group methods for the QR factored versions ((2.7), (2.8), (2.9)) of the pseudo-rigid body equations using these maps. For example, given algorithmic exponentials Exp_Q for $\mathrm{SO}(n)$ and Exp_R for $\mathrm{ST}_+(n)$ and a one step infinitesimal update map $\Xi\colon\ \mathrm{SO}(n)\times \mathrm{ST}_+(n)\times \mathrm{sym}_0(n)\times[0,\varepsilon)\to \mathrm{so}(n)\times \mathrm{st}_+(n)\times \mathrm{sym}_0(n)$ for the trivialized vector field (2.8), we can define the finite update map

$$U_{\Delta t}(q,r,s)=\big(q\,\mathrm{Exp}_Q(\xi_q),\mathrm{Exp}_R(\xi_r)r,s+\xi_s\big), \tag{3.6}$$

where $(\xi_q,\xi_r,\xi_s)=\Delta t\Xi(q,r,s,\Delta t)$.

Remark. In the physical case $n=3$, the Rodrigues formula provides a convenient closed form expression for the true exponential map of SO(3) (see, e.g., [**20**]); however, this map involves trigonometric functions, and hence may be slightly less efficient than the Cayley transform even in this low-dimensional setting. Unit quaternions have also been used in numerical simulations on SO(3).

Note that while centrally trivialized and momentum level set versions of the pseudo-rigid body equations are not defined in terms of the same variables as the full right-trivialized version (2.3), (approximate) trajectories of (2.7), (2.8), and (2.9) can easily be mapped onto (approximate) trajectories of the full right trivialized system using the maps

$$\begin{aligned}
&(q,r,\phi)\mapsto q\cdot(r,\phi)=(qr,q\phi q^T),\\
&(q,r,\sigma)\mapsto q\cdot\big(r,m(q)+\sigma\big)=(qr,m+q\sigma q^T), \quad \text{and}\\
&(q,r,s)\mapsto(qr,m+s).
\end{aligned}$$

The system (2.12) has a phase space that is *not* diffeomorphic to $\mathrm{SL}(n)\times\mathrm{sl}(n)$. The attitude matrix q is determined modulo the isotropy subgroup $\mathrm{SO}(n)_m$ by the equation $\mu=q^Tmq$. Given an (approximate) continuous trajectory $\big(r(t),m(t),\sigma(t)\big)$ of (2.12), we can 'reconstruct' an approximate trajectory of (2.3) by (approximately) solving the linear ODE

$$q^{-1}\dot q=\Omega=\mathcal{P}_{\mathrm{SO}}\big(\mathbb{L}_r(m+\sigma)\big)$$

for q, where $\mathcal{P}_{\mathrm{SO}}$ denotes projection onto the $\mathrm{so}(n)$ factor of $\mathrm{sl}(n)$. This process, known as reconstruction, is rarely carried out in practice. (See [**1**] for a detailed presentation of reduction and reconstruction.) However, as we shall discuss below, numerical implementation of reconstruction is closely related to the use of Lie group methods in integration of the system (2.12).

Angular momentum is conserved by any reasonable numerical scheme if the right trivialized equations (2.3) are used. The angular momentum function is simply projection onto the skew-symmetric component of the right-trivialized momentum; hence any momentum update consisting of addition of a symmetric matrix to the current momentum will preserve angular momentum. Since the angular momentum depends only on the momentum, not the configuration, in the right trivialization, our choice of configuration update does not effect angular momentum conservation. On the other hand, the angular momentum is a nonlinear function of both the configuration and momentum variables in the non-trivialized, left-trivialized, and centrally trivialized versions of the full equations, However, this limitation of the centered trivialization can be avoided by using the momentum level set equations (2.8) or (2.9), in which the angular momentum is eliminated from the dynamic variables and treated as a parameter. Since the configuration lies in the nonlinear group $\mathrm{SL}(n)$, while the momentum lies in the vector space $\mathrm{sl}(n)$, the right-trivialized equations and the momentum level set equations give us maximum flexibility where it is most needed. (See [**18, 19, 26, 27**] and references therein for other examples of momentum-preserving algorithms.)

One possible disadvantage of the level set equations (2.6) and (2.8) is the lack of a natural symplectic or Poisson structure on the momentum level set. If one wishes to use Hamiltonian schemes, this is obviously a serious limitation. (See, e.g. [**3, 6, 9, 18, 19**] and references therein for definitions and examples of symplectic numerical schemes.) The symplectically reduced phase space P_m is a homogeneous manifold, but not a Lie group; it is the direct product of the orbit $\mathcal{O}_m$ of m under the coadjoint action with the multiplicative group $\mathrm{ST}_+(n)$ and the additive group $\mathrm{sym}_0(n)$. Lie group methods can be applied to a homogeneous manifold M with a transitive G action by lifting the given vector field X on M to a family of vector fields Y_m on G that are mapped to X by the linearization of the group action, approximating the flow of Y_m using conventional Lie group methods, and then mapping the discrete trajectories in G back into M using the group action. (See, e.g. [**13, 14, 16, 23**].) Specifically, given $m \in M$, let $\widehat{\Phi}m \colon G \to M$ denote the map $\widehat{\Phi}m(g) := g \cdot m$. Let $Y \colon M \to \mathfrak{g}$ be a map satisfying $d\widehat{\Phi}mY(m) = X(m)$ for all $m \in M$ and define $Y_m(g) := Y(g \cdot m)$. Transitivity of the action guarantees that such a map Y exists. However, if some points have continuous isotropy, then the map Y is not unique. This nonuniqueness can be utilized in the construction of numerical integration schemes; see [**13–16**].) If, given a time step Δt, there is a neighborhood of the identity element e of G on which an update map $\widetilde{U}^m_{\Delta t}$ associated to Y_m is defined for any $m \in M$, then we can define the update map

$$U_{\Delta t}(m) := \widehat{\Phi}m\big(\widetilde{U}^m_{\Delta t}(e)\big)$$

for X on M.

We can apply Lie group methods as described above to the homogeneous manifold P_m, using, predictably, the coadjoint action on the coadjoint orbit factor $\mathcal{O}_m$. If $\widetilde{X}_{(h,m)}$ denotes the vector field for (2.9), then the induced vector fields $Y_{(r,\mu,\sigma)}$ satisfy

$$(Y_{(r,\mu,\sigma)} - \widetilde{X}_{(h,m)})(e, r, \sigma) \in \mathrm{so}(n)_\mu \times \{(0,0)\},$$

where $\mathrm{so}(n)_\mu = \{\xi \in \mathrm{so}(n) : [\xi, \mu] = 0\}$ denotes the isotropy algebra of μ. Given a family Ξ_μ of infinitesimal updates for the vector fields $Y_{(r,\mu,\sigma)}$, we obtain the update map

$$U_{\Delta t}(r, \mu, \sigma) = \big(\mathrm{Exp}_R(\xi_r)r, u^T\mu u, \sigma + \xi_\sigma\big), \quad \text{where} \quad u = \mathrm{Exp}_Q(\xi_q) \tag{3.7}$$

and $(\xi_q, \xi_r, \xi_\sigma) = \Delta t\Xi_\mu(q, r, \sigma, \Delta t)$. Note that while the centrally trivialized momentum level set system (2.9) determines a natural choice of lifted vector field, other choices are possible and may, in fact, yield superior approximations. If we use (2.9) to define our lifted vector field and track the 'side' variable q, rather than the reduced variable μ, choosing q_0 such that $\mu_0 = q_0^T m q_0$ and setting $q_{n+1} = q_n u_n$, we obtain the integration scheme

$$U_{\Delta t}(q, r, \sigma) = \big(q\,\mathrm{Exp}_Q(\xi_q), \mathrm{Exp}_R(\xi_r)r, \sigma + \xi_\sigma\big) \tag{3.8}$$

for (2.9), where $(\xi_q, \xi_r, \xi_\sigma) = \Delta t\Xi\big(r, m(q), \sigma, \Delta t\big)$. Thus, in this setting, the reconstruction procedure of Hamiltonian mechanics naturally arises in the implementation of a Lie group method on the symplectically reduced manifold. The interplay between reduction, symplectic integration, and reconstruction will be explored in future work.

References

1. R. Abraham and J. E. Marsden, *Foundations of mechanics*, 2nd ed., Benjamin–Cummings, Reading, MA, 1978.
2. S. Chandrasekhar, *Ellipsoidal figures of equilibrium*, Yale Univ. Press, New Haven, 1969.
3. P. J. Channell and C. Scovel, *Symplectic integration of Hamiltonian systems*, Nonlinearity **3** (1990), 231–259.
4. H. Cohen and R. G. Muncaster, *The theory of pseudo-rigid bodies*, Springer-Verlag, New York, 1988.
5. R. Dedekind, *Zusatz zu der vorstehenden Abhandlung*, J. Reine Angew. Math. **58** (1860), 217–228.
6. R. de Vogelaére, *Methods of integration which preserve the contact transformation property of the Hamiltonian equations*, Technical report, 1956.
7. G. L. Dirichlet, *Untersuchung über ein Problem der Hydrodynamik*, J. Reine Angew. Math. **58** (1860), 181–216.
8. F. Fasso and D. Lewis, *Stability properties of the Riemann ellipsoids*, Arch. Rational Mech. Anal. **158** (2001), 259–292.
9. J. M. Finn and A. J. Dragt, *Normal forms for mirror machine Hamiltonians*, J. Math. Phys. **20** (1979), 2649–2660.
10. G. H. Golub and C. F. Van Loan, *Matrix computations*, 2nd ed., The Johns Hopkins Univ. Press, Baltimore and London, 1989.
11. A. Iserles, H. Munthe–Kaas, S. Norsett, and A. Zanna, *Lie-group methods*, Acta Numerica **9** (2000), 215–365.
12. D. Lewis, *Conserving and approximately conserving algorithms*, Dynamics of Algorithms, IMA Vol. Math. Applications, vol. 118, Springer-Verlag, New York, 2000, pp. 31–54.
13. D. Lewis and N. Nigam, *Geometric integration on spheres and some interesting applications*, J. Comput. Appl. Math. **151** (2003), no. 1, 141–170.
14. ______, *A geometric integration algorithm with applications to micromagnetics*, Technical Report 1721, IMA Preprint Series, 2000.
15. D. Lewis, N. Nigam, and P. Olver, *Connections for general groups actions*, Commun. Contemp. Math. (to appear).
16. D. Lewis and P. Olver, *Geometric integration algorithms on homogeneous manifolds*, Found. Comput. Math. **2** (2002), 363–392.
17. D. Lewis and J. C. Simo, *Nonlinear stability of rotating pseudo-rigid bodies*, Proc. Roy. Soc. London Ser. A **427** (1990), 281–319.
18. ______, *Conserving algorithms for the dynamics of Hamiltonian systems on Lie groups*, J. Nonlinear Sci. **4** (1994), 253–299.
19. ______, *Conserving algorithms for the N-dimensional rigid body*, Fields Inst. Commun. **10** (1996), 121–139.
20. J. E. Marsden and T. S. Ratiu, *Introduction to mechanics and symmetry*, Springer-Verlag, New York, 1994.
21. H. Munthe–Kaas, *Runge-Kutta methods on Lie groups*, BIT **38** (1998), 92–111.
22. H. Z. Munthe–Kaas and B. Owren, *Computations in a free lie algebra*, R. Soc. Lond. Philos. Trans. Ser. A Math. Phys. Eng. Sci. **357** (1999), 957–981.
23. H. Z. Munthe–Kaas and A. Zanna, *Numerical integration of differential equations on homogeneous manifolds*, Foundations of Computational Mathematics (Rio de Janeiro, 1997), Springer, Berlin, 1997, pp. 305–315.
24. B. Owren and A. Marthinsen, *Integration methods based on canonical coordinates of the second kind*, Numer. Math. **87** (2001), 763–790.
25. B. Riemann, *Ein Beitrag zu den Untersuchungen über die Bewegung eines flüssigen gleichartigen Ellipsoides*, Abh. d. Königl. Gesell. der Wiss. zur Göttingen **9** (1860), 3–36.
26. J. C. Simo and D. D. Fox, *On a stress resultant geometrically exact shell model. I. Formulation and optimal parametrization*, Comput. Methods Appl. Mech Engrg. **72** (1989), 267–304.

27. J. C. Simo and L. Vu-Quoc, *On the dynamics in space of rods undergoing large motions—a geometrically exact approach*, Comput. Methods Appl. Mech. Engrg. **66** (1988), no. 2, 125–161.
28. ———, *The mechanics of an affinely-rigid body*, Internat. J. Theoret. Phys. **12** (1975), 271–296.
29. ———, *Affinely rigid body and Hamiltonian systems on* $\mathrm{GL}(n, R)$, Rep. Math. Phys. **26** (1988), 73–119.
30. A. Zanna and H. Z. Munthe-Kaas, *Generalized polar decompositions for the approximation of the matrix exponential*, SIAM J. Matrix Anal. Appl. **23** (2001)–(2002), 840–862.

MATHEMATICS DEPARTMENT, UNIVERSITY OF CALIFORNIA, SANTA CRUZ, SANTA CRUZ, CA 95064, USA.
E-mail address: lewis@math.ucsc.edu

Centre de Recherches Mathématiques
CRM Proceedings and Lecture Notes
Volume **39**, 2005

Towards a Variational Complex for the Finite Element Method

E. L. Mansfield and G. R. W. Quispel

ABSTRACT. Variational and symplectic integrators are now popular for mechanical systems, both because of their good long term stability and qualitative fit. Such integrators mimic or inherit the Lagrangian, respectively Hamiltonian, structure of the continuous model. A variational complex is a theoretical tool for the rigorous study of Lagrangian systems and their conservation laws. This article examines whether a formulation of a variational calculus for finite element methods, for an arbitrary finite element approximation scheme, is possible. The motivation is that this would allow a variational scheme to be written down for a given approximation model. Moreover, the stability and the conservation laws of such integrators could be studied without any need for individual, *ad hoc* arguments. A number of examples are considered, mainly one-dimensional, and conditions for a suitable complex derived.

1. Context and background

Variational principles are a central concept in the analysis of physically important models. The basic philosophy we take here is that numerical approximations which seek to mimic the underlying physics should have, of themselves, a variational structure to their formulation and analysis. There are several themes in classical and modern research that inform this article; the variational calculus, geometric integration, algebraic topology, and symbolic analysis. We consider these topics in turn.

1.1. Variational principles. The central tool in the mathematical analysis of variational principles is the variational calculus ([**32**, §§4.1, 5.4]). This calculus is also the correct setting for understanding the all-important theorem of Noether, which gives a conservation law for every variational symmetry.

A *variational calculus* is formulated in terms of a "locally exact differential complex". The most famous example of this concept is the grad–curl–div sequence

---

2000 *Mathematics Subject Classification.* Primary: 65M60.

ELM would like to thank the Institute for Advanced Studies at La Trobe University, Bundoora, Australia, for their support and hospitality; and Arieh Iserles for his encouragement. Her research is supported by EPSRC (UK) and the Leverhulme Trust. RQ's research is supported by the Australian Research Council.

This is the final form of the paper.

of operators. Exactness means that if one operator in the sequence sends an expression to zero, then that expression is in the image of the previous operator in the sequence, at least locally. So, if the curl of a vector field is zero, then locally it is the gradient of a function, and so forth. For this standard sequence, the extended sequence grad–curl–div–Euler–Lagrange–Helmholtz is still exact. This extended sequence is known as the *continuous variational complex.* With its attendant concepts and calculations, it answers questions such as,

(1) Is my equation an Euler–Lagrange equation[1], and if so, what is the Lagrangian?
(2) Is a given expression a total divergence, i.e. a conservation law?
(3) What are the conservation laws that arise from the symmetries of the Lagrangian?

It is now recognized that in order to obtain a stable numerical scheme, it is important to mimic this exactness.

> "In many contexts it is not enough that the numerical scheme be close to the original problem in a quantitative sense for it to inherit stability... the exactness properties of discrete differential complexes and their relation to differential complexes associated with the PDE are crucial tools in establishing the stability of numerical methods." [**3**]

Since exactness and variational principles are both important, it makes sense to investigate exact variational complexes for discretized systems. Such complexes are known for *difference* systems [**22, 25**]. Although there cannot be a "top down" analogue of the continuous variational complex, as there is no tangent structure on a discrete lattice, the difference variational calculus has theorems which are amazingly analogous to those of the continuous case.

1.2. Geometric integration. In recent years the area of numerical differential equations has undergone a quiet revolution called geometric integration [**17, 23, 27, 28**]. This arose from the confluence of two developments. The first is the increased awareness of the importance of geometric features of differential equations, not only symplectic structure, but volume-preservation, first integrals, symmetries, foliations and contact structure. The second development of course has been the exponential growth in computer speed over the last decades. Together these two developments have made geometric integration (the exact preservation of geometric features up to round-off error) both desirable and affordable.

> "The motivation for developing structure preserving algorithms for special classes of problems came independently from such different areas of research as astronomy, molecular dynamics, mechanics, theoretical physics and numerical analysis.... It turned out that the preservation of geometric properties of the flow not only produces an improved qualitative behaviour, but also allows for more accurate long-term integration than with general-purpose methods."[**17**]

[1]As the equation is precisely given. For equations only equivalent to an Euler–Lagrange equation, the problem is still open, although some particular cases have been solved [**1, 2, 7, 10, 11, 24**].

1.3. Numerical analysis and simplicial spaces. The influence of algebraic topology, in particular classical constructions of simplicial spaces, chains and cochains, boundary and coboundary operators, is of increasing interest to both physicists and numerical analysts [**5, 19, 31, 36, 37**]. The interplay of such notions with physical quantities and systems is increasingly being explored as a way to ensure that the correct geometry of a problem is encoded in the discretization. Indeed, some authors argue that the original models are already discrete, that is, formulated on the level of a small cell. The process of obtaining a continuum limit and then discretizing using an all-purpose numerical integrator is in fact *removing* the all-important geometry. Worse, this process can lead to a proliferation of "models" of the original, of no physical significance.

Just as one has a de Rham sequence for a smooth space which allows one to define and describe global features of gradient flows and conservation laws, so one has a matching simplicial complex for the triangulated space. The de Rham complex of a space captures the same global topological information of an orientable manifold, as the simplicial complex of its triangulation. Most importantly, one has a discrete Stokes' Theorem, allowing conservation laws to be exactly satisfied at the cellular level.

1.4. Can we use symbolic algebra to study numerical methods? Symbolic computation has been used with great success in obtaining symmetries and conservation laws of differential systems, and much progress has been made using these and other, strictly algebraic, techniques in on-line symbolic integrators. The potential use of symbolic computation to address the following questions is intriguing, and should lead to increased understanding of discrete systems in general.

- Can one design a numerical method, automatically, to inherit a variational principle and selected conservation laws?
- Can one obtain symmetries and hence conservation laws, automatically, of variational numerical methods?

These questions are beginning to be answered for finite difference models, (see [**22, 30**]). In particular, it is necessary to fully understand certain "no go" theorems, such as the Ge and Marsden result that symplecticity and energy can generally not both be conserved [**15**]. Their result holds when energy is the only integral; there is much that needs to be understood.

For 1-D problems, one can design schemes to inherit invariance under a given Lie symmetry group [**33**] although such schemes are necessarily high order for high dimensional groups. The theory of moving frames ([**12, 13, 29**]) will play a central role in any developments concerning the automatic derivation of such schemes for higher dimensional problems.

The computation of conservation laws for variational discrete systems ([**9, 16, 18, 21**]), with and without a discrete version of Noether's theorem, is also increasingly well understood. Even for smooth systems, however, the *efficient* computation of conserved densities is still being studied [**38**]. Meanwhile, the direct computation of symmetries for discrete systems has at least three different underlying philosophies, and hence theories (cf. [**20, 26, 34**]). Note that Noether's theorem can be applied only once you know the symmetries of the Lagrangian.

1.5. Road map of the article. In §2 we discuss the relationship between exact differential complexes and the Finite Element (FE) Method. We take as our starting point the article of D. Arnold [3], and discuss the relationship in terms of projection maps. In §3 we discuss the continuous variational complex and in particular, we address the question, what is the Euler–Lagrange (EL) operator? This discussion will be pivotal to motivating what follows. In §4 we look at the projections of Lagrangians as they are used in practice, and consider the question, What is the EL operator for an FE method? For one-dimensional problems, this turns out to be easily answered. However, the examples highlight what can go wrong, and in some cases, be fixed. This section concludes with a conjecture as to when a sensible EL system be obtained. In §5 we look at higher dimensional base spaces. Here, the EL operator depends on the mesh. In §6 we give a theoretical treatment of a variational complex for FE approximation schemes. This highlights the important problems still needing to to be addressed, which are summarized in §7.

2. Differential complexes and the finite element method

The fundamental differential complex is the de Rham complex, which for a three-dimensional base space is written as:

$$0 \to \mathbb{R} \to \Lambda^0 \xrightarrow{\mathrm{d}} \Lambda^1 \xrightarrow{\mathrm{d}} \Lambda^2 \xrightarrow{\mathrm{d}} \Lambda^3 \to 0. \tag{2.1}$$

The $\binom{3}{p}$-dimensional vector space Λ^p is the set of p-forms with coefficients being smooth functions in the co-ordinates of the three-dimensional base space (the independent variables). The linear operator d is the exterior derivative, which satisfies $\mathrm{d}^2 = 0$. One can think of Λ^p, $p > 0$, as being a vector field with $\binom{3}{p}$ components, and d as either the gradient, curl or divergence operator:

$$0 \to \mathbb{R} \to \mathcal{C}^\infty(X) \xrightarrow{\text{grad}} \mathcal{C}^\infty(X)^3 \xrightarrow{\text{curl}} \mathcal{C}^\infty(X)^3 \xrightarrow{\text{Div}} \mathcal{C}^\infty(X) \to 0,$$

but this is to rob the forms of much of their meaning. In the context of this article, a p-form is best thought of as the integrand of a p-dimensional integral. For example, inherent in the definition of the $\mathrm{d}x_i$, $\mathrm{d}x_i \wedge \mathrm{d}x_j$ and $\mathrm{d}x_i \wedge \mathrm{d}x_j \wedge \mathrm{d}x_k$ are the correct change of variable formulae learned in undergraduate calculus for such integrands ([6, p. 199]).

Remark 2.1. The precise correspondence between forms and vector fields or functions depends on the metric. Assuming the Euclidean metric, as we will throughout this article, the correspondence in 3-space is

$$\begin{aligned}
\Lambda^1 &: (f_1, f_2, f_3) \longleftrightarrow f_1\,\mathrm{d}x_1 + f_2\,\mathrm{d}x_2 + f_3\,\mathrm{d}x_3,\\
\Lambda^2 &: (f_1, f_2, f_3) \longleftrightarrow f_1\,\mathrm{d}x_2\mathrm{d}x_3 - f_2\,\mathrm{d}x_1\mathrm{d}x_3 + f_3\,\mathrm{d}x_1\mathrm{d}x_2,\\
\Lambda^3 &: f \qquad\qquad\;\; \longleftrightarrow f\mathrm{d}x_1\mathrm{d}x_2\mathrm{d}x_3 \equiv f\mathrm{d}\mathbf{x}.
\end{aligned}$$

In the following, we will use these interchangeably to ease the exposition.

Definition 2.2. Let X be a triangulated space, with the top-dimensional simplices denoted generically by τ. Let χ_τ be the characteristic function of τ. A *Finite Element approximation model* $\mathcal{F}_p$ of Λ^p is a set of p-forms, defined piecewise on the τ,

$$\mathcal{F}_p = \left\{ \sum_{\substack{\tau \\ i_1 < \cdots < i_p}} f_\tau^{i_1 \cdots i_p} \chi_\tau \,\mathrm{d}x_{i_1} \cdots \mathrm{d}x_{i_p} \;\middle|\; f_\tau^{i_1 \cdots i_p} : \tau \to \mathbb{R} \right\}$$

such that:

(1) $\mathcal{F}_p\big|_\tau$ is a finite dimensional real vector space. The coefficients of the basis elements on τ are called the **shape functions**, or function ansätze of the FE approximation, usually the same for all τ.

(2) The finite number of parameters on which each shape function $f_\tau^{i_1\cdots i_p}$ depends, can be obtained by specified linear functionals, one for each parameter. These functionals are called the **degrees of freedom** or **data** for the FE model.

(3) Each degree of freedom depends, in some sense, on a (sub)simplex of some dimension. Where a sub-simplex is shared by two top-level simplices in the triangulation, we assume the corresponding degrees of freedom for the shape functions in the different simplices agree.

Schemes which do not satisfy the third condition are said to be *non-conforming.* Strictly speaking, it is necessary to assign shared sub-simplices to just one τ in the sum above, in order to prevent double counting at those points.

Definition 2.3. The **projection map** $\Pi_p\colon \Lambda^p \to \mathcal{F}_p$ is the assignment to a given form, of a form with coefficients in the ansatz of the shape function, with those particular choices of the parameters such that the data of the function and its projection are equal.

Example 2.4. One of the simplest examples of an FE approximation space in 3-space is that of piecewise constant 3-forms. The projection operator from Λ^3 is:

$$\Pi_3(f\,\mathrm{d}\mathbf{x}) = \sum_\tau \alpha_\tau(f\,\mathrm{d}\mathbf{x})\chi_\tau\,\mathrm{d}\mathbf{x},$$

where

$$\alpha_\tau(f\,\mathrm{d}\mathbf{x}) = \frac{\int_\tau f\,\mathrm{d}\mathbf{x}}{\int_\tau \mathrm{d}\mathbf{x}},$$

that is, the normalized zeroth moment of the 3-form on τ.

Example 2.5. One of the simplest projection operators defined on $\Lambda^0 = \mathcal{C}^\infty$ is the usual piecewise linear interpolant defined by the values of the function at the vertices.

Following the discussion in [**3**], we make the following definition.

Definition 2.6. A **coherent** differential complex $\mathcal{F}^*$ of FE spaces, all relative to the same triangulation of the base space X, is defined as a set of choices of FE spaces $\mathcal{F}_i$ and projection maps Π_i such that the following diagram *commutes*,

$$\begin{array}{ccccccccccccc} 0 & \to & \mathbb{R} & \to & \Lambda^0 & \xrightarrow{\mathrm{d}} & \Lambda^1 & \xrightarrow{\mathrm{d}} & \Lambda^2 & \xrightarrow{\mathrm{d}} & \Lambda^3 & \to & 0 \\ & & & & \downarrow\Pi_0 & & \downarrow\Pi_1 & & \downarrow\Pi_2 & & \downarrow\Pi_3 & & \\ 0 & \to & \mathbb{R} & \to & \mathcal{F}_0 & \xrightarrow{\mathrm{d}} & \mathcal{F}_1 & \xrightarrow{\mathrm{d}} & \mathcal{F}_2 & \xrightarrow{\mathrm{d}} & \mathcal{F}_3 & \to & 0. \end{array} \tag{2.2}$$

We will motivate this definition at the end of §3.

Remark 2.7. In the literature, the spaces $\mathcal{F}_i$ are denoted as:

$$\mathcal{F}_0 = W_h, \quad \mathcal{F}_1 = Q_h, \quad \mathcal{F}_2 = S_h, \quad \mathcal{F}_3 = V_h,$$

where h is a real parameter indicating the "size" or "diameter" of the triangulation. Here we use an index to denote the dimension of the forms the space includes, for simplicity of the exposition.

Even though the respective projection operators are sometimes referred to as vertex, edge, face etc operators, it is not necessarily the case that the data for these spaces depends respectively only on function values on vertices, edges, faces and so on. It is also important to note that the functions in these spaces are defined piecewise on the top dimensional simplices. The exterior derivative in the lower sequence is usually taken to be defined only in the interior of the top dimensional simplices, because the forms are only piecewise smooth. When trying to prove theorems about such forms rigorously, concerning their local and global properties, the correct algebraic setting involves locally smooth forms together with an object which encodes how the mesh is put together. We will discuss this further in §§5,6. Using weak or distributional derivatives can lead to trouble for the applications in this article, see Example 4.2.

Example 2.8 (The zeroth order Raviart and Thomas scheme). Let τ be a 3-simplex with faces σ_i, edges e_i and vertices v_i. For a p-form ω defined on τ, the projection map takes ω to the piecewise linear p-form with the same zeroth moments on the p-dimensional sub-simplices of τ. So, letting a and b be constants, $\mathbf{a}$, $\mathbf{b}$ be constant vector fields, and writing the shape functions as vector fields,

Space	Dim	Shape	Data
$\mathcal{F}_0$	4	$a+\mathbf{b}\bullet\mathbf{x}$	$f \mapsto f(v_i)$
$\mathcal{F}_1$	6	$\mathbf{a}+\mathbf{b}\times\mathbf{x}$	$\omega \mapsto \int_{e_i}\omega$
$\mathcal{F}_2$	4	$\mathbf{a}+b\mathbf{x}$	$\omega \mapsto \int_{\sigma_i}\omega$
$\mathcal{F}_3$	1	a	$\omega \mapsto \int_{\tau}\omega$

where "Dim" denotes the dimension of the space per simplex τ. Stokes' theorem ([**6**, p. 212ff]),

$$\int_{\partial\rho}\omega = \int_{\rho}\mathrm{d}\omega,$$

where $\partial\rho$ is the boundary of the simplex ρ, implies that the diagram commutes. For example, if $f \in \Lambda^0 \equiv \mathcal{C}^\infty(X,\mathbb{R})$ and e is the edge running from vertex v_1 to v_2, then

$$\begin{aligned}\int_e \Pi_Q(\mathrm{d}f) = \int_e \mathrm{d}f = \int_{\partial e} f &= f(v_2) - f(v_1)\\ &= (\text{gradient of linear function with values}\\ &\qquad\qquad f(v_i) \text{ at } v_i)\cdot(v_2 - v_1)\\ &= \int_e \mathrm{d}\Pi_W(f).\end{aligned}$$

Since forms in the $\mathcal{F}_i$ with equal data are equal, $\Pi_Q(\mathrm{d}f) = \mathrm{d}\Pi_W(f)$.

Remark 2.9. The **de Rham** map is the pairing,

$$\langle\rho,\omega\rangle \mapsto \int_\rho \omega,$$

where ρ is a p-simplex, $\omega \in \Lambda^p$, extended linearly in both arguments. This map is used to show that the global topological information inherent in the de Rham and simplicial cohomology theories is equivalent for orientable manifolds ([**6**, p. 226]). In fact, the coherent FE differential complex given by the zeroth order Raviart and Thomas scheme, and the simplicial complex with real coefficients, are isomorphic.

By commutativity, that is, $\Pi_{p+1}\mathrm{d}(\omega) = \mathrm{d}\Pi_p(\omega)$, and Stokes' Theorem, the choice of data at the $\mathcal{F}_{p+1}$ level dictates to some extent the choice of data at the $\mathcal{F}_p$ level.

Example 2.10. Consider the one-dimensional problem,

$$\begin{array}{ccccccccc} 0 & \to & \mathbb{R} & \to & \Lambda^0 & \xrightarrow{\mathrm{d}} & \Lambda^1 & \to & 0 \\ & & & & \downarrow\Pi_0 & & \downarrow\Pi_1 & & \\ 0 & \to & \mathbb{R} & \to & \mathcal{F}_0 & \xrightarrow{\mathrm{d}} & \mathcal{F}_1 & \to & 0 \end{array}$$

where the projection Π_1 is to the piecewise constant 1-form

$$\omega\big|_e \mapsto \frac{\int_e \psi_e \omega}{\int_e \psi_e \,\mathrm{d}x}\,\mathrm{d}x$$

and the ψ_e are weight functions. Then since we require that the derivative of a 0-form in $\mathcal{F}_0$ lie in $\mathcal{F}_1$, the elements of $\mathcal{F}_0$ can be at most piecewise linear. Further, the projection map Π_0 on the edge e needs to involve the moment with respect to the weight function $\mathrm{d}\psi_e/\mathrm{d}x$, as well as the values of the 0-form at the vertices. Indeed, we have that

$$\begin{array}{ccc} u & \to & u_x\,\mathrm{d}x \\ \downarrow & & \downarrow \\ \sum_e (a_e x + b_e)\chi_e & \to & \sum_e a_e \chi_e \end{array}$$

where

$$a_e = \frac{\int_e \psi_e u_x \,\mathrm{d}x}{\int_e \psi_e \,\mathrm{d}x} = \frac{1}{\int_e \psi_e \,\mathrm{d}x}\left(\psi_e u|_{\partial e} - \int_e \psi_{e,x} u \,\mathrm{d}x\right).$$

The constants b_e are then chosen to make the elements in $\mathcal{F}_0$ piecewise continuous. While we have written the integrations to be over e, this is not necessary. There are examples where the intervals of integration for neighboring edges overlap to some small extent.

Example 2.11. Suppose the data for the projection map for Λ^p includes the first degree moments on the p-simplex σ. Let x denote a generic co-ordinate on X. Then since we have the identity

$$\int_\sigma x\mathrm{d}\omega = \int_\sigma \big(\mathrm{d}(x\omega) - \mathrm{d}x \wedge \omega\big) = \int_{\partial\sigma} x\omega - \int_\sigma \mathrm{d}x \wedge \omega$$

the data for Λ^{p-1} will involve the first order moments on the $p-1$ simplices, as well as the quantities $\int_\sigma \mathrm{d}x \wedge \omega$, which are the zeroth moments of the components of the form on p-dimensional simplices. One sees this in the data for the first order Raviart and Thomas scheme, where the shape functions are piecewise quadratic [**3**].

Theorem 2.12. *The de Rham complex, restricted to a top-level simplex, is* exact, *that is,*

$$\ker \mathrm{d}\big|_{\Lambda^p} = \operatorname{im} \mathrm{d}\big|_{\Lambda^{p-1}}. \tag{2.3}$$

This result is known as Poincaré's Lemma and is proved ([**6**, p. 224]) by producing a sequence of maps $h_p\colon \Lambda^p \to \Lambda^{p-1}$, such that

$$\mathrm{d}h_p + h_{p+1}\mathrm{d} = \mathrm{id}\big|_{\Lambda^p}.$$

The maps h_p are called *homotopy maps.* For then, if $\mathrm{d}\omega = 0$, we have $(\mathrm{d}h_p + h_{p+1}\mathrm{d})\omega = \mathrm{d}h_p\omega = \omega$. This shows that $\ker \mathrm{d} \subseteq \operatorname{im} \mathrm{d}$ while $\mathrm{d}^2 = 0$ shows the converse. In fact, whenever the base space X is diffeomorphic to the unit sphere, then the de Rham sequence on X is exact.

Theorem 2.13. *Let a coherent FE differential complex $\mathcal{F}^*$ be given, so that the diagram* (2.2) *commutes. If the upper sequence in* (2.2) *is exact, then so is the lower sequence;*

$$\ker \mathrm{d}\big|_{\mathcal{F}_p} = \operatorname{im} \mathrm{d}\big|_{\mathcal{F}_{p-1}}.$$

PROOF. For both upper and lower sequences we have that $\mathrm{d}^2 = 0$, so we need only produce homotopy operators for the lower sequence. If the homotopy operators for the upper sequence are h_i, then the homotopy operators for the lower sequence are $\Pi_i h_{i+1} \colon \mathcal{F}^{p+1} \to \mathcal{F}^p$, where we are using implicitly that $\mathcal{F}^p \subset \Lambda^p$ on every simplex. For

$$\begin{aligned}(\mathrm{d}\Pi_{i-1}h_i + \Pi_i h_{i+1}\,\mathrm{d})\omega &= \Pi_i(\mathrm{d}h_i + h_{i+1}\mathrm{d})\omega\\ &= \Pi_i\omega\\ &= \omega, \quad \text{as } \omega \in \mathcal{F}_i. \end{aligned}$$ □

When designing coherent FE differential complexes, the fact that they consist of finite dimensional vector spaces means that the following useful theorem applies:

Theorem 2.14. *If $\mathcal{F}^*$ is an exact complex, then* $\sum(-1)^n \dim(\mathcal{F}_n) = 0$.

Note that $\mathbb{R} \equiv \mathcal{F}_{-1}$ is included in this sum. Putting the above results together, we have that the alternating sum of the dimensions of the function spaces on each τ add to zero. Thus for the zeroth order Raviart and Thomas scheme, we have indeed that $1 - 4 + 6 - 4 + 1 = 0$. The theorem can be used to check that the correct amount of data at each level has been obtained. Of course, the converse is not true!

3. The variational complex for smooth forms

The variational complex is a theoretical tool for studying Lagrangian systems and their conservation laws. A detailed discussion with proofs can be found in Olver's text [**32**]. We define the basic objects for later reference, and consider in some depth the derivation of the continuous Euler–Lagrange operator. This is the map we wish to develop in the context of a FE approximation scheme, as a first step in asking whether a full variational complex for a given FE scheme exists.

The first half of the variational complex is similar to the de Rham complex, but with one major difference. The coefficient functions in Λ^p are now functions not just of the independent variables, but of the dependent variables and their derivatives. We assume that these functions are smooth functions with a finite number of arguments, but we do not specify any bounds in the degree of the derivatives. Moreover, the exterior derivative is taken to be a *total* exterior derivative, which we now denote by D. So, for example, we have

$$\mathrm{D}(yu_x\,\mathrm{d}x) = (yu_{xy} + u_x)\,\mathrm{d}y \wedge \mathrm{d}x + yu_{xz}\,\mathrm{d}z \wedge \mathrm{d}x.$$

Definition 3.1. Forms in the dependent variables and their derivatives, regarded as separate indeterminants, are known as *vertical* forms. The space of vertical p-forms is denoted by $\widehat{\Lambda}^p$. The vertical derivative $\widehat{\mathrm{d}}\colon \widehat{\Lambda}^p \to \widehat{\Lambda}^{p+1}$ operates in the natural way, but treats the independent variables as constants, for example,

$$\widehat{\mathrm{d}}\big((uu_x + 3xu_{xx})\mathrm{d}u\big) = u\,\mathrm{d}u_x \wedge \mathrm{d}u + 3\,\mathrm{d}u_{xx} \wedge \mathrm{d}u.$$

Theorem 3.2. [**32**, §5.4, p. 344] *The vertical complex*

$$\widehat{\Lambda}^0 \xrightarrow{\widehat{\mathrm{d}}} \widehat{\Lambda}^1 \xrightarrow{\widehat{\mathrm{d}}} \widehat{\Lambda}^2 \xrightarrow{\widehat{\mathrm{d}}} \cdots \tag{3.1}$$

is exact.

The second half of the variational complex are the vertical forms multiplied by the volume form on the base space, but with the "twist" that forms are defined only up to total divergence.

Definition 3.3. Let the total derivative with respect to the independent variable x_j, denoted by D_j, act linearly on the space of vertical forms, together with $\mathrm{D}_j(\omega\wedge\eta) = \mathrm{D}_j(\omega)\wedge\eta + \omega\wedge\mathrm{D}_j(\eta)$ and $\mathrm{D}_j\widehat{\mathrm{d}} = \widehat{\mathrm{d}}\mathrm{D}_j$. Then, for example, $\mathrm{D}_j\,\mathrm{d}u = \mathrm{d}u_{x_j}$, $\mathrm{D}_j\,\mathrm{d}x_k = 0$, and $\mathrm{D}_j(x_j\,\mathrm{d}u_{x_k}) = \mathrm{d}u_{x_k} + x_j\,\mathrm{d}u_{x_k x_j}$.

Definition 3.4. We say two vertical forms on an n-dimensional base space are equivalent if they differ by a total divergence, that is,

$$\omega \sim \eta \iff \omega - \eta = \mathrm{D}_1\zeta_1 + \cdots + \mathrm{D}_n\zeta_n. \tag{3.2}$$

The space of equivalence classes is the space of functional p-forms,

$$\Lambda^p_* = \widehat{\Lambda}^p/\mathrm{Div}(\widehat{\Lambda}^p)^n.$$

All elements of Λ^1_* on a one-dimensional base space can be represented as $f(x, u, u_x, \dots)\,\mathrm{d}u \wedge \mathrm{d}x$. For example,

$$xu_{xx}\,\mathrm{d}u_x = \mathrm{D}(xu_{xx}\,\mathrm{d}u) - \mathrm{D}(xu_{xx})\,\mathrm{d}u$$

and similarly for all $f\,\mathrm{d}u_{x\cdots x}$; note that $\mathrm{D}(\mathrm{d}u) = \mathrm{d}u_x$ etc. and $\mathrm{D}(\mathrm{d}x) = 0$.

Theorem 3.5. *The vertical derivative $\widehat{\mathrm{d}}$ is well defined on equivalence classes, and hence induces a map, the* variational derivative*,*

$$\delta\colon \Lambda^p_* \to \Lambda^{p+1}_*.$$

The variational complex in 3-space is written as

$$0 \to \mathbb{R} \to \Lambda^0 \xrightarrow{\mathrm{D}} \Lambda^1 \xrightarrow{\mathrm{D}} \Lambda^2 \xrightarrow{\mathrm{D}} \Lambda^3 \xrightarrow{\mathrm{E}} \Lambda^1_* \xrightarrow{\delta} \Lambda^2_* \xrightarrow{\delta} \cdots \tag{3.3}$$

where E is the Euler–Lagrange operator. The main theorem regarding this complex is the following.

Theorem 3.6 ([**32**, §5.4]). *If the base space is diffeomorphic to $\mathbb{R}^3$ then the complex* (3.3) *is exact.*

This means that an expression is precisely an Euler–Lagrange (EL) system if and only if the Helmholtz operator sends it to zero (it may be that a system is only equivalent to an EL system, in which case the test fails). An expression is precisely a total divergence if and only if the Euler–Lagrange operator sends it to zero. Moreover, there are operators called homotopy operators which allow you to answer questions such as, "if my system is an EL system, what is the Lagrangian?" and "if my system is a total divergence, what is it the divergence of?" Finally, the calculations and proofs for Noether's theorem are best given in the language of the variational complex.

The smooth Euler–Lagrange operator on Λ^3 is obtained in a two step process. First take the vertical exterior derivative, $\widehat{\mathrm{d}}$, and then take the projection π onto an equivalence class. For example, for the one-dimensional, degree 2 Lagrangian, $L = L(x, u, u_x, u_{xx})$,

$$\begin{aligned}\widehat{\mathrm{d}}L\,\mathrm{d}x &= \left(\frac{\partial L}{\partial u}\,\mathrm{d}u + \frac{\partial L}{\partial u_x}\,\mathrm{d}u_x + \frac{\partial L}{\partial u_{xx}}\,\mathrm{d}u_{xx}\right)\wedge \mathrm{d}x \\ &= \left(\frac{\partial L}{\partial u} - \frac{\mathrm{d}}{\mathrm{d}x}\frac{\partial L}{\partial u_x} + \frac{\mathrm{d}^2}{\mathrm{d}x^2}\frac{\partial L}{\partial u_{xx}}\right)\mathrm{d}u\wedge \mathrm{d}x + \mathrm{D}(A)\wedge \mathrm{d}x\end{aligned}$$

where

$$A = \left(\frac{\partial L}{\partial u_x} - \frac{\mathrm{d}}{\mathrm{d}x}\frac{\partial L}{\partial u_{xx}}\right)\mathrm{d}u + \frac{\partial L}{\partial u_{xx}}\,\mathrm{d}u_x$$

and then

$$\mathrm{E}(L) = \pi \circ \widehat{\mathrm{d}}(L\,\mathrm{d}x) = \left(\frac{\partial L}{\partial u} - \frac{\mathrm{d}}{\mathrm{d}x}\frac{\partial L}{\partial u_x} + \frac{\mathrm{d}^2}{\mathrm{d}x^2}\frac{\partial L}{\partial u_{xx}}\right)\mathrm{d}u\wedge \mathrm{d}x,$$

where $\pi(\omega) = \omega/\sim$, and $\omega \sim \psi \iff \omega - \psi = Dh$. The equivalence class of ω is written as $\int_X \omega$. Since by Stokes' Theorem we have

$$\int_X \mathrm{Div} h = \int_{\partial X} h,$$

the equivalence class can be thought of as "throwing away the boundary terms".

So the smooth Euler–Lagrange operator is the variational extremum for interior variations.

Important Remark 3.7. The motivation for studying coherent FE approximation schemes is that variational problems benefit from their use. This is shown in some detail by Arnold [**3**]. In 3-space, Lagrangians are 3-forms, which are made up of products of 0, 1, 2 and 3-forms. Arnold shows that provided the approximations of the components of a Lagrangian lie in the relevant FE approximation space of a *coherent* complex of such spaces, then the numerical approximation obtained will be stable; the exactness of a coherent complex implies that the functional analytic requirements of Brezzi's theorem, [**4**], for stability of mixed element schemes, will hold. The benefit of using a coherent scheme is that exactness is a relatively simple algebraic property, both to verify and to design.

4. Lagrangians and Euler–Lagrange equations for finite element schemes—the one-dimensional case

Comparing the complexes (2.2) and (3.3), the question, "Can we obtain a projection of the smooth variational complex to a Finite Element sequence and hence obtain similar or analogous results?" is a natural one. However, investigating the "top down" approach leads one to suspect that the answer is "no". First of all, for a wide range of projection operators,

$$\Pi(fg) \neq \Pi(f)\Pi(g), \quad \Pi(u_x) \neq \Pi(u)_x.$$

The left hand sides are how a "top down" version of a complex would project a Lagrangian. But the right hand sides are how a Lagrangian is projected in practice. Then, there is the problem of what is the meaning of the 100th derivative of a piecewise linear (quadratic, cubic...) approximation of u? Recall that there is no bound on the degree of differentiation in the arguments of Lagrangians. Finally, insisting on a commutative diagram leads quickly to strange looking results. For example, for the one-dimensional linear interpolation scheme, one obtains (using calculations that are explained in the sequel), that the projection of $\int 1\,\mathrm{d}u \wedge \mathrm{d}x$ must be $\sum \frac{1}{2}(x_{n+1}-x_n)\,\mathrm{d}u_n$.

On the other hand, finite element methods for variational problems are textbook material ([**8, 14**]). Hence the line of attack we take is to look at textbook examples using the mindset/technology/terminology of formal variational processes, and induce what the full analogue of the Finite Element variational complex should be.

In this section, we look at one-dimensional problems for simplicity. Hence, Lagrangians are 1-forms and the grid is a partition on $\mathbb{R}$. In the next section we will consider higher dimensional problems.

Example 4.1. Take the Lagrangian $L[u]=(u_x^2+2u)\,\mathrm{d}x$ which has the Euler–Lagrange equation,

$$\mathrm{E}(L)=\left(\frac{\partial}{\partial u}-\frac{\mathrm{d}}{\mathrm{d}x}\frac{\partial}{\partial u_x}\right)(L)=-2(u_{xx}-1).$$

The standard triangulation of $\mathbb{R}$ has vertices x_n and edges $e_n=(x_n,x_{n+1})$. The standard interpolation scheme with data $u_n=u(x_n)$ has projection operator

$$\begin{aligned}\Pi(u)|_{e_n}&=x\mapsto\frac{u_{n+1}-u_n}{x_{n+1}-x_n}x+\frac{u_nx_{n+1}-u_{n+1}x_n}{x_{n+1}-x_n}\\&=u^{e_n},\\\Pi(u)&=\sum u^{e_n}\chi_{e_n},\end{aligned}\tag{4.1}$$

where χ_A is the characteristic function for the set A.

The discrete Lagrangian is $\left(\Pi(u)_x^2-2\Pi(u)\right)\mathrm{d}x$ and hence

$$\mathcal{L}[u]=\int[\Pi(u)_x^2-2\Pi(u)]\,\mathrm{d}x=\sum_n\int_{e_n}\left((u^{e_n})_x^2-2u^{e_n}\right)\mathrm{d}x.$$

Performing the integrations yields

$$\mathcal{L}[u]=\sum_n\frac{(u_{n+1}-u_n)^2}{x_{n+1}-x_n}+(u_nx_{n+1}-u_{n+1}x_n)+(u_{n+1}x_{n+1}-u_nx_n).$$

The final group of terms telescopes and hence is in fact a boundary term, and ends by being discarded in the projection to the equivalence class.

Calculating the obvious analogue of the vertical derivative of $\mathcal{L}[u]$ yields

$$\sum_n\left(2\frac{u_{n+1}-u_n}{x_{n+1}-x_n}-x_n\right)\mathrm{d}u_{n+1}+\left(-2\frac{u_{n+1}-u_n}{x_{n+1}-x_n}+x_{n+1}\right)\mathrm{d}u_n.\tag{4.2}$$

Using the identity

$$\sum a_n\mathrm{d}u_{n+1}=\sum a_{n-1}\mathrm{d}u_n+\text{boundary terms}\tag{4.3}$$

yields the variational derivative, modulo boundary terms, of $\mathcal{L}[u]$ to be

$$\sum_n\left(2\frac{u_n-u_{n-1}}{x_n-x_{n-1}}-x_{n-1}-2\frac{u_{n+1}-u_n}{x_{n+1}-x_n}+x_{n+1}\right)\mathrm{d}u_n.$$

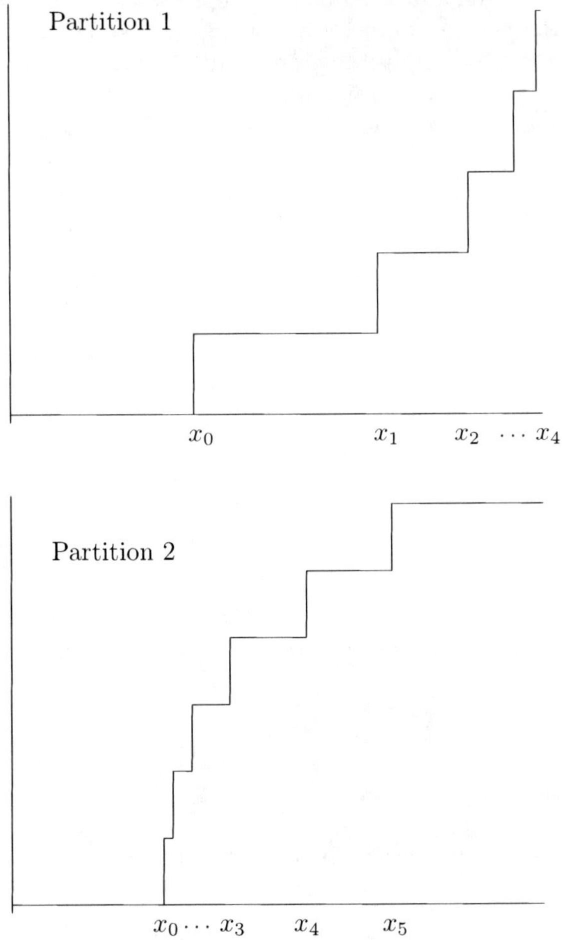

FIGURE 1. The solution to $\alpha_{n+1} - \alpha_n = \kappa \in \mathbb{R}$ for two different partitions.

Since each variation du_n is independent, each coefficient must be zero (at least in the function space ℓ_2), and this then yields the discrete Euler–Lagrange equation. Integrating twice gives

$$u_n = \tfrac{1}{2}x_n^2 + \kappa x_n + \left(u_0 - \tfrac{1}{2}x_0^2 - \kappa x_0\right)$$

which is the exact solution at each node.

Not all choices of Finite Element are suited to variational methods. We look now at two examples to show the problems that can occur. The first example shows that having too few degrees of freedom can lead to problems. We then show how the problem can be addressed.

Example 4.2. Consider the projection to piecewise constant functions,

$$\Pi(u) = \sum_n \alpha_n \chi_{e_n},$$

where

$$\alpha_n = \frac{1}{x_{n+1} - x_n} \int_{x_n}^{x_{n+1}} u \, \mathrm{d}x.$$

For $L[u] = \frac{1}{2}u_x^2 \, \mathrm{d}x$ we have $\Pi(u)_x = \sum(\alpha_n - \alpha_{n-1})\delta(x - x_n)$ taking the weak or distributional meaning of the derivative, and thus

$$\frac{1}{2}\int u_x^2 \, \mathrm{d}x \mapsto \frac{1}{2}\sum_n (\alpha_n - \alpha_{n-1})^2.$$

The weak derivative is needed to avoid a null discrete Lagrangian[2]. The variational derivative of the discrete Lagrangian with respect to the moments α is

$$\sum_n (\alpha_n - \alpha_{n-1})(\mathrm{d}\alpha_n - \mathrm{d}\alpha_{n-1}). \tag{4.4}$$

Equation (4.4) is equivalent to $\sum_n (2\alpha_n - \alpha_{n-1} - \alpha_{n+1})\mathrm{d}\alpha_n$, modulo boundary terms. The discrete Euler–Lagrange equation is then $2\alpha_n - \alpha_{n-1} - \alpha_{n+1} = 0$ or $\alpha_{n+1} - \alpha_n = \kappa$, $\kappa \in \mathbb{R}$.

Figure 1 shows the solution to this equation for two different partitions. By changing the partition, we can make the piecewise constant function α resemble a wide variety of functions! Note that α_n is the average value of u in $[x_n, x_{n+1}]$ and so the graph of α resembles that of u in some sense. The correct answer to the continuous Euler–Lagrange problem is $u(x) = ax + b$ for constants a and b. This has zeroth moments

$$\alpha_n = \frac{1}{x_{n+1} - x_n} \int_{x_n}^{x_{n+1}} (ax + b) \, \mathrm{d}x = \frac{1}{2}a(x_{n+1} + x_n) + b$$

and thus $\alpha_{n+1} - \alpha_n = \frac{1}{2}a(x_{n+2} - x_n)$. This is constant only for the regular partition, so that the discrete EL equation is correct only in that case.

This approximation scheme yields even worse problems for other Lagrangians, including unsolvable EL equations such as $1/\alpha_n = 0$!

By taking a piecewise linear approximation for a degree 1 Lagrangian, we avoid the need for a distributional or weak derivative. This can be achieved, still using the zeroth moments α_n and α_{n+1} for u, on (x_n, x_{n+1}) and (x_{n+1}, x_{n+2}) respectively, to create a piecewise linear approximation for u on (x_n, x_{n+2}). This yields

$$\Pi(u)_n = x \mapsto A_n x + B_n \tag{4.5}$$

where

$$\begin{aligned} A_n &= 2\left(\frac{\alpha_n - \alpha_{n+1}}{x_n - x_{n+2}}\right), \\ B_n &= \left(\frac{x_{n+1} + x_{n+2}}{x_{n+2} - x_n}\right)\alpha_n - \left(\frac{x_{n+1} + x_n}{x_{n+2} - x_n}\right)\alpha_{n+1}. \end{aligned} \tag{4.6}$$

Hence we get a piecewise linear approximation on the partition

$$\cdots x_{n-4},\ x_{n-2},\ x_n,\ x_{n+2},\ x_{n+4}, \cdots.$$

Doing the same calculations as before, the resulting EL equation "integrates" to

$$\frac{\alpha_n - \alpha_{n+1}}{x_n - x_{n+2}} = \kappa, \quad \kappa \in \mathbb{R}$$

---

[2]We note that products of such derivatives is problematic and the conditions we specify in the sequel will in fact avoid them.

which is correct.

This and other examples lead one to the conclusion that there must be at least as many data per top-level simplex as the order of the Euler–Lagrange equation, that is, double the order of the Lagrangian.

Unfortunately, this condition is still insufficient for success, as the next example shows.

Example 4.3. Suppose we take in each edge $e_n = [x_n, x_{n+1})$ the piecewise linear approximation

$$\Pi(u)\big|_{e_n} = x \mapsto u_n + u_n'(x - x_n)$$

constructed from the data

$$u_n = u(x_n), \qquad \alpha_n = \frac{1}{x_{n+1} - x_n} \int_{x_n}^{x_{n+1}} u\,\mathrm{d}x.$$

Then

$$u_n' = 2\frac{\alpha_n - u_n}{x_{n+1} + x_n}.$$

Substituting this into any first order Lagrangian and taking the variational derivatives, one for the moments u_n and one of the moments α_n, leads to two Euler–Lagrange equations in two unknowns; u_n and α_n. No recurrence relation is obtained, there is no one "solution" to the global problem depending on two constants of integration linking all the moments; in short, the result is nonsense. The problem is that the degrees of freedom in the one edge are unrelated, they are not shifts of each other in some sense.

In Examples 4.2 and 4.3, approximation schemes were given for functions, that is, for Λ^0, without any reference to what the approximation might be for Λ^1 that would be coherent with it, in the sense of Definition 2.6. When trying to ponder what these might be, it seems that in fact there isn't one! For example, using the zeroth moment for Λ^0 presupposes that the first degree moment is used for Λ^1. This in turn implies that not only $u(x_n)$ but $u(x_{n+1})$ is needed in the approximations of functions in each edge. This would then lead to the discrete Euler–Lagrange equations being recurrence relations of the correct order.

The improved approximation scheme in Example 4.2, given in Eqs: (4.5) and (4.6), does in fact form part of a coherent scheme. The projection of the 1-form $u(x)\,\mathrm{d}x$ is to the piecewise constant 1-form with the value

$$\int_{x_n}^{x_{n+2}} u(x)\psi_n(x)\,\mathrm{d}x,$$

where

$$\psi_n(x) = \begin{cases} \frac{2(x-x_n)}{(x_{n+2}-x_n)(x_{n+1}-x_n)}, & x_n < x < x_{n+1}, \\[2ex] \dfrac{2(x_{n+2}-x)}{(x_{n+2}-x_n)(x_{n+2}-x_{n+1})}, & x_{n+1} < x < x_{n+2} \end{cases}$$

in the interval (x_n, x_{n+2}).

Conjecture 4.4. Suppose that a one-dimensional Lagrangian of order m is given together with a coherent Finite Element differential complex, and that at least twice as many degrees of freedom are involved in the projected Lagrangian as the order of the Lagrangian. Then the Euler–Lagrange system is a recurrence system whose solution depends on $2m$ constants and converges to the solution of the continuous Euler–Lagrange equations in some well defined sense.

Our final example in this section shows that Noether's Theorem can be applied, at least where the group acts on the dependent variables. This takes advantage of the fact that the theorem is known for Lagrangian difference systems (cf. [**18, 22**]), and the clear analogy of these to the discrete EL equations obtained here. We show only a simple case.

Example 4.5. Consider the scaling invariant Lagrangian,

$$L[u] = \frac{1}{2}\left(\frac{u_x}{u}\right)^2 \, \mathrm{d}x.$$

Using the standard piecewise linear interpolation, we have

$$\int L[u]\, \mathrm{d}x \mapsto \sum_n L_n = \sum_n \frac{(u_{n+1}-u_n)^2}{2u_{n+1}u_n(x_{n+1}-x_n)}.$$

Now the discrete EL equation is

$$\mathrm{E}(L) = \partial_{u_n} L_n + S^{-1}\partial_{u_{n+1}} L_n, \tag{4.7}$$

where S is the shift map, $x_n \mapsto x_{n+1}$, $u_n \mapsto u_{n+1}$. The scaling symmetry $u \mapsto \mu u$ "translates" to the discrete Lagrangian as:

$$(u_n\partial_{u_n} + u_{n+1}\partial_{u_{n+1}})L_n = 0 \tag{4.8}$$

which can be checked directly. This last is by analogy with the characteristic form of the variational symmetry. Now, (4.7) and (4.8) together give a first integral of the EL equation:

$$u_n\partial_{u_n} L_n = \kappa, \qquad \kappa \in \mathbb{R}.$$

This yields

$$\left(\frac{u_{n+1}}{u_n}\right)^2 - \kappa(x_{n+1}-x_n)\left(\frac{u_{n+1}}{u_n}\right) - 1 = 0$$

or

$$u_{n+1} = H_n u_n, \qquad H_n \sim \left(\mp 1 + \tfrac{1}{2}\kappa(x_{n+1}-x_n)\right).$$

If $(x_{n+1}-x_n) \sim x/n$ this integrates to

$$u_n = \left(1 + \frac{\kappa x}{2n}\right)^n u_0 \sim u_0 \exp(\kappa x/2)$$

which is the correct result.

5. Higher dimensional calculations

The question, "what is the Euler–Lagrange operator for a Lagrangian defined on a mesh" comes down to the question, "which forms have integrals depending only on their boundary values?" Inside a simplex, such forms are, of course, a total divergence. But piecewise projected forms are not necessarily smooth across the boundaries of a simplex. For forms defined piecewise on a mesh, the analogue of a divergence is known as a "coboundary of a cochain". We next explain this concept.

While we restrict ourselves to the two-dimensional case for the sake of a simple exposition, the definitions and arguments are all adaptable to higher dimensions. We take a two-dimensional mesh to be a triangulation with vertices v, oriented edges e and oriented faces σ.

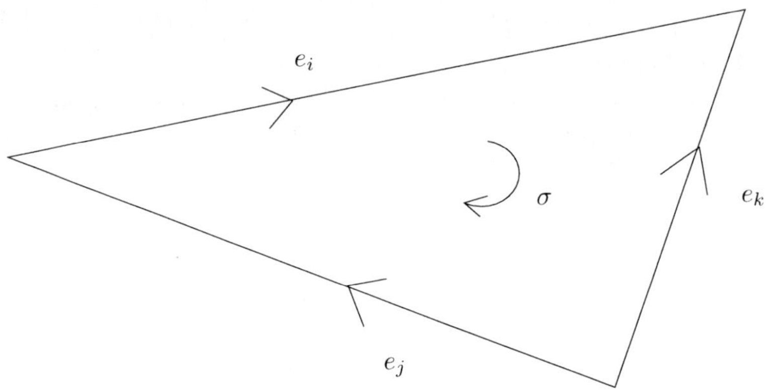

FIGURE 2. A 2-simplex and its boundary.

Definition 5.1. For any set S, let $\langle S\rangle_R$ denote the linear space with elements of S considered as a basis, and with coefficients in a ring R. We define *cochains* as follows:

- 0 cochains are linear maps $\langle\{v_i\}\rangle_R \to R$,
- 1 cochains are linear maps $\langle\{e_i\}\rangle_R \to R$,
- 2 cochains are linear maps $\langle\{\sigma_i\}\rangle_R \to R$.

The set of p-cochains will be denoted by $\mathcal{C}^p$. The ring R will be $\mathbb{R}$ at first, but will be extended later on.

Definition 5.2. Given the 1-cochain $F\colon \langle\{e_i\}\rangle_R \to R$, define the *coboundary* of F to be the 2-cochain $\delta F\colon \langle\{\sigma_i\}\rangle_R \to R$ by $\delta(F)(\sigma) = F(\partial\sigma)$ where $\partial\sigma$ is the sum of edges constituting the boundary of the face. We note that edges are signed in this sum according to whether their orientation matches that induced on the boundary by the orientation of the face. The map δ is then extended linearly to act on the space of 1-cochains, and is called the *coboundary operator*.

Example 5.3. For σ as in Figure 2, the coboundary map is $(\delta F)(\sigma) = F(e_i + e_j - e_k) = F(e_i) + F(e_j) - F(e_k)$.

If you are using an interpolation scheme, all the data lie on the vertices. The set of faces is then a set of ordered triples of indices and the set of edges is a set of ordered pairs of indices; see Figure 3. The ordering gives the orientation. The coboundary of a map F on the face (ijk) is easily seen to be

$$(\delta F)(ikj) = F(kj) + F(ji) + F(ik).$$

That is, coboundaries are cyclic sums on edges.

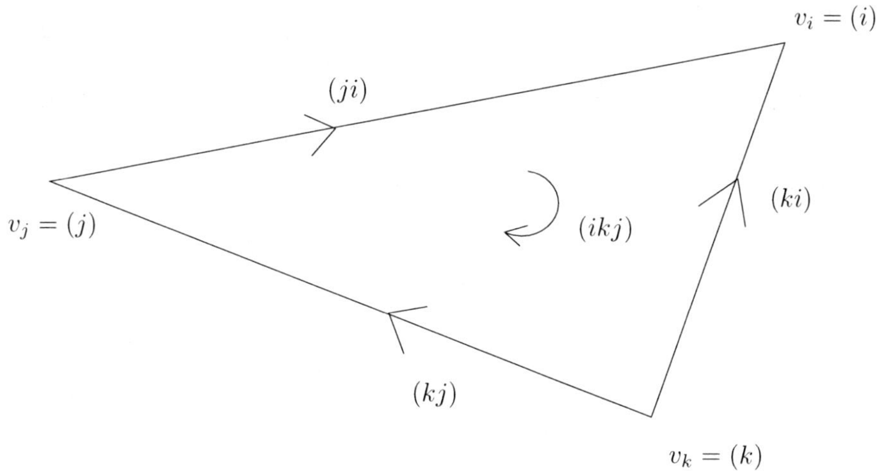

FIGURE 3. A 2-simplex and its sub-simplices labelled in ordered index notation.

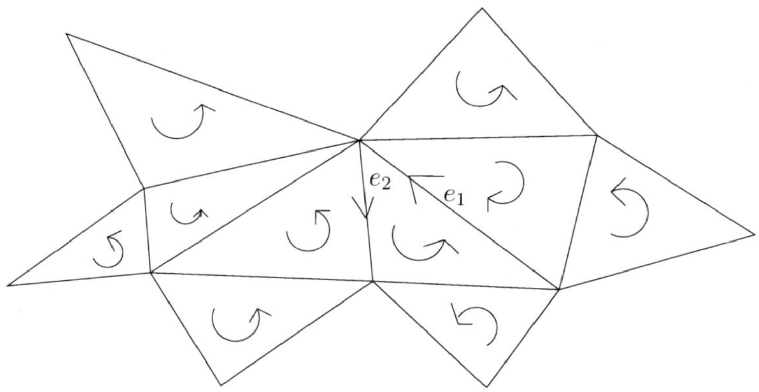

FIGURE 4. A mesh showing faces with their orientations. When integrating a coboundary, the sums over all the internal edges, for example e_1 and e_2, will cancel.

Definition 5.4. For a p-cochain F, and a p-surface Y composed of p-simplices ρ_i, $i = 1, \dots, N$, define the integral of F over Y as

$$\int_Y F = \sum_{k=1}^{N} (-1)^{[\rho_k, Y]} F(\rho_k),$$

where

$[\rho_k, Y] = 0$ if the orientation of ρ_k matches that induced by Y,
$[\rho_k, Y] = 1$ otherwise.

Theorem 5.5 (Discrete Stokes' Theorem). *If the dimension of X is n and F is an $(n-1)$-cochain, then*

$$\int_X \delta(F) = \int_{\partial X} F.$$

Indeed, we have that

$$\int_X \delta(F) := \sum_\tau (-1)^{[\tau,X]} \delta F(\tau) = \sum_{\sigma\in\partial X} (-1)^{[\sigma,X]} F(\sigma) =: \int_{\partial X} F.$$

For in the left hand sum, every internal edge will appear twice, with opposite signs, leaving only the sum over the external edges; see Figure 4. In fact, there is a version of the discrete Stokes' Theorem valid at every dimension, not just the top one (and with the same proof) but a proper discussion of this topic requires a knowledge of the Poincaré duality between homology and cohomology of a triangulation.

The astute reader will have noticed that coboundaries of 1-cochains derive from maps on edges, while coherent FE complexes have all forms defined throughout. By using the de Rham map defined in Remark 2.9, the two concepts can be "married" as follows.

Definition 5.6. Let $\omega \in \Lambda^p$, and let the p-simplices be denoted as σ. Define the $\mathbb{R}$-valued p-cochain $\boldsymbol{\omega}$, induced by ω, by

$$\boldsymbol{\omega}\colon \langle\{\sigma\}\rangle_{\mathbb{R}} \to \mathbb{R}, \quad \sum a_i\sigma_i \mapsto \sum a_i \int_{\sigma_i} \omega.$$

We will also denote $\boldsymbol{\omega}$ by $\int \omega$, and write

$$\int : \Lambda^p \to \mathcal{C}^p.$$

The map $\int$ is well-defined on $\mathcal{F}^n$ where n is the dimension of the base space X. Piecewise defined n-forms whose global integrals depend only on their boundary values are those whose induced cochains are coboundaries; $\boldsymbol{\omega} = \delta\boldsymbol{\eta}$. For then we have that

$$\int_X \omega = \int_X \boldsymbol{\omega} = \int_X \delta\boldsymbol{\eta} = \int_{\partial X} \boldsymbol{\eta}, \tag{5.1}$$

where $\int_X \omega$ is the standard integral from calculus, and $\int_X \boldsymbol{\omega}$ is given in Definition 5.4.

The fact that the forms in $\mathcal{F}^n$ are not smooth, and indeed may not even be defined on internal faces, means that globally, the smooth Stokes' Theorem does not hold, and we are forced to use the discrete version on the induced forms.

We are interested in the case where we have a Lagrangian depending on arbitrary functions and their derivatives. This projects to an FE Lagrangian depending on arbitrary, or unevaluated, degrees of freedom. For these forms, the above arguments all go through, but now the coefficients are functions of these unevaluated degrees of freedom; note that the degrees of freedom are labelled by the (sub)simplices of the mesh.

In the n-dimensional case where the n-form is a Lagrangian density, the projection of $\int_X L[u]\,\mathrm{d}\mathbf{x}$ used in practice is

$$\sum_{\tau}(-1)^{|\tau|}\int_{\tau}\Pi(L[u]\,\mathrm{d}\mathbf{x}), \tag{5.2}$$

where $|\tau| = 0$ if τ has the anti-clockwise orientation, and $|\tau| = 1$ for the reverse orientation. We thus take the induced n-cochain of $\Pi(L[u]\,\mathrm{d}\mathbf{x})$ to be

$$\tau \mapsto \int_{\tau}\Pi(L[u]\,\mathrm{d}\mathbf{x}).$$

In order to prove, rigorously, theorems concerning exactness and to obtain (eventually) formulae for conservation laws, it is helpful to introduce the following algebraic device to ease the notation. We first define a particular cochain which is reminiscent of the characteristic function of a simplex and plays a similar role.

Definition 5.7. Let the cochain which takes the value 1 on the simplex σ and zero elsewhere be called $\boldsymbol{\chi}_{\sigma}$. Then the set of p-cochains with real coefficients is $\mathcal{C}^p(\mathbb{R}) = \langle \boldsymbol{\chi}_{\sigma} \mid \sigma \in X^p\rangle_{\mathbb{R}}$, where we use X^p to denote the set of pth simplices in the triangulation. Changing the coefficients to R, where R is to be specified, we have

$$\mathcal{C}^p(R) \cong R \otimes \mathcal{C}^p(\mathbb{R}),$$

where $\otimes$ is the tensor product[3]. Then for example, the 0-cochain which takes the value a_n at x_n is written as $\sum a_n \otimes \boldsymbol{\chi}_{x_n}$ or more simply as $\sum a_n \boldsymbol{\chi}_{x_n}$, while the integral over X is then the signed sum of the coefficients, $\int_X \sum_{\tau} a_{\tau}([u]) \otimes \boldsymbol{\chi}_{\tau} = \sum_{\tau}(-1)^{[\tau,X]}a_{\tau}([u])$. The tensor notation gives explicitly the value of a cochain on each simplex in the form of a sum. This will enable us to write down the variational derivative of an induced cochain in the next section.

Note that the induced cochain of $\chi_{\tau}\mathrm{d}\mathbf{x}$, where χ_{τ} is the characteristic function for τ, is $(\int_{\tau}\mathrm{d}\mathbf{x}) \otimes \boldsymbol{\chi}_{\tau}$.

The simplicial theory of cochains and coboundaries is attractive. It allows us to use results and intuition from classical work on triangulations in algebraic topology, and it generalizes to n dimensions. However, using it to obtain closed form expressions for the EL equations defined on an arbitrary mesh seems difficult. Much better is to use the coboundary operator in matrix form, and to take advantage of the well-developed theory concerning the incidence matrices of meshes. This is a development for the future.

For regular meshes, it is simple to obtain a closed form expression for the EL equations. Collate triangles into a regular "square" lattice, for example, as 1, 2 or 3 in Figure 5. Since

$$\sum_{\text{triangles}} = \sum_{\text{squares}}$$

the boundary terms are then of the form

$$\sum(\mathrm{Shift}_1 - id)(\cdots) + (\mathrm{Shift}_2 - id)(\cdots).$$

[3]If V and W are real vector spaces of dimension n and m and with bases v_i and w_j respectively, then $V \otimes W$ is a real vector space of dimension nm and basis $v_i \otimes w_j$, satisfying $c(v \otimes w) = (cv) \otimes w = v \otimes (cw)$ for c a constant. In physics texts the symbol $\otimes$ is usually omitted in expressions of elements.

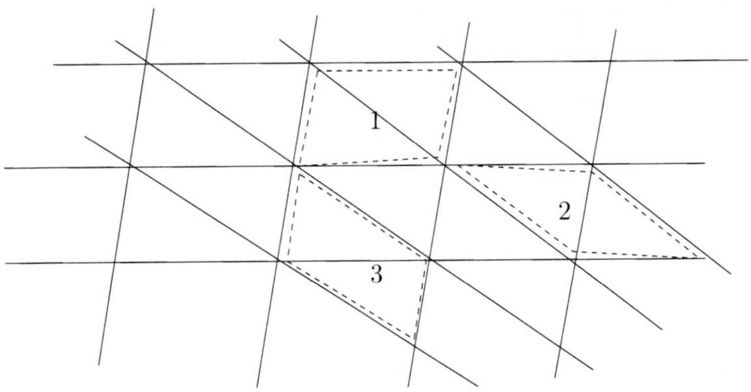

FIGURE 5. In a regular mesh, triangles can be collated into "squares".

This can be used in the same way as the identity (4.3) was used in the one-dimensional case. Euler–Lagrange equations then become difference systems, and can be studied using methods for those [**22**]. In particular, Noether's theorem is known for such systems (see references listed in §1.4).

6. A variational complex for coherent finite element schemes

We can now put together an FE variational complex, based on a coherent FE approximation scheme, which incorporates how the calculations are naturally done. But first, we look more closely at the algebraic structure of the de Rham complex, to motivate the construction of the complex we eventually study in Theorem 6.12.

The de Rham complex Λ^*, (2.1), is a *graded algebra* ([**6**, p. 197]). This means that forms can be added, multiplied by constants, and multiplied together, in this case using the wedge product;

$$\wedge\colon \Lambda^p \times \Lambda^q \to \Lambda^{p+q}, \quad (\alpha, \beta) \mapsto \alpha \wedge \beta.$$

The degree of the form gives the grading. Finally, the de Rham complex is a *differential graded algebra* since it is closed under the exterior derivative (i.e. the operator sends the set of all forms to itself). In addition, there is a metric dependent Hodge star operator, $*$, which sends, in n-space, p-forms to $(n-p)$-forms.

Lagrangians for physical problems in 3 dimensions are 3-forms which are composed of products of 0, 1, 2 and 3-forms. For example, $(u_x^2 + u_y^2 + u_z^2)\,\mathrm{d}\mathbf{x}$ is in fact $(*\mathrm{d}u) \wedge \mathrm{d}u$, and since $u \in \Lambda^0$, $\mathrm{d}u \in \Lambda^1$ and $*\mathrm{d}u \in \Lambda^2$, we have $(*\mathrm{d}u) \wedge \mathrm{d}u \in \Lambda^3$. A potential function $V(u)$ in a Lagrangian is in fact a three-form $*V(u) = V(u)\mathrm{d}\mu$, where $\mathrm{d}\mu$ is the volume form for the given metric. It can be seen that the construction of Lagrangians takes advantage of the differential graded algebra structure of Λ^*. Moreover, these 3-forms have the dependent variables and their derivatives appearing explicitly in the coefficient functions. One way to think of this is that the dependent variables are "unevaluated". And indeed when solving the EL equations, we regard them as variables to be determined.

On the other hand, a coherent FE complex is not an algebra, because the wedge product of elements in $\mathcal{F}^*$ need not be in $\mathcal{F}^*$. Looking at the discrete Lagrangians derived in §3, we see that in fact they are not in $\mathcal{F}^3$ as it stands, but in a larger space, the 3-forms in the algebra that $\mathcal{F}^*$ generates. Moreover, just as in the continuous case, we need the degrees of freedom to be unevaluated, so that we may speak of forms with coefficients that depend on arbitrary dependent variables and their derivatives. Such coefficient functions take the form, in the simplest one-dimensional case,

$$f\left(x, \Pi(u), \left(\Pi(u)\right)_x, \left(\Pi(u)\right)_{xx}, \dots\right),$$

where all such coefficients are smooth functions of a *finite* number of arguments.

Definition 6.1. For $\mathcal{F}^*$ be a coherent FE approximation complex, let $\widetilde{\mathcal{F}}^*$ denote the algebra generated by $\mathcal{F}^*$.

Theorem 6.2. *The algebra $\widetilde{\mathcal{F}}^*$ is a differential graded algebra. Moreover,*

$$0 \to \mathbb{R} \to \widetilde{\mathcal{F}}^0 \xrightarrow{\mathrm{d}} \widetilde{\mathcal{F}}^1 \xrightarrow{\mathrm{d}} \widetilde{\mathcal{F}}^2 \xrightarrow{\mathrm{d}} \widetilde{\mathcal{F}}^3 \xrightarrow{\mathrm{d}} \cdots \xrightarrow{\mathrm{d}} \widetilde{\mathcal{F}}^n \tag{6.1}$$

is exact in every n-dimensional simplex.

PROOF. Elements of $\widetilde{\mathcal{F}}^*$ can be written as *finite* sums and wedge products of elements in $\mathcal{F}^*$. Since the exterior derivative operator d is linear and an antiderivation, that is, obeys the product rule (up to signs), and $\mathcal{F}^*$ is closed under d, then so will $\widetilde{\mathcal{F}}^*$ be closed. Local exactness of (6.1) follows using the same homotopy maps as in the proof as the continuous case, but where instead of u, v, ... appearing in the coefficients, we have $\Pi(u)$, $\Pi(v)$, ... appearing. The homotopy operators H in the continuous case are maps which satisfy $\mathrm{id} = H\mathrm{d} + \mathrm{d}H$ for every specialization of every unevaluated dependent variable, and this is precisely what we need for the present result (see the discussion in [**32**, p. 343]). Finally, we note that the top-dimensional simplices satisfy the domain requirements for the continuous theorem to hold. □

Remark 6.3. The Hodge star operator also does not map $\mathcal{F}^*$ to itself and therefore needs to be modified. A first attempt might be to use $\Pi \circ *$, but this fails to satisfy the other properties of the star operator, such as being its own inverse (up to a sign). A better idea, propounded by several authors ([**31, 37**]) is that the discrete Hodge operator is defined in terms of a bilinear mapping between the mesh and its Poincaré dual. A full analogy of the variational complex involving these concepts is an important development for the future.

The second ingredient in a variational complex is the vertical complex (3.1), consisting of forms in the dependent variables. In our case, these dependent variables will be the unevaluated degrees of freedom in the approximation scheme. The exterior derivative of a Lagrangian with respect to the degrees of freedom in the model is a vertical form; for example (4.2) in Example 4.1, or (4.4) in Example 4.2.

In order to fix the notation, we will denote all the degrees of freedom by α_σ^j where σ is a simplex, and j being the labelling on the different degrees of freedom per simplex.

Definition 6.4. We first define vertical forms on a top-dimensional simplex, which is a form in the $\{\mathrm{d}\alpha^j_\sigma\}$, that is, a *finite* sum

$$\sum_{j_1<j_2<\cdots<j_p} f^{k_1\cdots k_p}_{j_1\cdots j_p}\,\mathrm{d}\alpha^{k_1}_{j_1}\ldots\mathrm{d}\alpha^{k_p}_{j_p},$$

where the coefficients f are smooth functions with a *finite* number of arguments, in the α^j_σ and the mesh variables. The space of such p-forms is denoted $\mathcal{V}^p$. The exterior derivative $\widehat{\mathrm{d}}\colon \mathcal{V}^p \to \mathcal{V}^{p+1}$ operates in the natural way, but treats any mesh variables (the "independent variables") as constants; for example,

$$\begin{aligned}\widehat{\mathrm{d}}[(x_n\alpha^{j-1}_n\alpha^j_n)^2\,\mathrm{d}\alpha^j_n] &= 2(x_n\alpha^{j-1}_n\alpha^j_n)(x_n\alpha^{j-1}_n\,\mathrm{d}\alpha^j_n + x_n\alpha^j_n\,\mathrm{d}\alpha^{j-1}_n)\wedge\mathrm{d}\alpha^j_n\\ &= 2x_n^2\alpha^{j-1}_n(\alpha^j_n)^2\,\mathrm{d}\alpha^{j-1}_n\wedge\mathrm{d}\alpha^j_n,\end{aligned}$$

since $(\mathrm{d}\alpha^j_n)^2 = 0$.

Definition 6.5. Globally, the vertical p-forms are obtained by taking a copy of $\mathcal{V}$ for each simplex, and composing the tensor product, $\widehat{\mathcal{F}}^p = \mathcal{V}^p \otimes \mathcal{C}^n$, where n is the dimension of the base space. Then the vertical derivative is given by

$$\widehat{\mathrm{d}}(v\otimes\boldsymbol{\omega}) = (\widehat{\mathrm{d}}v)\otimes\boldsymbol{\omega}.$$

In other words, global vertical forms are n-cochains with coefficients in $\mathcal{V}$.

Theorem 6.6. *The complex*

$$\widehat{\mathcal{F}}^0 \xrightarrow{\widehat{\mathrm{d}}} \widehat{\mathcal{F}}^1 \xrightarrow{\widehat{\mathrm{d}}} \widehat{\mathcal{F}}^2 \xrightarrow{\widehat{\mathrm{d}}} \widehat{\mathcal{F}}^3 \xrightarrow{\widehat{\mathrm{d}}} \cdots \tag{6.2}$$

is exact in every top-dimensional simplex.

The proof of this theorem is the same as for the corresponding smooth vertical complex ([**32**, p. 344]), since the cochain part of the expressions are unaffected by the vertical derivative operator.

The next step is to consider equivalence classes of vertical forms, where equivalence is up to forms whose integrals depend only on their boundary values. To do this, we use the induced action of the coboundary operator discussed in the previous section, but on the spaces $\widehat{\mathcal{F}}^p$.

Example 6.7. Let us look at the simplest 1−d case, using the new notation we are developing. Using the standard interpolation approximation given in Eq. (4.1), the projection of $\frac{1}{2}u_x^2\,\mathrm{d}x$ is

$$\sum_n \frac{1}{2}\left(\frac{u_{n+1}-u_n}{x_{n+1}-x_n}\right)^2 \chi_{e_n}\,\mathrm{d}x$$

and the induced cochain is

$$\boldsymbol{\omega} = \sum_n \frac{1}{2}\frac{(u_{n+1}-u_n)^2}{x_{n+1}-x_n}\otimes\boldsymbol{\chi}_{e_n},$$

where we have used the fact that the induced cochain of $\chi_{e_n}\,\mathrm{d}x$ is $(x_{n+1}-x_n)\boldsymbol{\chi}_{e_n}$ (see Definition 5.7). Then (omitting the $\otimes$ symbol)

$$\widehat{\mathrm{d}}\boldsymbol{\omega} = \sum_n \left(\frac{u_{n+1}-u_n}{x_{n+1}-x_n}\right)(\mathrm{d}u_{n+1}-\mathrm{d}u_n)\boldsymbol{\chi}_{e_n}.$$

Definition 6.8. Two vertical forms are said to be *equivalent* if their induced cochains differ by a coboundary. We write the equivalence class of a form ω as $\pi(\omega)$. The space of equivalence classes is the space of Finite Element *functional* p*-forms* (compare Definition 3.4),

$$\mathcal{F}^p_* = \widehat{\mathcal{F}}^p/\delta(\mathcal{V}^p \otimes \mathcal{C}^{n-1}).$$

Remark 6.9. Our notation follows that of the smooth case as given by Olver [**32**]. There, the de Rham complex was denoted by Λ^*, the vertical complex by $\widehat{\Lambda}^*$ and the complex of functional forms by Λ^*_*. Since Λ^* is an algebra, there is no need to introduce $\widetilde{\Lambda}^*$.

Example 6.7 (cont.). An arbitrary 0-cochain is $\boldsymbol{\eta} = \sum a_n \boldsymbol{\chi}_{x_n}$ and its coboundary is $\delta\boldsymbol{\eta} = \sum a_n \delta\boldsymbol{\chi}_{x_n}$. Now $\delta\boldsymbol{\chi}_{x_n} = \boldsymbol{\chi}_{e_n} - \boldsymbol{\chi}_{e_{n+1}}$ if x_n is an internal vertex. It is not hard to see that

$$\delta\boldsymbol{\eta} = \sum_n (a_{n+1} - a_n)\boldsymbol{\chi}_{e_n}.$$

To obtain $\pi\widehat{\mathrm{d}}\boldsymbol{\omega}$, we can add whatever coboundary we like—the point is to find a good coboundary to obtain the most useful representative of the vertical derivative. So, if we add the coboundary of

$$\boldsymbol{\eta} = -\sum_n \left(\frac{u_{n+1} - u_n}{x_{n+1} - x_n}\right) \mathrm{d}u_n \boldsymbol{\chi}_{x_n},$$

we will get that

$$\pi\widehat{\mathrm{d}}\boldsymbol{\omega} \sim -\sum_n \left(\frac{u_{n+1} - u_n}{x_{n+1} - x_n} - \frac{u_n - u_{n-1}}{x_n - x_{n-1}}\right) \mathrm{d}u_n \boldsymbol{\chi}_{e_n}.$$

Despite the abstraction of "equivalence relations", the operations we perform are all those occurring naturally in practical problems.

Theorem 6.10. *The vertical exterior derivative is well defined on equivalence classes, that is, if* $\omega_1 \sim \omega_2$ *then* $\widehat{\mathrm{d}}\omega_1 \sim \widehat{\mathrm{d}}\omega_2$. *Thus* $\widehat{\mathrm{d}}$ *induces a map* $\mathrm{d}^* : \mathcal{F}^p_* \to \mathcal{F}^{p+1}_*$ *which satisfies* $(\mathrm{d}^*)^2 = 0$ *and* $\pi \circ \widehat{\mathrm{d}} = \mathrm{d}^* \circ \pi$.

Definition 6.11. The Euler–Lagrange operator for a coherent FE approximation scheme is

$$\mathcal{E}L = \pi \circ \widehat{\mathrm{d}} \circ \int.$$

Recall that the map $\int$ takes a form to its induced cochain (Definition 5.6).

Theorem 6.12. *The Finite Element variational complex, shown here for a three-dimensional base space,*

$$0 \to \mathbb{R} \to \widetilde{\mathcal{F}}_0 \xrightarrow{\mathrm{d}} \widetilde{\mathcal{F}}_1 \xrightarrow{\mathrm{d}} \widetilde{\mathcal{F}}_2 \xrightarrow{\mathrm{d}} \widetilde{\mathcal{F}}_3 \xrightarrow{\mathcal{E}L} \mathcal{F}^1_* \xrightarrow{\mathrm{d}^*} \mathcal{F}^2_* \xrightarrow{\mathrm{d}^*} \cdots \tag{6.3}$$

is locally exact.

PROOF. It suffices to show that the following diagram commutes, and the bottom row is exact. The result then follows from standard algebraic arguments (see for example [**22**, §4.1]).

$$\begin{array}{ccccccccc} & \xrightarrow{\mathrm{d}} & \widetilde{\mathcal{F}}^3 & \xrightarrow{\widehat{\mathrm{d}}\circ\int} & \widehat{\mathcal{F}}^1 & \xrightarrow{\widehat{\mathrm{d}}} & \widehat{\mathcal{F}}^2 & \xrightarrow{\widehat{\mathrm{d}}} & \cdots \\ & & \big\downarrow \pi\circ\int & & \big\downarrow \pi & & \big\downarrow \pi & & \\ 0 & \longrightarrow & \mathcal{F}^0_* & \xrightarrow{\mathrm{d}^*} & \mathcal{F}^1_* & \xrightarrow{\mathrm{d}^*} & \mathcal{F}^2_* & \xrightarrow{\mathrm{d}^*} & \cdots \end{array}$$

That the diagram commutes follows from Theorem 6.10. That the bottom row is exact to the right of $\mathcal{F}^0_*$ onwards, follows from the fact that it is a projection of an exact sequence, (6.2). Thus we need only show exactness at $\mathcal{F}^0_*$. In other words, we need only show that $\mathrm{d}^*\big|_{\mathcal{F}^0_*}$ is an injection. This follows firstly from the fact that any function which maps to zero under the vertical derivative must involve only the independent variables, and is therefore locally a divergence. Secondly, any cochain mapping to a coboundary under the vertical derivative is already a coboundary. □

7. Discussion and open problems

In this article, we have laid the foundation for a rigorous analogue of the variational complex for Finite Element schemes. In order for this to be of practical benefit in the design and analysis of schemes that inherit both a variational structure and specified conservation laws, the following open problems and questions need to be addressed.

(1) Incorporate the Poincaré dual mesh model of the Hodge star operator.
(2) Develop a computationally useful notion of the EL equations in terms of the incidence matrix of the mesh.
(3) Derive formulae for discrete Noether's theorem relative to a given mesh.

References

1. I. M. Anderson and T. E. Duchamp, *Variational-principles for 2nd-order quasi-linear scalar equations*, J. Differential Equations **51** (1984), 1–47.
2. I. M. Anderson and G. Thompson, *The inverse problem of the calculus of variations for ordinary differential-equations*, Mem. Amer. Math. Soc. **98** (1992), no. 473.
3. D. Arnold, *Differential complexes and numerical stability*, Proc. of the International Congress of Mathematicians (Beijing, 2002), vol. I, Higher Ed. Press, Beijing, 2002, pp. 137–157, `http://www.ima.umn.edu/~arnold/papers/icm2002.pdf`.
4. F. Brezzi, *On the existence, uniqueness and approximation of saddle point problems arising from Lagrange multipliers*, RAIRO Anal. Numer. **8** (1974), 129–151.
5. J. A. Chard and V. Shapiro, *A multivector data structure for differential forms and equations*, Math. Comput. Simulation **54** (2000), 33–64.
6. Y. Choquet–Bruhat, C. DeWitt–Morette, and M. Dillard–Bleick, *Analysis, manifolds and physics*, North-Holland, Amsterdam, 1982.
7. M. Crampin, G. E. Prince, W. Sarlet, and G. Thompson, *The inverse problem of the calculus of variations: Separable systems*, Acta Appl. Math. **57** (1999), 239–254.
8. A. J. Davies, *The finite element method: A first approach*, Clarendon Press, Oxford, 1980.
9. V. Dorodnitsyn, *Noether-type theorems for difference equations*, Appl. Numer. Math. **39** (2001), 307–321.
10. J. Douglas, *Solution of the inverse problem of the calculus of variations*, Trans. Amer. Math. Soc. **50** (1941), 71–128.
11. M. Fels, *The inverse problem for fourth order scalar differential equations*, Trans. Amer. Math. Soc. **348** (1996), 5007-5029.
12. M. Fels and P. J. Olver, *Moving Coframes.* I, Acta Appl. Math. **51** (1998), 161–213.
13. ———, *Moving coframes.* II, Acta Appl. Math. **55** (1999), 127–208.
14. R. H. Gallagher, *Finite element analysis: Fundamentals*, Prentice-Hall Inc., New Jersey, 1975.

15. Ge Zhong and J. E. Marsden, *Lie–Poisson Hamilton–Jacobi theory and Lie–Poisson integrators*, Phys. Lett. A **133** (1988), 134–139.
16. Ü. Göktaş and W. Hereman, *Computation of conserved densities for nonlinear lattices*, Physica D **123** 425–436 (1998).
17. E. Hairer, C. Lubich, and G. Wanner, *Geometric numerical integration*, Structure-preserving Algorithms for Ordinary Differential Equations, Springer Ser. Comput. Math., vol. 31. Springer-Verlag, Berlin, 2002.
18. W. Hereman, J. A. Sanders, J. Sayers, and J. P. Wang, *Symbolic computation of conserved densities, generalized symmetries, and recursion operators for nonlinear differential-difference systems*, this volume.
19. R. Hiptmair, *Finite elements in computational electromagnetism*, Acta Numer. **11** (2002), 237–340.
20. P. E. Hydon, *Symmetries and first integrals of ordinary difference equations*, Proc. Roy. Soc. A **456** (2000), 2835–2855.
21. ———, *Conservation laws of partial difference equations with two independent variables*, J. Phys. A **34** (2001), 10347–10355.
22. P. E. Hydon and E. L. Mansfield, *On a calculus of variations for difference systems*, Found. Comput. Math. (to appear).
23. A. Iserles, H. Munthe–Kaas, S Norsett, and A. Zanna, *Lie group methods*, Acta Numerica **9** (2000), 215-365.
24. M. Juras, *The inverse problem of the calculus of variations for sixth- and eighth-order scalar ordinary differential equations*, Acta Appl. Math. **66** (2001), 25–39.
25. B. Kuperschmidt, *Discrete Lax equations and differential-difference calculus*, Astérisque **123** (1985).
26. D. Levi, S. Tremblay, and P. Winternitz, *Lie symmetries of multidimensional difference equations*, J.Phys. A **34** (2001), 9507–9524.
27. R. I. McLachlan and G. R. W. Quispel, *Six lectures on the geometric integration of* ODEs, Foundations of Computational Mathematics (R. A. DeVore et al, eds.), Cambridge Univ. Press, Cambridge, 2001, pp. 155-210.
28. ———, *Splitting methods*, Acta Numer. **11** (2002), 341–434.
29. E. L. Mansfield, *Moving frames and differential algebra*, Differential Algebra and Related Topics, (Li Guo, P. J. Cassidy, W. F. Keigher, and W. Y. Sit, eds.), World Scientific Press, Singapore, 2002, pp. 257–279.
30. E. L. Mansfield and P. E. Hydon, *Towards approximations which preserve integrals*, Proc. ISSAC (B. Mourrain, ed.), ACM Publishing, 2001, pp. 217–222.
31. C. Mattiussi, *An analysis of finite volume, finite element, and finite difference methods using some concepts from algebraic topology*, J. Comput. Phys **133** (1997), 289–309.
32. P. J. Olver, *Applications of Lie groups to differential equations*, 2nd ed., Springer Verlag, New York, 1993.
33. ———, *Geometric foundations of numerical algorithms and symmetry*, Appl. Algebra Engrg. Comm. Comput. **11** (2001), 417–436.
34. G. R. W. Quispel and R. Sahadevan, *Lie symmetries and the integration of difference equations*, Phys. Lett. A. **184** (1993), 64–70.
35. P. A. Raviart and J. M. Thomas, *A mixed finite element method for 2nd order elliptic problems*, Mathematical aspects of Finite Element Methods, Proc. Conf., Consiglio Naz. delle Ricerche (C.N.R.) (Rome, 1975), Lecture Notes in Math., vol. 606, Springer, Berlin, 1977, pp. 292–315.
36. W. Schwalm, B. Moritz, M. Giona, and M. Schwalm, *Vector difference calculus for physical lattice models*, Phys. Rev. E **59** (1999), 1217–1233.
37. E. Tonti, *On the formal structure of physical theories*, Istituto di Matematica del Politecnico di Milano, Milan, 1975; `http://www.dic.univ.trieste.it/perspage/tonti/papers.htm`
38. T. Wolf, *A comparison of four approaches to the calculation of conservation laws*, European J. Appl. Math. **13** (2002), 129–152.

Institute of Mathematics, Statistics and Actuarial Science, University of Kent, Canterbury, CT2 7NF, U.K.
E-mail address: E.L.Mansfield@kent.ac.uk

Department of Mathematics, and The ARC Centre of Excellence for Mathematics and Statistics of Complex Systems, La Trobe University, Victoria 3086, Australia.
E-mail address: R.Quispel@latrobe.edu.au

Centre de Recherches Mathématiques
CRM Proceedings and Lecture Notes
Volume **39**, 2005

Models of Resonantly Driven Motion of Motor Proteins in 2D Potentials

J. Middleton and J. A. Tuszynski

ABSTRACT. This paper discusses the mathematical description of motor protein motion along protein filaments. We have included the presence of 2D potentials in the stochastic mechanism of motor protein propagation. We derive a Fokker–Planck equation for this case and show its qualitative and quantitative agreement with experimental data.

1. Introduction

Motor proteins are ubiquitous in the cell and are essential for a wide range of cellular functions, mainly transport and force generation. They come in different forms but share a common structural characteristic which is a catalytic core that is able to bind and hydrolyze ATP to release the chemical energy stored in it to be used for mechanical work [**8**]. Motor proteins carry out their functions in the cell through their association with protein polymer filaments which form a matrix of structural support throughout the cellular cytoplasm, undergoing significant rearrangement during different stages of the cell's life cycle. This matrix is termed the cytoskeleton. The constituent protein filaments of the cytoskeleton are microtubules (MT's), actin filaments, and intermediate filaments.

Two of the principal motor proteins that attach to MTs are kinesin and dynein. While kinesin moves towards the plus end of a MT, dynein is negative end directed. Each of these proteins consists of a globular head region and an extended coiled-coil tail section as shown schematically in Fig. 1. The study of motor proteins has shown that the essential components for force generation are located within the globular head. Models of motor protein movement can essentially be divided into those with diffusion and those with a power stroke [**12**]. The diffusion models require an oscillating potential that is presumably driven by a conformation change of the motor-MT bond. Diffusion occurs in a flat potential state and once the potential reverts to its asymmetric form, its geometry is such that forward propagation of the motor protein is favored [**1, 2, 6**]. A review of such schemes may be

2000 *Mathematics Subject Classification.* Primary: 54C40, 14E20; Secondary: 46E25, 20C20.
Support information for the second author.
This research was supported by grants from NSERC and the MITACS-MMPD project.
This is the final form of the paper.

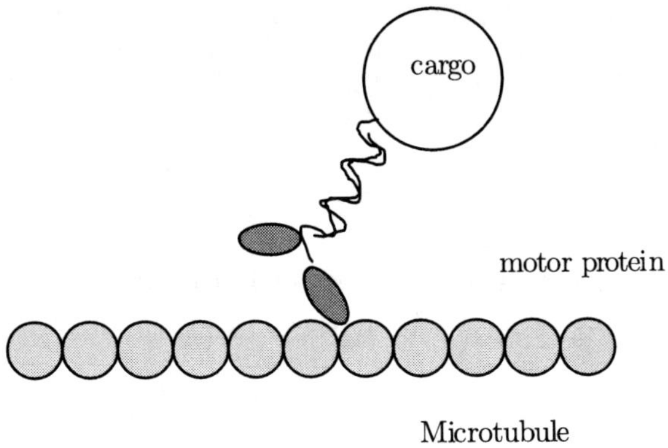

FIGURE 1. A motor protein is shown 'walking' along a MT protofilament.

found in Ref. [**9**]. In the power stroke models by contrast, it is the motor protein whose structure changes. In such models one imagines the protein to stretch and bind at a second location before relaxing to its original conformation when the 'back leg' releases its grip on the MT to start the process anew once additional ATP has arrived. The motor protein has two or more distinct states where at least one conformational change occurs and is driven by ATP hydrolysis as has been experimentally demonstrated for myosin [**15**]. Phosphorylation of the motor protein may lead to subsequent conformational changes. Thus the protein may be viewed as walking along the MT powered by ATP. Recent attempts at modeling the motor protein behavior use isothermal ratchets in which the motor is subjected to an external potential that is periodic and asymmetric. In addition, a fluctuating stochastic force due to thermal oscillations $F(t)$ is acting on the motor. A review article on this topic [**9**] distinguishes three subclasses of such models, namely:

(i) A Langevin-based approach [**4**] with the equation of motion given by

$$\xi \frac{dx}{dt} = -\partial_x W(x) + F(t) \tag{1.1}$$

where x is the direction along the protofilament axis, ξ is the friction coefficient, $W(x)$ is the potential due to the fiber and $F(t)$ is the fluctuating force whose time average is zero, i.e. $\langle F(t)\rangle = 0$. This fluctuating force is typically related to the kinetics of ATP binding to the motor protein and a subsequent hydrolysis of ATP into ADP that provides the excess energy allowing the protein to unbind from the filament.

(ii) An approach in which the potential fluctuates in time [**1**]

$$\xi \frac{dx}{dt} = -\partial_x W(x,t) + f(t) \tag{1.2}$$

where the Gaussian (white) noise term $f(t)$ satisfies the fluctuation-dissipation theorem.

(iii) A generalized model with several internal states of the particle which is described by the Langevin equation that depends on the state $i = 1, \ldots, N$, i.e.

$$\xi_i \frac{dx}{dt} = -\partial_x W_i(x) + f_i(t). \tag{1.3}$$

However, a more convenient and elegant approach is through the Fokker–Planck formalism [14] for the probability distribution function $P(x,t)$ that describes the motion of the motor in a statistical manner. Here, $D^{(1)}$ is the drift coefficient and $D^{(2)}$ the diffusion coefficient.

$$\frac{\partial P}{\partial t} = \left[-\frac{\partial}{\partial x} D^{(1)}(x) + \frac{\partial^2}{\partial x^2} D^{(2)}(x) \right] P(x,t) \tag{1.4}$$

So far, models of the above types have been fairly successful in representing the gross features of the experimental data obtained [16] but many questions still remain unanswered. It is worth mentioning that an entirely different group of models involves the formalism of chemical kinetics [10].

2. A 2-D potential

The main focus of this paper is placed on the Fokker–Planck formalism of Brownian motion. We propose a mechanism that presents new behavior in molecular transport being based on a random walk in two dimensions (as opposed to the one-dimensional walks of previously developed models) with time periodic external forcing. We show that for certain two-dimensional structures resonant behavior can be seen for an external periodic forcing, as opposed to the monotonic dependence on forcing frequency seen in the one-dimensional case. To extend the periodically forced rachet to two dimensions the form of periodic forcing must be considered. Previous models used a time periodic forcing, of a sinusoidal type with zero average. In the present model we hypothesize that the periodic forcing involved is due to the torque generated by the neck-linker region of the motor [13] and it occurs as a result of ATP hydrolysis. Kinesin molecules walk on the surface of a microtubule whose 8 nm periodicity along the axis should be reflected by a periodic potential that has local minima that reflect the location of binding sites on the outer surface of the beta-tubulin monomer. Thus the 2D potential employed in our model is a map of the microtubule surface. The periodicity arises from the repetitive structure of the polymer, and yet in the y-direction we are not using a periodic potential even though the microtubule surface wraps around on itself. This could be explained through the lack of flexibility of the linker chain to move in the lateral direction and is supported by observations that motor proteins move primarily in the radial direction.

Although the exact formula for the forces involved may vary and will be ultimately determined from experimental data, the force components used for the simulations reported here are

$$F_x = a_x(1 + \sin(-\omega t)) \quad F_y = -a_y \cos(-\omega t). \tag{2.1}$$

The postulated potentials due to the filament are shown in Figs. 2–4. Figure 2 shows a potential which is given by a double well profile along the y-axis (superposed on a periodic well profile along the x-axis). The potential satisfies the natural boundary conditions, i.e. $\Phi(x,y) \to 0$ as $y \to \pm\infty$, and periodic boundary conditions in the x-direction, i.e., $\Phi(x,y) = \Phi(x + 2\pi, y)$ Figs. 3–4 represent shifts of the wells in the lower half of the potential by $\pi/2$, π , and $3\pi/2$, respectively. The reason for using these potentials is to explore the possible effects of resonant phenomena in Brownian walk models in motor protein motion.

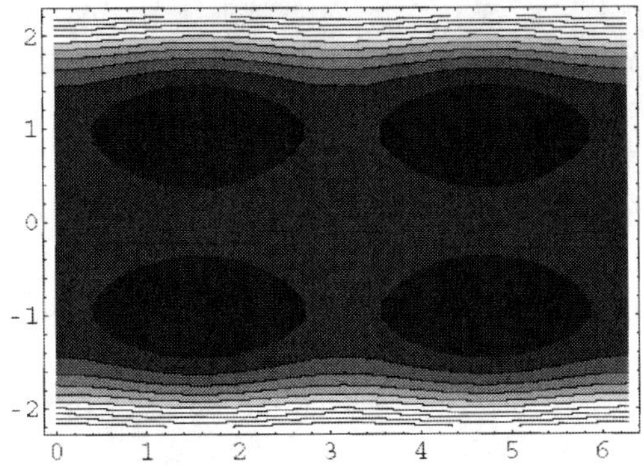

FIGURE 2. A sample two-dimensional potential to test the torques given by (2.1). This potential displays 2-fold reflection symmetry as well as discrete 2π translational symmetry.

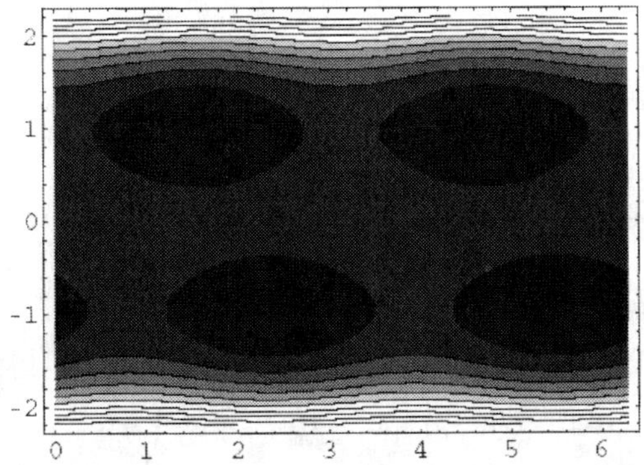

FIGURE 3. A sample two-dimensional potential to test the torques given by (2.1). The lower half of the potential in Fig. 2 is shifted to the right by $\pi/2$. This potential displays a discrete 2π translational symmetry.

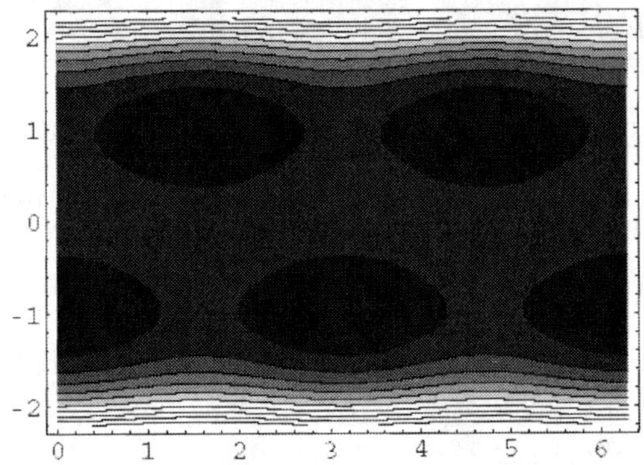

FIGURE 4. A sample two-dimensional potential to test the torques given by (2.1) . The lower half of the potential in Fig. 2 is shifted to the right by $3\pi/2$. This potential displays a discrete $2\ \pi$ translational symmetry.

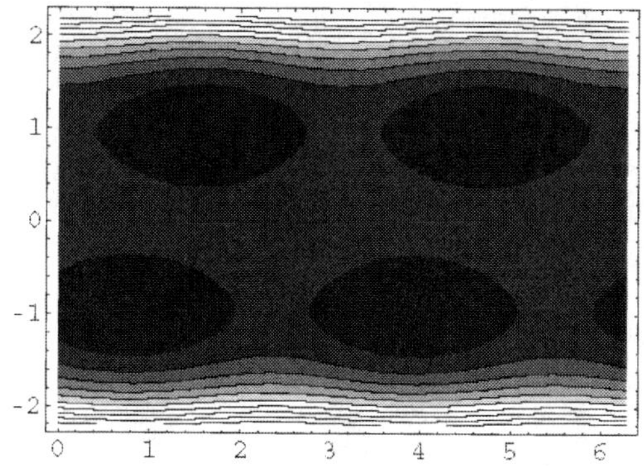

FIGURE 5. A sample two-dimensional potential to test the torques given by (2.1) . The lower half of the potential in Fig. 2 is shifted to the right by π. This potential displays a discrete 2 π translational symmetry.

3. Solution of the Langevin equation

To solve for the average velocity of the particle under different parameters we integrate the Langevin equation for several different parameter sets as shown in Fig. 6 in terms of the average velocity as a function of the frequency of the periodic forcing at a fixed temperature. From top to bottom, these curves correspond to motion in the potentials shown in Figs. 2–5, to be referred to hereafter as potentials 1 through 4. A resonant-like dependence of the average velocity on the frequency of forcing reminiscent of stochastic resonance in bistable systems [**7,11**] is seen. There is an asymptotic value of the average velocity for very small frequencies, and then the velocity increases with frequency until it peaks for an intermediate frequency value.

As the frequency of forcing is increased further the average velocity drops down even below very low frequency values as the system becomes frustrated. The resonant curves are highly sensitive to the symmetry of the potential. While all the potentials possess discrete 2 π translational symmetry along the x-axis, only potential 1 possesses reflection symmetry about the x-axis. Potential 4 also possesses the reflection symmetry about the x-axis in conjunction with an axial π-translation. The reason that the average velocity is the lowest for motion in potential 1 is that the resonant motion in the y-direction is largely decoupled from motion in the x-direction. The periodic forcing back and forth in the y-direction couples with the noise optimally for a given frequency to give regular jumping back and forth in the y-direction. If the amplitude in the x-direction is small enough then there will be hopping back and forth between adjacent minima in the y-direction with occasional jumping forward. The forward motion in the x-direction is due mostly to the offset or non-zero average x-component of the force. For potential 4, however, the adjacent minima are no longer lined up and the resonant hopping is different from the previous case.

A sample trajectory for motion in potential 4 is shown in Fig. 7 for a frequency of 0.1, with the x-position shown by the upper curve and the y-position shown by the lower one. It can be seen that for every switch in the y-direction there

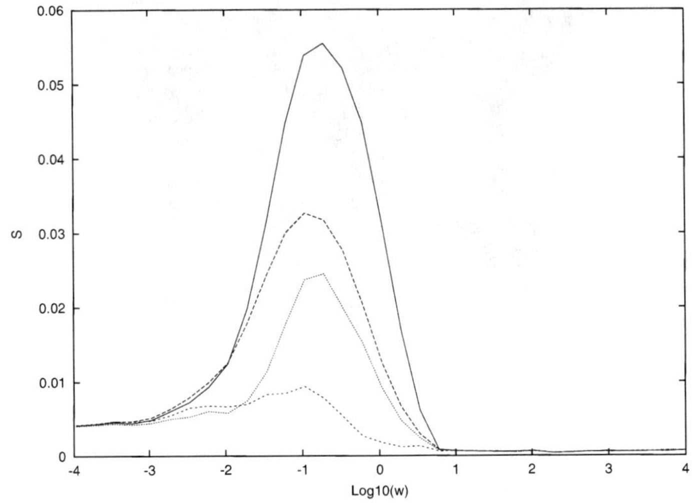

FIGURE 6. From bottom to top, the average velocity, or current, curves obtained by simulating Brownian motion in the potentials shown in Figs. 2–3–5, and 4, respectively.

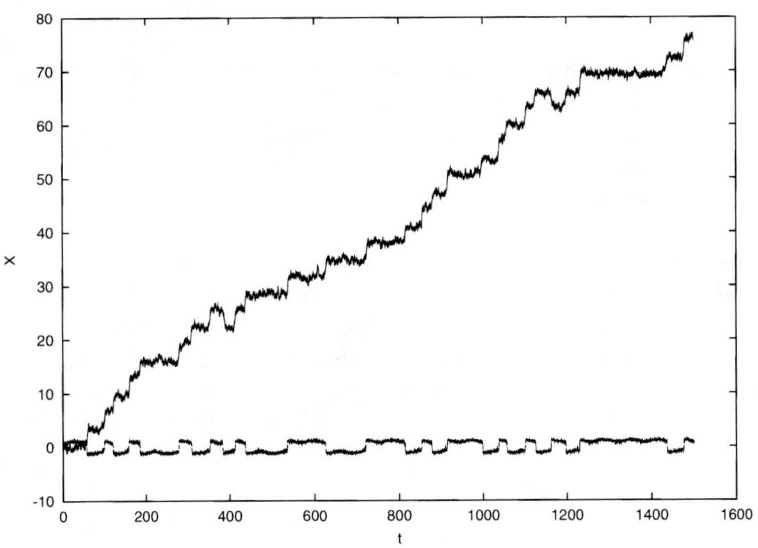

FIGURE 7. Sample trajectory from Langevin equation with the 2D potential shown in Fig. 5. The upper trajectory is the x-position, along the microtubule axis. The lower trajectory is the y-position, away from the microtubule surface. Note that nearly every switching in the y-position corresponds to a step forward of the x- position.

is a corresponding step forward. This displays a 1 : 1 tight coupling. A sample trajectory for motion in potential 1 for the same frequency is shown in Fig. 8. It can be seen that there are many switches in the y-direction which have no corresponding step forward, and hence the resonance in this case is loosely coupled.

It is interesting to examine the remaining two curves for relative shifts of the lower half of the potential of $\pi/2$ and $3\pi/2$. These potentials, while not having a peak velocity as high as they do for the completely aligned anti-potential 4, have higher peak velocities than does the potential 1. They do now, however, share the same peak potential; the curve with the higher peak velocity of the two is that of potential 3, depicted in Fig. 4. Coincidentally this is the potential that shares some similarities with the microscopic potential calculated using molecular dynamics simulations [**5**]. The fact that these two potentials with different symmetries have different velocity frequency curves indicates that the model potential has a polarity, or a preferred direction of motion. This is an important quality of microtubules which appears to be essential for unidirectional motor protein transport.

The nature of the torque, or force that the kinesin experiences is more accurately described as a position-dependent quantity rather than a time-dependent one. This is a drawback of our model. If the neck-linker region is exerting a torque on the non-anchored motor domain, it will do so until the latter binds in the forward position. If the kinesin has not swept out its full head-over-head motion before the prescribed period of the forcing is through, it will not begin a new cycle of forcing but will do so when the forward head binds to the microtubule and its ATP is hydrolyzed. Note that in systems characterized by stochastic resonance a regularity or predictability of motion is achieved for the optimal frequency. For example, a particle in a bistable double well potential jumps back and forth predictably for the optimal frequency. If a motor protein in the above model fell into a resonant state is could step forward at a predictable and regular rate and become locked in this resonant state.

4. Linearized Fokker–Planck equation approach

In this section we discuss an approach to solve the problem of resonant motion in a two-dimensional potential using a probability distribution and its Fokker–Planck equation. Writing the Fokker–Planck operator as a sum of its time independent and dependent parts we have

$$L_{FP}(x,t) = L_{FP}(x) + L_{ext}(x,t) \tag{4.1}$$

Using the external, time-dependent forcing from Eq. (2.1)

$$F_{ext}(x,t) = a_x \sin(\omega t)\widehat{x} + a_y \cos(\omega t)\widehat{y} \tag{4.2}$$

the external operator corresponding to it is

$$L_{ext} = -\frac{\partial}{\partial x} a_x \sin(\omega t) - \frac{\partial}{\partial y} a_y \cos(\omega t) \tag{4.3}$$

where a and ω are, respectively, the amplitude and frequency of the external forcing. The solution of the Fokker–Planck equation can similarly be split into time dependent and independent parts. A convenient choice for the time independent part is the stationary solution of the time independent part of the Fokker–Planck operator:

$$W(x,t) = W_{st}(x) + w(x,t). \tag{4.4}$$

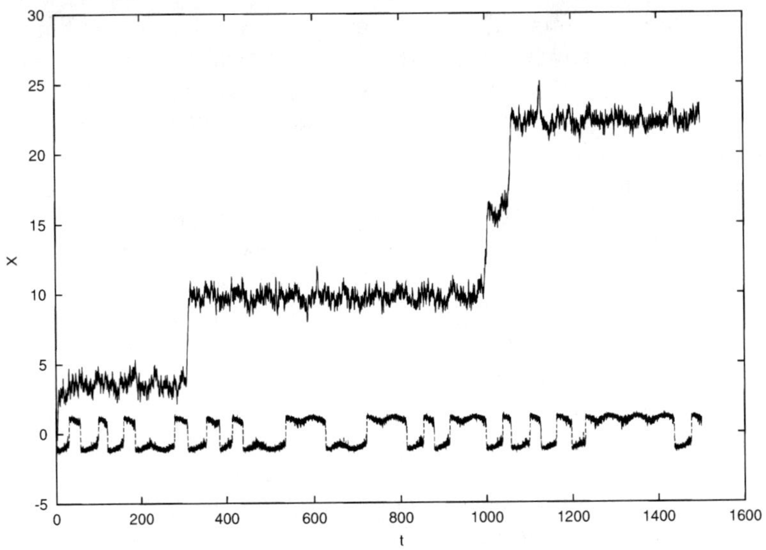

FIGURE 8. Sample trajectory from Langevin equation with the 2D potential shown in Fig. 2. The upper trajectory is the x-position, along the microtubule axis. The lower trajectory is the y-position, away from the microtubule surface. Note that there are many switchings in the y-position without corresponding steps forward of the x-position.

The Fokker–Planck equation then becomes

$$\frac{\partial}{\partial t}W(x,t) = L_{FP}(x)W_{st}(x)+L_{FP}(x)w(x,t)+L_{ext}W_{st}(x)+L_{ext}(x,t)w(x,t). \tag{4.5}$$

The first term on the right is zero, because this is equal to the time derivative of the stationary solution. If the magnitudes, a_x and a_y, of the external periodic forcing are small it might be possible that the change this makes to the stationary solution is small as well. Therefore, the last term on the right hand side of the previous equation can also be neglected in the linear response limit. The equation for the time dependent correction then becomes

$$\frac{\partial}{\partial t}w(x,t) = L_{FP}(x)w(x,t) + L_{ext}W_{st}(x) \tag{4.6}$$

whose formal solution is

$$\begin{aligned} w(x,t) &= -a_x \int_{-\infty}^{t} e^{L_{FP}(x)\cdot(t-t')} L_{ext}(x,t)W_{st}(x)\,dt' \\ &= -a_x \int_{-\infty}^{t} e^{L_{FP}(x)\cdot(t-t')} \sin(\omega t)\frac{\partial}{\partial x}W_{st}(x)\,dt' \\ &\quad - a_y \int_{-\infty}^{t} e^{L_{FP}(x)\cdot(t-t')} \cos(\omega t)\frac{\partial}{\partial y}W_{st}(x)dt' \end{aligned} \tag{4.7}$$

For even the simplest systems this equation is very difficult to solve in closed form, so we will turn to approximations and numerical methods. The analytical approximation involves taking the integrand of this equation and expanding it in terms of a series of the eigenfunctions of the Fokker–Planck equation. The numerical technique used to calculate these eigenfunctions is based on discrete singular convolution methods [**18, 19**].

The stationary solution is itself an eigenfunction of the Fokker–Planck operator, with eigenvalue zero, and so, were it not for the spatial derivatives in the integrand, the exponential operator would just act on the stationary solution and give back the stationary solution with a multiple of one in front. The derivative operators, however, project the stationary solution onto the other eigenfunctions and thus a series expansion into the different modes is necessary. If we assume that an arbitrary function can be expanded in terms of the eigenfunctions of L_{FP}we have:

$$f(x) = \sum_i c_i \phi_i(x) \tag{4.8}$$

with coefficients given by

$$c_i = \int e^{\Phi(x)} \phi_i(x) f(x)\, dx \tag{4.9}$$

Using

$$\frac{\partial}{\partial x} W_{st}(x) = \sum_i c_i^x \phi_i(x) \tag{4.10}$$

and

(4.11)

$$\frac{\partial}{\partial y} W_{st}(x) = \sum_i c_i^y \phi_i(x) \tag{4.12}$$

and substituting into equation (4.7), we have:

$$\begin{aligned} w(x,t) = &-a_x \sum_i c_i^x \phi_i(x) e^{-\lambda_i t} \int_{-\infty}^{t} \sin(\omega t') e^{\lambda_i t'}\, dt' \\ &- a_y \sum_i c_i^y \phi_i(x) e^{-\lambda_i t} \int_{-\infty}^{t} \cos(\omega t') e^{\lambda_i t'}\, dt' \end{aligned} \tag{4.13}$$

With the solutions of the time integrals; equation (4.13) becomes

$$w(x,t) = \sum_i b_i^x(t) \phi_i(x) + \sum_i b_i^y(t) \phi_i(x) \tag{4.14}$$

with the coefficients given by

$$b_i^x(t) = -a_x c_i^x \left(\frac{\lambda_i \sin(\omega t) - \omega \cos(\omega t)}{\lambda_i^2 + \omega^2} \right), \tag{4.15}$$

$$b_i^y(t) = -a_y c_i^x \left(\frac{\lambda_i \cos(\omega t) - \omega \sin(\omega t)}{\lambda_i^2 + \omega^2} \right). \tag{4.16}$$

The probability current, which is ultimately the quantity of interest to us, is given by

$$L_{FP}(x,t) W(x,t) = -\nabla S(x,t). \tag{4.17}$$

Since $\Phi(\vec{x}) \to \infty$ as $y \to \pm\infty$, the current in this direction will average out to be zero. We could, potentially, have flow in the x direction, given by:

$$S_x(\vec{x},t) = \left(a_x \sin(\omega t) - \frac{\partial \Phi(\vec{x})}{\partial x}\right) W(\vec{x},t) - D\frac{\partial}{\partial x} W(\vec{x},t). \tag{4.18}$$

Using equation (4.4), this expression becomes:

$$\begin{aligned} S_x(\vec{x},t) = &\left(-\frac{\partial \Phi(\vec{x})}{\partial x}\right) W_{st}(\vec{x}) - D\frac{\partial}{\partial x} W_{st}(\vec{x}) + a_x \sin(\omega t) W_{st}(\vec{x}) \\ &+ \left(a_x \sin(\omega t) - \frac{\partial \Phi(\vec{x})}{\partial x}\right) w(\vec{x},t) - D\frac{\partial}{\partial x} w(\vec{x},t). \end{aligned} \tag{4.19}$$

The first term in brackets on the right hand side of the latter equation is just the stationary current density, which is zero due to the symmetry of the stationary probability density and of the potential. The remaining expressions give us the current density as a function of time. This quantity will be oscillatory in time, but we can average these out over one period to see if there is any net motion. The net average current is then

$$S_x(\vec{x},t) = \frac{2a}{\pi} W_{st}(\vec{x}) + \sum_i a_x \left(\overline{b_i^x(t)} + \overline{b_i^y(t)}\right) \tag{4.20}$$

with

$$\overline{b_i^x(t)} = -\frac{2ac_i^x \lambda_i}{\pi(\lambda_i^2 + \omega^2)} \tag{4.21}$$

$$\overline{b_i^y(t)} = -\frac{2ac_i^y \lambda_i}{\pi(\lambda_i^2 + \omega^2)}. \tag{4.22}$$

As mentioned earlier, this linearization technique has problems generating the same average velocity curves obtained through numerical solution of the corresponding Langevin equation. The source of this problem is probably due to nonlinear effects dominating the process. An alternative technique to solve for the asymptotic behavior of the probability density is given in Wei [**18, 19**]. This technique involves taking the formal solution of the Fokker–Planck equation and Taylor expanding the exponential term in powers of the Fokker–Planck operator.

5. Random Walk in a one-dimensional projection of the mechano-chemical phase plane

In this section we show that the sigmoidal force velocity curves obtained from chemical kinetics modeling [**8, 10, 16**] can also be obtained from a random walk in a two-dimensional phase space in which the potential is featureless in the physical position direction. The 2D potential used in the model proposed here has one variable that represents the position along the microtubule axis, and the complementary variable represents the internal conformational state of the motor. Our approximation leading to a 1D potential would be achieved by constraining the motion in 2D along the minimal potential path. To quantify the average velocity of a point particle moving in a periodic potential we turn again to the Fokker–Planck equation.

The Langevin equation for a point particle moving in a potential field in the over-damped case is

$$\gamma\dot{x} = \left(-f'(x) + F + \Gamma(t)\right) \tag{5.1}$$

where $f(x)$ is the potential the particle is moving in, F is an externally imposed force, and $\Gamma(t)$ is Gaussian white noise. The corresponding Fokker–Planck equation is:

$$\frac{\partial W}{\partial t} = \frac{1}{\gamma}\left(f' - F + D\frac{\partial}{\partial x}\right)W = -\frac{\partial S}{\partial x} \tag{5.2}$$

where S is the probability current. The stationary solution is obtained by setting the time derivative of W to zero. This gives a probability current that is uniform in space whose solution is given by

$$\gamma S = (F - f')W - D\frac{\partial W}{\partial x}. \tag{5.3}$$

The solution to this equation is given by [**14**]

$$W(x) = e^{-V(x)/D}\left[N - \gamma(S/D)\int_0^x e^{V(x')/D}\,dx'\right] \tag{5.4}$$

where $V(x) = f(x) - Fx$ is the total potential energy from both the periodic structure and the external force. We must show periodicity for the above solution and show that this solution is normalizeable within one period. We begin by looking at the behavior of the first integral on the right hand side of equation (5.4) under integer multiple periodic translations

$$\begin{aligned} &\int_0^{x+2n\pi} e^{V(x')}\,dx' \\ &= \int_0^{2\pi} e^{V(x')/D}dx' + \cdots + \int_{2\pi(n-1)}^{2\pi n} e^{V(x')/D}\,dx' + \int_{2\pi n}^{2\pi n+x} e^{V(x')/D}\,dx' \\ &= I + Ie^{-2\pi F/D} + \cdots + Ie^{-2\pi(n-1)F/D} + \int_0^x e^{V(x')}dx' e^{-2\pi F/D} \\ &= I\frac{1-e^{-2\pi F/D}}{1-e^{-2\pi F/D}} + e^{-2\pi F/D}\int_0^x e^{V(x')/D}\,dx' \end{aligned} \tag{5.5}$$

with I given by

$$I = \int_0^{2\pi} e^{V(x)/D}\,dx \tag{5.6}$$

Inserting equation (5.5) into equation (5.4) gives

$$\begin{aligned} W(x+2n\pi) = {}& e^{-V(x)/D}\left[N - \frac{\gamma SI}{D(1-e^{-2\pi F/D})}\right]e^{-2\pi nF/D} \\ &+ e^{-V(x)/D}\left[\frac{\gamma SI}{D(1-e^{-2\pi F/D})} - \gamma\frac{S}{D}\int_0^x e^{V(x')/D}\,dx'\right]. \end{aligned} \tag{5.7}$$

We can then see that the stationary solution satisfies periodicity if the first term in the second pair of brackets is equal to the normalization constant, N, or, equivalently if the first term in brackets is equal to zero, i.e.,

$$\gamma SI = DN(1 - e^{-2\pi F/D}). \tag{5.8}$$

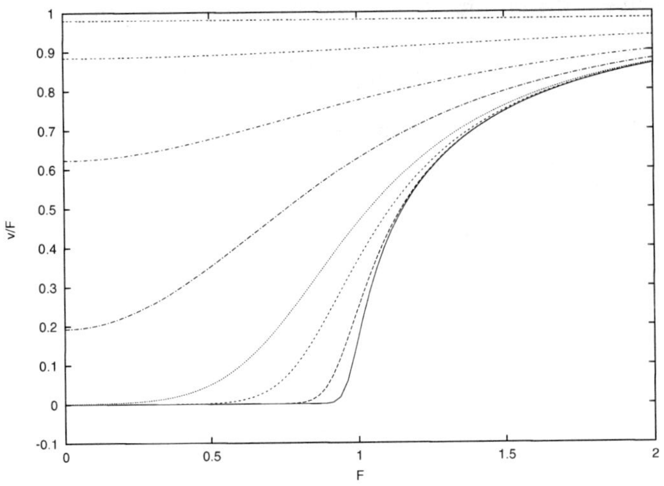

FIGURE 9. Mobility curves for a cosine periodic potential [**14**]. The curves, from top to bottom, correspond to temperatures of 5, 2, 1, 0.5, 0.2, 0.1, 0.03, and 0.01.

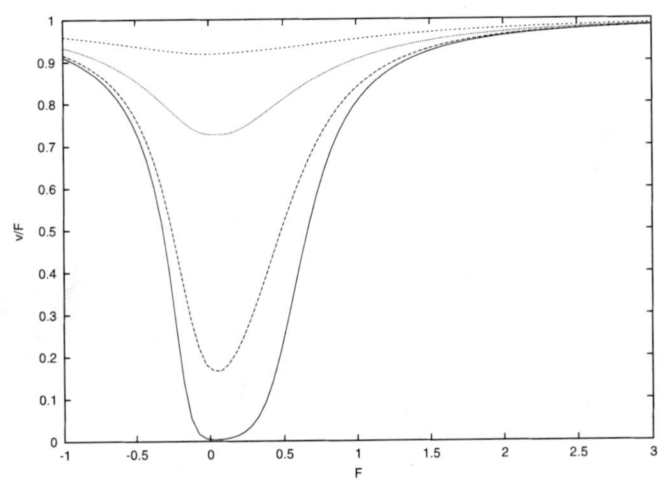

FIGURE 10. Mobility curves for a sawtooth periodic potential [**14**]. The curves, from top to bottom, correspond to temperatures of: 1, 0.5, 0.2, and 0.1.

To solve for the normalization constant we integrate the stationary probability density over one period

$$\int_0^{2\pi} W(x)\, dx = 1. \tag{5.9}$$

Using Eq. (5.9) in conjunction with Eq. (5.8) and the fact that the average velocity is the integral of the current density over one period, i.e., $\langle v \rangle = 2\pi S$, we

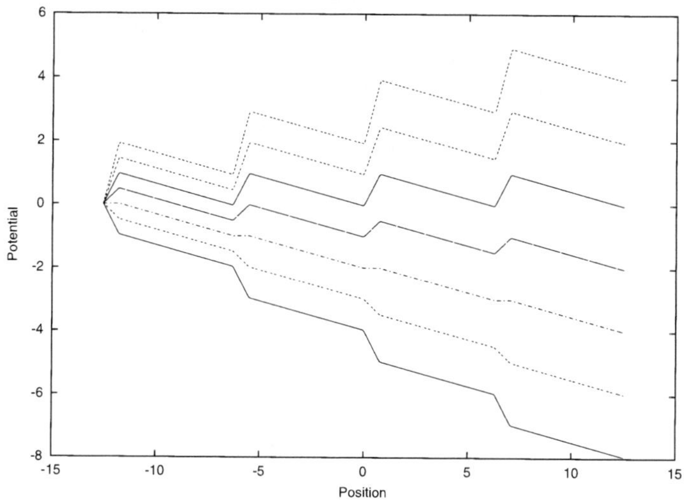

FIGURE 11. The overall potential for the one-dimensional description of the mechano-chemical random walk, for varying load forces. From top to bottom, the potentials correspond to loading forces of: $8/\pi$, $6/\pi$, $4/\pi$, $2/\pi$, 0, $-2/\pi$, and $-4/\pi$.

obtain an analytic expression for the average velocity of the particle:

$$\gamma\langle v\rangle = \frac{2\pi D(1-e^{-2\pi F/D})}{\left(\int_0^{2\pi} e^{-V(x)/D}\,dx' - (1-e^{-2\pi F/D})\int_0^{2\pi} e^{-V(x)/D}\int_0^{x} e^{-V(x')/D}\,dx'\,dx\right)}. \tag{5.10}$$

To examine the influence of external forces on periodic potentials we must choose the form of the potentials to be used. Figures 9–10 show the mobility times the damping constant, $\gamma\mu = v/F$, for the cosine potential and sawtooth potential respectively, as a function of external force. The sawtooth potential used for Fig. 10 is given by:

$$f(x) = \begin{cases} 2/\pi, & 0 < x < \pi/2, \\ -2/3\pi, & \pi/2 < x < 2\pi. \end{cases} \tag{5.11}$$

Figure 10 also displays the asymmetry of the potential. The reason for the greater response of the system to negative forces over their positive counterpart is because the particle has to overcome a gentler slope in this direction to reach the next well. Another consideration when choosing the relative fraction that each space takes up in the period is how long, on average the motor protein spends in chemical transition relative to its diffusive walk forward.

Figure 11 shows the sawtooth potential for varying forces. When the load force is zero, the net slope of the potential is negative so that particles would flow in the forward direction. The fraction that the physical space occupies for the potential in Fig. 11 is 1/8, or on the normalized period length of 2π it occupies a length of $\pi/4$.

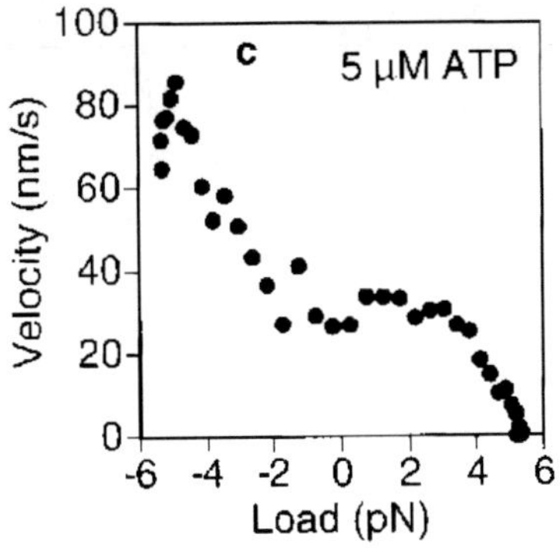

FIGURE 12. Experimental force-velocity data for kinesin with an ATP concentration of $5\mu m$ from Coppin, et al. [**3**].

The stalling force occurs when the net slope of the potential is zero [**16**], being $4/\pi$ in Fig. 11.

For our calculations we used the same parameters as those used by Fox and Choi [**6**]. At any point in time the largest moving component of kinesin is just one of the motor domains, which have a diameter of approximately $R_0 = 2.94$ nm each. The viscosity of the medium is slightly higher than water, at about 1 x 10^{-3} kg$/ms$, and thus the damping coefficient, γ, is equal to $6\pi\eta R_0 = 5.54 \times 10^{-11}$ kg$/s$. The diffusion coefficient is then $D = 2kT/\gamma = 7.2 \times 10^{-11} m^2/s$. The force-velocity curves for these parameters are shown in Fig. 13 . The top curve in Fig. 13 is for a sawtooth potential which has straight edges in both the physical and the chemical spaces. To explore the possibility that there night be several intermediate states along the particles' excursion through chemical space, numerous small barriers were added to this part of the potential. Figure 14 illustrates this by showing the potential without any barriers and also with three barriers. Reassuringly, these curves are very similar qualitatively to those produced by the chemical kinetics models previously mentioned [**8, 10, 16**].

Conclusions

The curves obtained by the random walk model qualitatively reproduce the experimental data, (see Fig. 12) for positive load forces up until smaller negative loads [**3**]. Once the forward pull, or negative load, on the motor proteins increases too much, another upward turn in the curve is seen. Neither the random walk model nor the chemical kinetics models can effectively reproduce this behavior. This shortcoming of the chemical kinetics models can be understood in view of the fact that the reactions proposed are events or processes in series. The increased negative load causes the forward rate constants of the reactions it influences to increase. For very large values of forward rate constant the mean time to pass from one state to the next becomes negligible. The remaining rate constants, however, are unaffected by the load and this remains a challenge for the mathematical modeling effort.

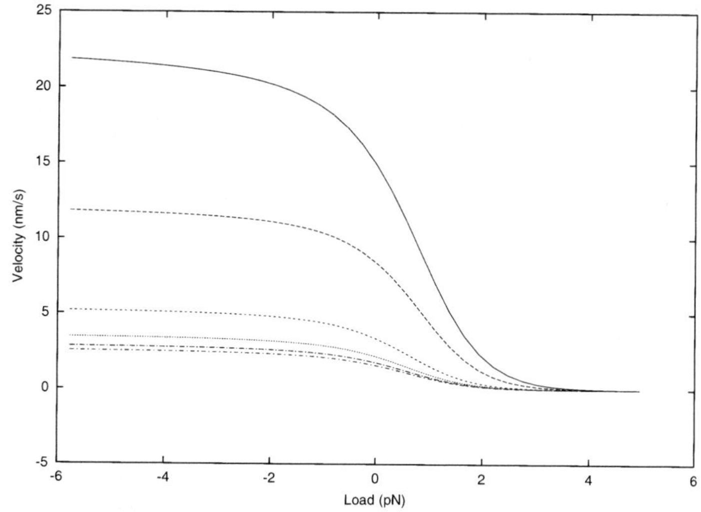

FIGURE 13. Velocity as a function of force for the random walk model with varying number of barriers in the chemical phase space. From top to bottom, the number of barriers are: 0, 1, 2, 3, 4, and 5. The data was obtained, assuming a stalling force of $7pN$.

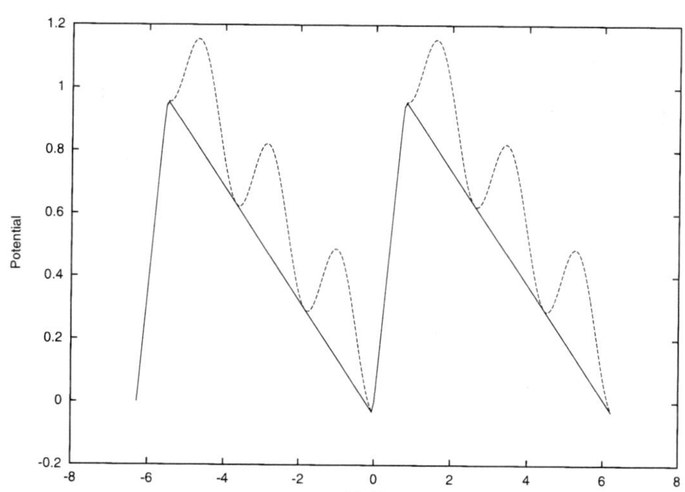

FIGURE 14. Sawtooth potential used to obtain force-velocity relationships in Fig. 13. Shown are potentials with no barriers in the chemical phase space, and with three barriers in the chemical phase space. Each sawtooth represents the spatial exentent of one dimer, and the superposed cosine potential represents the local chemical potential.

References

1. R. D. Astumian and M. Bier, *Fluctuation driven ratchets: Molecular motors*, Phys. Rev. Lett. **72** (1994), 1766–1769.
2. R. D. Astumian and I. Derenyi, *A chemically reversible brownian motor: Applications to kinesin and ncd*, Biophys. J **77** (1999), 993–1002.
3. C. M. Coppin, J. T. Finer, J. A. Spudich, and R. D. Vale, *Detection of* sub − 8 − nm *movements of kinesin by high-resolution optical-trap microscopy*, Proc. Nat. Acad. Sci. USA **93** (1996), 1913–1917.
4. C. R. Doering, W. Horsthemke, and J. Riordan, *Nonequilibrium fluctuation-induced transport*, Phys. Rev. Lett. **72** (1994), 2984–2987.
5. H. P. Erickson, *Co-operativity in protein-protein association*, J. Mol. Biol. **206** (1989), 465–474.
6. R. F. Fox and M. H. Choi, *Rectified brownian motion and kinesin motion along microtubules*, Phys. Rev. E **63** (2001), 1–12.
7. L. Grammaitoni, P. Haenggi, P. Jung, and V. F. Marchesoni, *Stochastic resonance*, Rev. Modern Phys. **70** (1998), 223–287.
8. J. Howard, *Mechanics of motor proteins and the cytoskeleton*, Macmillan, New York, 2001.
9. F. Jülicher, A. Adjari, and J. Prost, *Modeling molecular motors*, Rev. Modern Phys. **69** (1997), 1269–1281.
10. D. Keller and C. Bustamante, *The mechanochemistry of molecular motors*, Biophys.
11. R. Landauer, *Motion out of noisy states*, J. Stat. Phys. **53**(1988), 233–248.
12. S. Leibler and D. A. Huse, *Porters versus rowers: A unified stochastic model of motor proteins*, J. Cell Biol. **121** (1993), 1357–1368.
13. A. Mogilner, A. J. Fisher, and R. J. Baskin, *Structural changes in the neck linker of kinesin explain the load dependence of the motor's mechanical cycle*, J. Theoret. Biol. **211** (2001), 143–157.
14. H. Risken, *The Fokker–Planck equation*, Springer-Verlag, Berlin, 1989.
15. K. M. Ruppel, M. Lorenz, and J. A. Spudich, *Myosin structure or function: A combined mutagenesis-crystallography approach*, Curr. Opin. Struct. Biol. **5** (1995), 181–186.
16. K. Svoboda and S. M. Block, *Force and velocity measured for single kinesin molecules*, Cell **77** (1994), 773–784.
17. G. Woehlke, A. K. Ruby, C. L. Hart, B. Ly, N. Hom–Booher, and R. D. Vale, *Microtubule interaction site of the kinesin motor* Cell **90** (1997), 207–216.
18. G. W. Wei, *Solving quantum eigenvalue problems by discrete singular convolution*, J. Phys. B **33** (2000), 343–352.
19. ———, *Discrete singular convolution for the solution of the Fokker–Planck equations*, J. Chem. Phys. **110** (1999), 8930–8942.

Department of Physics, University of Ottawa, 150 Louis Pasteur, Ottawa, Ontario, K1N 6N5, Canada
E-mail address: `jmidd620@science.uottawa.ca`

Departement of Physics, University of Alberta, Edmonton, Alberta, T6G 2J1, Canada.
E-mail address: `jtus@phys.ualberta.ca`

Centre de Recherches Mathématiques
CRM Proceedings and Lecture Notes
Volume **39**, 2005

Determination of Approximate Symmetries of Differential Equations

J. Bonasia, François Lemaire, Greg Reid, Robin Scott, and Lihong Zhi

ABSTRACT. There has been considerable progress in the theory and computer implementation of symbolic computation algorithms to automatically determine and exploit exact symmetries of exact differential equations. Such programs usually apply a finite number of exact differentiations and eliminations to the overdetermined linearized equations for the unknown symmetries (the symmetry defining equations), to complete them to certain involutive or standard forms. The symmetry properties can be determined from these involutive forms.

In many applications, however, the differential equations describing a model are only approximately known. For example they may contain parameters that are only known approximately. Symbolic methods are unstable if applied to the symmetry defining equations directly, and indirect techniques (e.g. replacing approximate parameters by symbolic ones) might not be practical in cases with many parameters. Discussion and examples are given of such difficulties.

A new generation of symbolic-numeric methods is described and applied to the problem of determining symmetries of differential equations.

We introduce a class of differential-elimination methods which uses Numerical Linear Algebra, and in particular the Singular Value Decomposition, to perform the elimination process on the symmetry defining equations. Our approach uses symbolic differentiations but not symbolic eliminations. Substitution of an appropriate random point in the independent variables, followed by numerical projection is used to test for the conditions of completion to a projective involutive form. We prove that this form is equivalent to the involutive form of the Cartan–Kuranishi theory of partial differential equations.

Our method is applied to determining symmetry properties of 50 ODE from the collection in Kamke's book.

2000 *Mathematics Subject Classification.* Primary: 34M15; Secondary: 68W30.

Support from NSERC and ORDCF is gratefully acknowledged by the first author.

The authors thank Edgardo Cheb–Terrab for providing and giving help on the test set of Kamke ODE. We thank Rob Corless for helpful information about the SVD and backward error analysis. We are grateful for very helpful discussions at the conference that prompted us to seek an interpretation and calculation of nearby ODE with large symmetry groups that led to Section 6 of the paper.

This is the final form of the paper.

1. Introduction

The continuous (exact) point symmetries of a differential equation with independent variables x and dependent variables y are the group of transformations:

$$\tau : \widehat{x} = X(x, y), \quad \widehat{y} = Y(x, y) \tag{1.1}$$

leaving invariant its solution set S:

$$\tau(S) = S. \tag{1.2}$$

Applications of symmetries to differential equations include determination of invariant solutions, mappings from intractable to tractable equations (e.g. non-linear to linear), and the determination of conservation laws. They have become a major tool for determining analytic features and solutions of differential equations [**2, 15**].

The (exact) symmetries of a differential equation are usually not known a priori and have to be determined. By associating the differential equation with a submanifold $\mathcal{M}_S$ of a *Jet space*, condition (1.2) is replaced with

$$\mathrm{pr}(\tau)(\mathcal{M}_S) = \mathcal{M}_S \tag{1.3}$$

where the group has been prolonged to act on derivatives up to the order of the differential equation.

The Infinitesimal Lie symmetries are the linearized form of such symmetries about the identity transformation:

$$\widehat{x} = X(x, y; \varepsilon) = x + \varepsilon\xi(x, y) + \mathcal{O}(\varepsilon^2), \quad \widehat{y} = Y(x, y; \varepsilon) = y + \varepsilon\eta(x, y) + \mathcal{O}(\varepsilon^2). \tag{1.4}$$

In particular, the $\mathcal{O}(\varepsilon)$ terms of condition (1.3) leads to a linear homogeneous system of over-determined partial differential equations on the function ξ and η [**2, 15**], which we call the symmetry defining system.

In this paper we concentrate on symmetries of second order ordinary differential equations of the form

$$y_{xx} = g(x, y, y_x). \tag{1.5}$$

Extensive studies have been made of the exact symmetry and equivalence properties of this class of equations starting with Lie and continuing with the work of Tresse, Cartan and others [**16**].

The Symmetry Defining System for (1.5) is:

$$\begin{aligned} &-g_y\eta + g(-3y_x\xi_y - 2\xi_x) - g_x\xi + g_{y_x}(-y_x\eta_y + y_x^2\xi_y + y_x\xi_x - \eta_x) \\ &+ \eta_{x,x} + y_x^2\eta_{y,y} - y_x^3\xi_{y,y} - 2y_x^2\xi_{x,y} + 2y_x\eta_{x,y} - y_x\xi_{x,x} = 0, \quad \xi_{y_x} = 0, \; \eta_{y_x} = 0, \end{aligned} \tag{1.6}$$

There are many computer algebra programs for automatically generating such symmetry defining systems (see the review article by Hereman [**12**]). They can be large and highly over-determined. Indeed such systems can be simplified to canonical forms by a finite number of differentiations and eliminations. This process is a differential generalization of Gröbner bases [**3, 14, 18, 22, 24**]. The canonical forms can then often be used for symbolic explicit integration of the system and a number of symbolic computation approaches have been based on this method [**31**].

Such symbolic differential-elimination algorithms tend to explode in memory on some non-trivial examples, reflecting their underlying exponential complexity. Moreover they are not guaranteed to deal with models with floating point numbers in them and this motivated us to develop a symbolic-numeric differential elimination method.

An underlying principle of our method, is our strong emphasis on geometry. In particular we emphasize jet space geometry, the geometry of differential systems [**11, 17, 18, 24, 29**]. In comparison to the symbolic differentiation-elimination approaches, where symbolic or algebraic manipulations of the equations are emphasized, our approach focuses on the solutions of the system regarded as algebraic equations. This is the jet space picture of a differential equation, in which one regards all the appearing variables (derivatives etc.) as formal unknowns on an equal footing.

Our approach is based on combining techniques from numerical analysis and symbolic computation. This work falls under the new area of hybrid symbolic-numeric computation. Symbolic elimination algorithms (e.g. the Gaussian elimination and Buchberger algorithms) can be executed with systems with floating point arithmetic. However, the output is not well defined (for example the Buchberger algorithm relies on using a field and floating point numbers do not form a field). Moreover, due to round-off errors, the output is almost always unstable with respect to small changes in the coefficients of the input systems.

Our approach uses only symbolic differentiation, and avoids symbolic elimination, in an attempt to address the significant expression swell seen in many symbolic elimination algorithms. In contrast, symbolic differentiation is relatively cheap. Techniques from Numerical Linear Algebra [**10**], and in particular the Singular Value Decomposition (SVD) are used. It was first outlined briefly in [**30**].

Our approach is motivated by the geometric differential elimination methods that arose in the approaches of Cartan and Kuranishi. This work has also been influenced by [**23**] in which an approach is proposed for ordinary differential systems that generate prime differential ideals. That approach uses symbolic differentiations, and avoids nonlinear eliminations, instead using linear eliminations based on substitution of a random point.

First we introduce (exact) Jet geometry, and the symbolic (exact) method underlying our completion method. Theorems are given on the relation between our method and the classical Cartan–Kuranishi method. The implementation of this method which uses the Singular Value Decomposition is discussed.

We use 50 ODE taken from Kamke [**13**] as a test bed for our symbolic-numeric methods. This is a classic reference work for ODE having significant analytic properties (e.g. being exactly solvable). Finally in Section 6 we outline an analytic interpretation of our method, that characterizes it in terms of nearby ODE admitting large symmetry groups.

Our work is part of a series in which we are developing the theory and applications of symbolic-numeric methods for general systems of differential equations. An early example of this series [**20**] has been that of constant coefficient linear homogeneous PDE in particular as used in the problem of determining a camera's orientation from reference data (the *camera pose problem*, an important problem in computer vision). In that paper, we showed experimentally that the methods presented here, could stably predict orientations, even in singular configurations which had been difficult for other methods in the literature. In [**19**], we apply the new methods of Numerical Algebraic Geometry given in [**27**] to the problem of completion of nonlinear systems of ordinary differential equations with constraints. Research is underway on how to efficiently combine such nonlinear methods with the linear ones developed in this article.

Compared to the large volume of results on exact symmetries of differential equations very little work has been done on approximate symmetries. A method was proposed by Baikov, Gazizov and Ibragimov [**1**] for the investigation and exploitation of approximate symmetries (also see [**8, 9**]). This method depends on being able to identify a small symbolic parameter in the original differential equation, and then constructs Lie symmetry vector fields by a symbolic perturbation approach up to a given order in the small parameter. Our approach does not depend on identifying a small parameter, and is more thoroughly numeric, using tools from Numerical Analysis rather than the perturbation techniques. Moreover the methods of Ibragimov et al do not appear to be able to identify nearby differential equations with large symmetry groups.

2. Symmetries and introductory examples

We illustrate some of the difficulties associated with symbolic methods when applied to cases containing floating point numbers. In particular consider:

$$y_{xx} + 6.708203932yy_x + 5y^3 = 0 \tag{2.1}$$

which will be used to motivate the need to have truly symbolic-numeric methods for determining approximate features of systems.

2.1. Defining system for symmetries of $y_{xx} + 6.708203932yy_x + 5y^3 = 0$. Lie point symmetries of (2.1) are transformations of form $(x, y) \mapsto (\widehat{x}, \widehat{y}) = \big(x + \xi(x, y)\varepsilon + \mathcal{O}(\varepsilon^2), y + \eta(x, y)\varepsilon + \mathcal{O}(\varepsilon^2)\big)$. They generate a Lie algebra of vector fields $\xi(x, y)\partial/\partial x + \eta(x, y)\partial/\partial y$ which follows the flow of the group. The symmetry defining system of (2.1) is:

$$\begin{gathered}
\xi_{yy} = 0, \\
\eta_{yy} - 2.0\xi_{xy} + 13.41640787y\xi_y = 0, \\
2.0\eta_{xy} - 1.0\xi_{xx} + 15.0y^3\xi_y + 6.708203928y\xi_x + 6.708203932\eta = 0, \\
\eta_{xx} - 5.0y^3\eta_y + 10.0y^3\xi_x + 6.708203932y\eta_x + 15.0y^2\eta = 0.
\end{gathered} \tag{2.2}$$

where subscript notation has been used for partial derivatives. This system has been produced by the Maple command `DEtools[odepde]` and results from polynomial decomposition of (1.6) with $g = -6.708203932yy_x - 5y^3$ with respect to powers of y_x.

The exact differential elimination packages `RifSimp` and `diffalg` are based on theories that assume that the input systems to these `Maple` packages have coefficients from a field. Since floating point numbers do not form a field (e.g. the associative property fails), computations with floating point numbers with these packages are not guaranteed to produce a theoretically well-founded result.

2.2. Direct use of floating point numbers in exact differential elimination methods. The package `diffalg` will process floating point numbers in un-altered form in its internal computations, and when applied to system (2.2) yields:

$$\xi_x = 0.0, \quad \xi_y = 0.0, \quad \eta = 0.0 \tag{2.3}$$

which represents a 1 parameter translation symmetry in x for (2.1). The `Maple` package `RifSimp` will by default, convert floating point numbers into their exact rational approximations, in the input system, and this strategy is discussed next.

2.3. Rational substitution strategy. A common strategy, for symbolic programs is to replace input floating point numbers with rational numbers. Executing this strategy with system (2.2) (so that 6.708203928 is replaced with $6.708203928 = 6708203928/10^9$ etc.) both the packages `RifSimp` and `diffalg` output the same simplified system:

$$\xi_x = 0, \quad \xi_y = 0, \quad \eta = 0, \tag{2.4}$$

which clearly has a one parameter solution space. In the formation of the defining system (2.2) small round-off errors have been introduced (e.g. 6.708203928 appears in one coefficient and 6.708203932 in another). So an alternative strategy is to make the rational replacement $6.708203932 = 6708203932/10^9$ in the original ODE. Applying this strategy both `RifSimp` and `diffalg` output the following simplified form of the defining system:

$$\xi_{xx} = 0, \quad \xi_y = 0, \quad \eta = -y\xi_x. \tag{2.5}$$

Integration easily shows that $\xi = ax+b$, $\eta = -ay$. These correspond to the obvious symmetries of translation in x and a mutual scaling in $(x, y) \mapsto (x/c, cy)$ possessed by the input ODE. As the above example illustrates the symbolic simplification methods are not continuous with respect to small changes in the coefficients. The method can, for this example, be regarded as more successful since it found a larger group of symmetries of (2.1) than the previous method.

2.4. Symbolic substitution strategy. A common strategy by symmetry and computer algebra researchers to address difficulties faced when using floating point numbers is to replace such quantities with symbolic parameters (or in general with unspecified functions).

In particular when this is done with the above example, it is embedded in the one parameter class of ODE:

$$y_{xx} + \alpha y y_x + 5y^3 = 0. \tag{2.6}$$

Application of `RifSimp` and `diffalg` which can be called using the convenient interface `DEtools[casesplit](sys, option)` with `option = rif` or `option = diffalg` yields:

Case 1 ($\alpha^2 \neq 45$).

$$\xi_{xx} = 0, \quad \xi_y = 0, \quad \eta = -y\xi_x \tag{2.7}$$

and

Case 2 ($\alpha^2 = 45$).

$$\begin{aligned} \eta &= \frac{2}{15}\xi_{xxxy} + 2y^2\xi_{xy} - y\xi_x - \frac{2}{15}y\alpha\xi_{xxy} + \frac{1}{15}\alpha\xi_{xx} - \frac{1}{3}\alpha\xi_y y^3, \\ \xi_{xxxx} &= \frac{30y^2}{\alpha}\xi_{xxy} - \frac{30y}{\alpha}\xi_{xxx}, \\ \xi_{yy} &= 0. \end{aligned} \tag{2.8}$$

Further algorithms can determine that the dimension of the solution space for Case 2 is eight dimensional. Thus ODE (2.6) has an 8 dimensional Lie group of symmetries when $\alpha^2 = 45$ and a 2 dimensional group when $\alpha^2 \neq 45$. ODE with larger symmetry groups are amenable to a wider class of analytical techniques. For example, second order ODE are linearizable [**16**] if and only if they have an eight dimensional group. Hence the ODE (2.1) is very close to a linearizable ODE.

2.5. Discussion of difficulties with symbolic approaches. As the above examples illustrate, care has to be taken with symbolic approaches. They are not designed for using floating point numbers and can produce unstable results when used with numeric coefficients. Strategies such as rational replacement need to be used with care, since round-off errors can mean that the rational approximation does not inherit desired properties of the original system. Further, the more general approach of symbolic replacement, although powerful in certain situations, can be impractical due to the greater complexity of the calculations involved.

In summary it is useful to explore approaches that more thoroughly integrate symbolic and numeric methods, and such methods need to consider *close-by* systems. For example it would be desirable for a symbolic-numeric approach to be able to determine that the case $\alpha = \sqrt{45} \approx 6.708203932$ was close by with a desirably large (8) dimensional symmetry group characterizing a linearizable ODE.

The linearity and homogeneity of the symmetry defining equations motivated us to consider the related problem of finding rank deficient matrices $\tilde{A}$ near a matrix A. This linear algebra problem is well known to be solved using the Singular Value Decomposition (SVD). For more details on SVD, see [**10, 28**]. Given a tolerance ε, the SVD of A can determine the lowest rank (equivalently largest dimension for $\mathrm{Null}(A)$) that the family of matrices $\tilde{A}$ satisfying $\|\tilde{A} - A\|_2 \leq \varepsilon$ can have. The SVD gives information on how close rank deficient matrices are to A.

The objective of the work initiated in this article is to stably seek close-by systems admitting large symmetry groups in the presence of round off errors from the original system.

3. Jet space and involution

In this section we describe the symbolic algorithm underlying our symbolic-numeric method. It depends heavily on the methods of Jet Space Geometry.

The most basic step in the theory of differential elimination algorithms is to replace differential equations by algebraic ones, by regarding derivatives as formal variables.

A polynomially nonlinear qth order system with n independent variables $x = (x_1, x_2, \ldots, x_n)$ and m dependent variables $(\Psi^1, \Psi^2, \ldots, \Psi^m)$ is represented as a polynomial in $x, u, \underset{1}{u}, \underset{2}{u}, \ldots$. Equivalently and more formally it is represented as an element of the differential ring $\mathbb{F}[x, u, \underset{1}{u}, \underset{2}{u}, \ldots]$ where the field $\mathbb{F}$ is $\mathbb{R}$, $\mathbb{C}$ or $\mathbb{Q}$ and

$$u = \underset{0}{u} \leftrightarrow (\Psi^1, \Psi^2, \ldots, \Psi^m),$$

$$\underset{1}{u} \leftrightarrow \frac{\partial \Psi^l}{\partial x_j},$$

$$\underset{2}{u} \leftrightarrow \frac{\partial^2 \Psi^l}{\partial x_j \partial x_k},$$

$$\ldots.$$

The formal total derivative is:

$$D_{x_j} = \frac{\partial}{\partial x_j} + \sum_l u^l_{x_i} \frac{\partial}{\partial u^l} + \cdots . \tag{3.1}$$

For sake of notational brevity, it has become customary to use the same variable names as the names of the variables in the original physical formulation of the differential system. We follow this convention in our paper.

Thus we consider systems of total derivative order q, of form $R^1 = 0, \ldots, R^s = 0$, or more concisely $R = 0$, where

$$R^k \colon J^q \to \mathbb{C}, \quad J^q = \mathbb{C}^{N_q}, \tag{3.2}$$

and $N_q = n + m\binom{q+n}{q}$, is the number of jet variables of order less than or equal to q.

The spirit of the geometric approach to differential equations is that it is concerned with the (jet) variety of the system:

$$\mathcal{R} = V(R) := \{(x, u, \underset{1}{u}, \ldots, \underset{q}{u}) \in J^q : R^k(x, u, \underset{1}{u}, \ldots, \underset{q}{u}) = 0\}, \tag{3.3}$$

where $\underset{r}{u}$ represents the rth order derivatives. If all of the equations in the system have derivative of order exactly q then a single symbolic prolongation of the system is taken to be:

$$D(R) := \{(x, u, \underset{1}{u}, \ldots, \underset{q+1}{u}) \in J^{q+1} : R^k = 0, D_{x_i} R^k = 0\}. \tag{3.4}$$

If there are equations in R with derivative order less than q then derivatives of those equations are appended to the system to form DR. This process is continued until no undifferentiated equations of lower order remain. The prolongations of a system can be obtained more efficiently by omitting the obviously equivalent equations resulting from the commutativity of partial differentiation. For example, given $\partial R^1/\partial x_1$, $\partial R^1/\partial x_2$, in the first prolongation, $\partial/\partial x_1 \partial/\partial x_2 R^1$ only needs to be listed once in the second prolongation. These and many more refined efficiency improvements are discussed in [**24**].

The corresponding geometric operation to elimination is that of geometric projection. A single geometric projection is defined as:

$$E(R) := \{(x, u, \underset{1}{u}, \ldots, \underset{q-1}{u}) \in J^{q-1} : R^k(x, u, \underset{1}{u}, \ldots, \underset{q-1}{u}, \underset{q}{u}) = 0\}. \tag{3.5}$$

We call the systems: ER, E^2R, ..., E^qR projected systems. They are of respective orders: $q-1, q-2, \ldots, 0$.

The symbol of a qth order system is the linearized highest order part of the system, which is given by the matrix:

$$\operatorname{Symbol} R := \frac{\partial R}{\partial \underset{q}{u}}. \tag{3.6}$$

The symbol of a qth order system $R = 0$ is involutive if

$$\operatorname{rank} \operatorname{Symbol}(DR) = \Sigma_{j=1}^n j \beta_j^{(q)} \tag{3.7}$$

and the $\beta_j^{(q)}$ are the dimensions of certain subspaces of the Null Space of the Symbol of R. Details of this test are given in [**17, 24**].

Definition 1 (Involutive system). *A differential system of order q is said to be involutive* [**17,24**] *in a δ-regular system of coordinates if it passes the elimination test: $(E \circ D)R = R$, has involutive symbol, and satisfies the constant rank conditions given below. Almost all coordinate systems are δ-regular, and can be achieved by a random (linear) change of coordinates if necessary.*

Definition 2 (Constant Rank Conditions). *Let $R(x, \underset{q}{v}) = 0$ be a qth order system with independent variables x and jet variables $\underset{q}{v} = (u, \underset{1}{u}, \ldots, \underset{q}{u})$ corresponding to dependent variables and their derivatives. Then this system satisfies the constant rank conditions at $(x^0, \underset{q}{v}^0) = (x^0, u^0, \underset{1}{u}^0, \ldots, \underset{q}{u}^0) \in V(R) \subseteq J^q$ if there exist nonzero constants α, β such that*

$$\operatorname{rank} \frac{\partial R(x, \underset{q}{v})}{\partial(x, \underset{q}{v})} = \alpha = \operatorname{rank} \frac{\partial R}{\partial \underset{q}{v}}, \quad \operatorname{rank} \frac{\partial R}{\partial \underset{q}{u}} = \beta \tag{3.8}$$

in a neighbourhood of $(x^0, \underset{q}{v}^0)$ in the usual Euclidean norm. We call $(x^0, \underset{q}{v}^0)$ a Euclidean point.

Involution enables a local existence and uniqueness theorem (the Cartan–Kähler Theorem) to be stated for the system [**4**]. It has been used to determine consistent initial conditions for the numerical solution of systems.

In this article we confine ourselves, unless otherwise stated, to linear homogeneous systems of PDE. The constant rank conditions are then satisfied automatically except at a lower ($< n$) dimensional set of non-Euclidean points in the space of independent variables. The more complicated situation, encountered in non-linear systems, of decomposition into components of different dimensions does not occur. As a special case of the Cartan–Kuranishi prolongation theorem [**4**] we have:

Theorem 3.1 (Cartan–Kuranishi prolongation theorem). *At regular points a linear homogeneous system becomes involutive after a finite number of projections and prolongations. That is, given an input system $R = 0$ there exist $C_r, \ldots, C_2, C_1$ with $C_j = E$ or $C_j = D$ for each j:*

$$C_r, \ldots, C_2 C_1(R) \tag{3.9}$$

is an involutive system of order greater than or equal to q except at a lower ($< n$) dimensional set of non-Euclidean points in the space of independent variables.

The approach which will be described here does not use symbolic elimination (the usual approach of exact differential elimination methods such as [**24**]). Instead our approach constructs prolonged systems $D^k R$ and finds an l which for some k, yields $E^l D^k R$ as an involutive system. In other words we always apply prolongations first. To prove that such a k, l exist we need the following easily derived results concerning dimensions and permuting D and E.

Theorem 3.2 (Monotonicity properties of D and E). *Given a system R the following properties hold:*

$$R \subseteq S \implies DR \subseteq DS \quad \text{and} \quad ER \subseteq ES; \tag{3.10}$$

$$EDR \subseteq V(R), \quad EDR \subseteq DER; \tag{3.11}$$

$$\dim E^{l+1} D^k R \leq \dim E^l D^k R, \quad \dim E^{l+1} D^{k+1} R \leq \dim E^l D^k R. \tag{3.12}$$

Proof. The property $R \subseteq S \implies DR \subseteq DS$ and $ER \subseteq ES$ follows directly from the definitions of D and E.

Note that $EDR = \{(x, u, \underset{1}{u}, \ldots, \underset{q}{u}) \in J^q : R(x, u, \underset{1}{u}, \ldots, \underset{q}{u}) = 0, D_{x^j} R = 0\} \subseteq V(R)$, so we have the first property in (3.11). So given $(x, u, \underset{1}{u}, \ldots, \underset{q}{u}) \in EDR$ then

$(x, u, \underset{1}{u}, \ldots, \underset{q}{u}) \in V(R)$, so $(x, u, \underset{1}{u}, \ldots, \underset{q-1}{u}) \in ER$. Therefore for some $\underset{q+1}{w}$ we have $(x, u, \underset{1}{u}, \ldots, \underset{q}{u}, \underset{q+1}{w}) \in DER$. Thus $EDR \subseteq DER$ and the second property of (3.11) is proved.

We have $\dim E^{l+1}D^kR \leq \dim E^lD^kR$ since $\dim ES < \dim S$ for any system S. Let $S = D^kR$, then since $EDS \subseteq V(S)$ from (3.11) we have $\dim EDS \leq \dim S$. By repeated application of this property we obtain $\dim E^{l+1}DS \leq \dim E^lS$. Consequently we have the monotonicity property $\dim E^{l+1}D^{k+1}R \leq \dim E^lD^kR$. □

Theorem 3.3 (Projected involutive systems). *For a given linear homogeneous system of* PDE *there exists a non-negative integer l, and an integer $0 \leq k \leq l$ such that E^kD^lR is involutive except at a lower ($< n$) dimensional set of non-Euclidean points in the space of independent variables.*

Proof. Applying Theorem 3.1, there exist $C_1, \ldots, C_r$ where each C_j is either D or E such that $C_r \cdots C_2C_1(R)$ is an involutive system. Let $k = \#j$ such that $C_j = D$ and $l = \#j$ such that $C_j = E$ then $0 \leq l \leq k$ since the total order of the system is greater than q (i.e. $q + k - l \geq q$).

From the repeated application of the permutation rules (3.11) we have

$$E^lD^kR \subseteq C_r, \ldots, C_2C_1R. \tag{3.13}$$

We will prove that $E^lD^kR = C_r, \ldots, C_2C_1R$. Suppose by contradiction that $E^lD^kR \neq C_r, \ldots, C_2C_1R$ and let $w \in C_r, \ldots, C_2C_1R \backslash E^lD^kR$. Since $w \in C_r, \ldots, C_2C_1R$ by the Cartan–Kähler theorem [4] there exists a local analytic solution passing through w. However such a solution must also satisfy E^lD^kR. But $w \notin E^lD^kR$, contradicting the existence of a local solution. Hence:

$$E^lD^kR = C_r, \ldots, C_2C_1R \tag{3.14}$$

is an involutive system. □

The symbolic algorithm on which our symbolic-numeric method is based is to determine the k, l whose existence is given above, such that E^lD^kR is involutive. However the classical tests for involution, for example the classical elimination test:

$$\dim EDE^lD^kR = \dim E^lD^kR \tag{3.15}$$

together with the classical involutive symbol test, depend on systems which are not in the form E^sD^rR. Our need to develop tests for involution based solely on systems of the form E^sD^rR motivates us to define a projectively involutive system in the following manner.

Definition 3 (Projectively involutive system). *A differential system E^lD^kR is said to be projectively involutive if it passes the projective elimination test:*

$$\dim E^{l+1}D^{k+1}R = \dim E^lD^kR \tag{3.16}$$

and passes the projective involutive symbol test

$$\operatorname{rank} E^lD^{k+1}R = \sum_j j\beta_j \tag{3.17}$$

where the $\beta_j \equiv \beta_j^{(q+k-l)}$ are the characters for the $(q+k-l)$th order system E^lD^kR in a δ-regular system of coordinates.

Theorem 3.4. *A system is projectively involutive if and only if it is involutive.*

PROOF. Suppose that E^lD^kR is projectively involutive.
Then the properties (3.11) and (3.13) imply:

$$\dim E^{l+1}D^{k+1}R \leq \dim EDE^lD^kR \leq \dim E^lD^kR \tag{3.18}$$

and consequently from (3.16) that

$$\dim EDE^lD^kR = \dim E^lD^kR. \tag{3.19}$$

So the elimination test for classical involution is satisfied.

Next we want to show that the projective involutive symbol test (3.17) implies the classical involutive symbol test (3.7). Rewriting the ranks in terms of codimensions we wish to show that

$$N_{q+k+1-l} - \dim E^lD^{k+1}R = \sum_j j\beta_j \rightarrow N_{q+k+1-l} - \dim DE^lD^kR = \sum_j j\beta_j. \tag{3.20}$$

In δ-regular coordinates, $N_{q+k+1-l} - \dim DE^lD^kR \geq \sum_j j\beta_j$, so we only need to show that $N_{q+k+1-l} - \dim DE^lD^kR \leq \sum_j j\beta_j$, or equivalently that $N_{q+k+1-l} \leq \dim DE^lD^kR + \sum_j j\beta_j$. Now

$$N_{q+k+1-l} = \dim E^lD^{k+1}R + \sum_j j\beta_j \leq \dim DE^lD^kR + \sum_j j\beta_j \tag{3.21}$$

by using the permutation properties of dimensions and consequently the symbol is involutive.

Suppose that E^lD^kR is involutive. Then $\dim EDE^lD^kR = \dim E^lD^kR$. Also $\dim E^{l+1}D^{k+1}R \leq \dim EDE^lD^kR = \dim E^lD^kR$.

Repeated application of the properties (3.10), (3.11) yields:

$$E^lD^{k+1}R \subseteq DE^lD^kR. \tag{3.22}$$

Then DE^lD^kR is also involutive since any prolongation of an involutive system is also involutive [**17**].

We will prove that $E^lD^{k+1}R = DE^lD^kR$. Suppose by contradiction that $E^lD^{k+1}R \neq DE^lD^kR$ and let $w \in DE^lD^kR \backslash E^lD^{k+1}R$. Since $w \in DE^lD^kR$ by the Cartan–Kähler Theorem [**4**] there exists a local analytic solution passing through w. However such a solution must also satisfy $E^lD^{k+1}R$. But $w \notin E^lD^{k+1}R$ contradicting the existence of a local solution. Hence:

$$E^lD^{k+1}R = DE^lD^kR. \tag{3.23}$$

Consequently applying E to (3.23) gives

$$E^{l+1}D^{k+1}R = EDE^lD^kR. \tag{3.24}$$

Since $EDE^lD^kR = E^lD^kR$ (3.24) implies $\dim E^{l+1}D^{k+1}R = \dim E^lD^kR$ and so the projective elimination test is satisfied. Also $\operatorname{rank} E^lD^{k+1}R = \operatorname{rank} DE^lD^kR$ from (3.24), so E^lD^kR is projectively involutive. □

In summary a qth order system is prolonged (differentiated) until a projection of the prolonged system (of order $\geq q$) satisfies the projected elimination test. If any of these projections also satisfy the projected involutive symbol test, then we have found an involutive system (without loss we can choose the minimum order ($\geq q$) such system).

4. Symbol-numeric differential elimination method

The method is outlined for linear homogeneous systems with independent variables z. A linear homogeneous differential system of order q can be written in the matrix form:

$$\underset{q}{B}(z)\underset{q}{v} = \mathbf{0}, \quad \underset{q}{v} = \begin{pmatrix} \underset{q}{u} \\ \cdot \\ \cdot \\ \underset{1}{u} \end{pmatrix}. \tag{4.1}$$

For example, consider again the determining system from our simple introductory example (2.2). When we write this system in matrix form with $z = (x, y)$, rounding the floating point numbers for compactness of presentation, it becomes:

$$\begin{bmatrix} 1 & 0 & 0 & 0 & 0 & 0 & 0 & 0 & 0 & 0 & 0 & 0 \\ 0 & 1 & -2 & 0 & 0 & 0 & 13.416y & 0 & 0 & 0 & 0 & 0 \\ 0 & 0 & 0 & 0 & 0 & 1 & 0 & -5.0y^3 & 10.0y^3 & 6.708y & 0 & 15.0y^2 \\ 0 & 0 & 0 & 2 & -1 & 1 & 15.0y^3 & 0 & 6.708y & 0 & 0 & 6.708 \end{bmatrix} \underset{2}{v} = \begin{bmatrix} 0 \\ 0 \\ 0 \end{bmatrix}$$

where $\underset{2}{v}$ is the transpose of the vector $[\xi_{yy}, \eta_{yy}, \xi_{xy}, \eta_{xy}, \xi_{xx}, \eta_{xx}, \xi_y, \eta_y, \xi_x, \eta_x, \xi, \eta]$.

Successive prolongations (differentiations) of the system yield $R, DR, D^2R, \ldots$, and a sequence of corresponding linear homogeneous matrix systems:

$$\underset{q}{B}(z)\underset{q}{v} = \mathbf{0}, \quad \underset{q+1}{B}(z)\underset{q+1}{v} = \mathbf{0}, \quad \underset{q+2}{B}(z)\underset{q+2}{v} = \mathbf{0}, \ldots. \tag{4.2}$$

For linear homogeneous systems our hybrid symbolic-numeric approach is to choose a random point $z = z^0 = (x^0, y^0)$ in the space of independent variables, and substitute it into the sequence of systems above. This yields a sequence of constant matrix systems:

$$\underset{q}{B}(z^0)\underset{q}{v} = \mathbf{0}, \quad \underset{q+1}{B}(z^0)\underset{q+1}{v} = \mathbf{0}, \quad \underset{q+2}{B}(z^0)\underset{q+2}{v} = \mathbf{0}, \ldots. \tag{4.3}$$

For each prolongation, D^kR, the projected systems E^lD^kR, $l = 0, 1, \ldots, k$, are numerically constructed, by replacing the symbolic projection operator E with a numeric projection operator $\widehat{E}$ resulting in the family of systems $\widehat{E}^l\big(D^k(R)\big)$.

It is this projection operator $\widehat{E}$ which we use in the numerical implementation of the involution test sketched below. We first find the singular value decomposition (SVD) of $\underset{q+k}{B}(z^0)$ at a random point z^0:

$$\underset{q+k}{B}(z^0) = U\Sigma V^t \tag{4.4}$$

using one of the available numerical packages (in our case the NAG library). Here U and V are unitary matrices. Σ is a diagonal matrix whose diagonal entries are real decreasing non-negative numbers. The approximate rank r is the number of singular values bigger than a fixed tolerance. Deleting the first r rows of V^t yields an approximate basis for the Null Space of $\underset{q+k}{B}(z^0)$. This yields an estimate for $\dim(D^kR)$. To estimate $\dim\big(\widehat{E}(D^kR)\big)$ the components of the approximate basis for D^kR corresponding to the highest order ($(q+k)$th order) derivatives are deleted. This projected basis yields an approximate spanning set for $\widehat{E}(D^kR)$. Proceeding in the same way, deleting components corresponding to highest order jet variables from the approximate spanning set just obtained, yields an approximate

TABLE 1. $\dim \widehat{E}^l D^k R$ for (2.1)

	$k=0$	$k=1$	$k=2$
$l=0$	8	8	8
$l=1$	6	$\underline{8}$	8
$l=2$		6	8
$l=3$			6

spanning set for $\widehat{E}^2(D^k R)$, and then for $\left(\widehat{E}^3(D^k R)\right)$, etc. Application of the SVD to each of these approximate spanning sets yields the approximate dimensions of $E(D^k R), E^2(D^k R), E^3(D^k R), \ldots$, required for the application of the approximate projected elimination test (3.16).

To execute the projected involutive symbol test (3.7) bases for the spaces $E^l D^k R$ are first constructed as above. Then the subspaces of their symbols are extracted by using the SVD based Subspace Intersection Algorithm described in [**10**, p. 602]. The integer from the left hand side of this test is easily approximated using the methods discussed above. The approximation of the $\beta_k^{(q)}$ is more complicated. In this case, the symbol, or rather approximate bases for each of the subspaces corresponding to the $\beta_k^{(q)}$ can be determined by further use of the SVD based Subspace Intersection Algorithm [**10**].

5. Examples

5.1. $y_{xx}+6.708203932yy_x+5y^3 = 0$. When the symbolic-numeric method is applied to (2.2) we get the dimensions shown in Table 1.

We have calculated slightly more than needed here, by calculating an additional projection. This enables us to calculate the symbol dimensions via:

$$\dim\left(\text{Symbol}\, E^l(D^k R)\right) = \dim E^l(D^k R) - \dim E^{l+1}(D^k R). \tag{5.1}$$

When $\dim\left(\text{Symbol}\, E^l(D^k R)\right) = 0$, or equivalently when the symbol has full rank (a case that occurs in this paper), it is easily shown that the symbol of $E^l(D^k R)$ is involutive [**17**]. It follows that the projected involutive symbol test amounts to testing:

$$\dim E^l(D^k R) = \dim E^{l+1}(D^k R). \tag{5.2}$$

We seek the smallest k such that there exists an $l = 0, \ldots, k$ with $\widehat{E}^l D^k R$ approximately involutive (choosing the largest such l if there are several such values for the given k). Passing the projected elimination test (3.16), amounts to finding the first column with an equal entry in the next column entry with a neighboring equal entry diagonally downwards to its right with both entries being on or above the main diagonal $k = l$. This first occurs for $k = 0$ and $l = 0$. However in this case the version of the involutive symbol test in the form (5.2) is not passed.

At the next prolongation $k = 1$, the approximate projected elimination test is passed when $l = 0$, $l = 1$. Examining the dimensions of the symbol shows that the involutive symbol test is passed when $l = 1$ (the underlined entry in Table 1). Consequently EDR is approximately involutive, and we expect approximately an 8 dimensional solution space and symmetry group. This result coincides with the dimension expected for the case $\alpha = \sqrt{45}$ whose approximate value is 6.708203932.

TABLE 2. $\dim \widehat{E}^l D^k R$ for (5.3)

	$k=0$	$k=1$	$k=2$	$k=3$	$k=4$	$k=5$
$l=0$	8	8	6	3	2	2
$l=1$	6	8	6	3	2	2
$l=2$		6	6	3	2	2
$l=3$			4	3	2	2
$l=4$				3	$\underline{\mathbf{2}}$	2
$l=5$					2	2
$l=6$						2

This example shows that the algorithm was able to determine the dimension of the symmetry group of the close-by equation $y_{xx} + \sqrt{45} y y_x + 5y^3 = 0$. In Section 6 we give some discussion on this topic.

5.2. $\boldsymbol{y_{xx} + 7.1 y y_x + 5y^3 = 0}$**.** Applying the symbolic-numeric method to the case $\alpha = 7.1$ in (2.6)

$$y_{xx} + 7.1 y y_x + 5y^3 = 0 \tag{5.3}$$

yields the results in Table 2.

Applying the same process as in Section 5.1, we conclude that $\widehat{E}^l D^k R$ is approximately involutive for $k = 4$ and $l = 4$ (the underlined entry in Table 2), and we expect approximately a 2 dimensional solution space and symmetry group. This result is consistent with the one we found by the symbolic substitution strategy, namely that in the case $\alpha \neq \sqrt{45}$ there should be a 2 dimensional symmetry group.

5.3. Test examples from the Kamke collection. Kamke [**13**] provided a large collection of ODE with interesting features (usually amenable to some analytical solution technique). This set has become popular as a suite for testing the effectiveness of ODE symbolic solving software. Many of the ODE have symmetries, and systematic methods using invariants of the symmetries can be used to assist in integrating them. Thus they are also used as a suite to bug-check and test software to reduce the overdetermined systems for the symmetries of the ODE.

We took 50 ODE from Kamke's collection, and generated the overdetermined systems of PDE for their symmetries using the `Maple`'s `DEtools[odepde]` command. Integer values were substituted for any parameters occurring in the ODE since our method does not currently allow for parameters.

The 50 overdetermined systems were symbolically reduced (there were no floating point quantities in the systems), without difficulty by the `Maple` differential elimination package `RifSimp`. Using the `RifSimp` function `initialdata`, the number of parameters in the solution spaces of the systems (i.e. the dimensions of their symmetry groups) were determined. Additionally the dimensions of the projected spaces both for the system and its symbol were symbolically constructed at involution to provide exact data to test the output of our symbolic-numeric method.

One of the methods used to detect numerical instability was violation of the monotonicity properties given by Theorem 3.2 in the tables of dimensions for the projected systems and their symbol spaces.

Using a tolerance of 10^{-9}, and performing calculations with `Maple`'s Digits := 15, we applied our symbolic-numeric method to the 50 examples.

In the initial versions of the program 45 of the 50 ODE had their symmetry dimensions successfully and stably determined. The remaining calculations for ODE 24, 26, 27, 28, 33 in Table 3 exhibited instability. All these ODE have high degree monomial terms leading us to conjecture that the instability was due to either very large (or very small) entries in the matrices after substitution of the random points with values greater than 1 (less than 1). This would disguise linear dependencies and independencies in the system.

This motivated us to substitute random points of modulus 1 in the complex exponential form $\exp(i\theta_j)$, and in some cases random real numbers in a sufficiently small real interval about 1. These changes removed the instability observed in the remaining cases and gave results consistent with those determined by the (exact) symbolic differential elimination algorithms.

6. Analytic interpretation of the symbolic-numeric method

Exact methods applied to approximate problems are generally unstable (e.g. three lines intersecting exactly at a point lose this property when their coefficients are replaced by approximate numbers). Backward error approaches seek to address such problems by embedding them in a family C of nearby problems characterized by parameters or functions, so that the solution procedure converges to a nearby exact representative in C (e.g. the least squares solution of the system). We outline such an approach for approximate symmetries for a specific input ODE $y'' = f(x, y, y')$ which will be described in more detail in a forthcoming work.

Fix a tolerance $\varepsilon > 0$ for the symmetry defining system and a point $z_0 = (x_0, y_0, y_0')$ in the region in which we wish to determine approximate symmetries. We embed our specific ODE in the class of ODE of form $y'' = g(x, y, y')$ with g analytic at $z = z_0$:

$$(6.1)\quad y'' = g(x, y, y') = g(z_0) + g_x(z_0)(x - x_0) + g_y(z_0)(y - y_0) + g_{y'}(z_0)(y' - y_0') + \cdots .$$

The Symmetry Defining System for general g is given in (1.6). Since only first partial derivatives of g appear in (1.6), we initially describe the class by the Taylor coefficients up to first order (i.e. by $\mathbf{b} = (g(z_0), g_x(z_0), g_y(z_0), g_{y'}(z_0))$) for general g and by the corresponding numerical vector of values $\mathbf{a}$ for the specific input f).

Thus, using our Linear Algebra approach, the symmetry defining system is written in the form of a matrix system $\underset{2}{B}(\mathbf{b})\underset{2}{v} = 0$ with $\mathbf{b} = \mathbf{a}$ initially as above. Here and in what follows the dependence on the point z^0 is not displayed. The prolongations are carried out symbolically using the general class g. Constant matrix systems result from substituting numerical vector $\mathbf{a}$ of the Taylor coefficients of f for $\mathbf{b}$ at each prolongation. The vector $\mathbf{b}$ at a given prolongation consists of the Taylor coefficients of g up to the maximum order of derivative of g appearing in the prolonged matrix system.

We now describe the determination of a nearby symmetry system supposing that a prolongation of order s has been reached, writing $\underset{s}{B}(b)\underset{s}{v} = 0$ as $B(\mathbf{b})v = 0$ for brevity and $\mathbf{b}$ consists of the Taylor coefficients of g at $z = z^0$ up to order p say.

Then the SVD of $B(\mathbf{a})$ is the decomposition $B(\mathbf{a}) = U\Sigma V^t$. The SVD gives a map ψ_ε from $B(\mathbf{a})$ to a matrix $B^{(1)} = \psi_\varepsilon(B(\mathbf{a})) = U\widetilde{\Sigma}V^t$ where $\widetilde{\Sigma}$ is the matrix obtained by deleting all singular values of magnitude less than the input tolerance ε. Specifically $B^{(1)}$ is a matrix with largest dimension null space (and rank r say) such that $\|B(\mathbf{a}) - B^{(1)}\|_2 = \sigma_{r+1} < \varepsilon$. However $B^{(1)}v = 0$ is not necessarily a symmetry

TABLE 3. Statistics for application of Symbolic-Numeric Method to 50 Kamke ODE where k, l are integers such that $E^l D^k R$ is approximately involutive. $\dim(\mathcal{G})$ is the dimension of the symmetry group of the ODE calculated using the symbolic numeric results.

	ODE	k	l	$\dim(\mathcal{G})$
1	$y'' - y^2 = 0$	5	5	2
2	$y'' - 6y^2 - x^4 = 0$	7	7	0
3	$y'' - 6y^2 - x = 0$	7	7	0
4	$y'' - 6y^2 + 4y = 0$	6	6	1
5	$y'' + y^2 + 2x + 3 = 0$	7	7	0
6	$y'' - 2y^3 - yx + 1 = 0$	5	5	0
7	$y'' - y^3 = 0$	4	4	2
8	$y'' - 2y^3 + 4yx - 2 = 0$	5	5	0
9	$y'' + 4 + 2yx + 3y + y^3 = 0$	5	5	0
10	$y'' + 4 + 2y^2 + 3y + y^3 = 0$	4	4	1
11	$y'' + x^4y^9 = 0$	4	4	1
12	$y''y - 1 = 0$	4	4	2
13	$y''y = 0$	1	0	8
14	$y''y - x^2 = 0$	4	4	1
15	$2(1 + x^2)y'' - xy'^2(x + 4y') + 2(x + y')y' - 2y = 0$	4	4	0
16	$y''' - (1 - x^3y')^3 = 0$	2	2	1
17	$y'' - y'^2/(y + 4) - y'/(x + 2) = 0$	1	0	8
18	$y' - y = 0$	0	0	∞
19	$y'''' + 9y = 0$	2	2	6
20	$y'' - 3y' - y^2 - 2y = 0$	6	6	1
21	$y'' - 7y' - y^3 + 12y = 0$	4	4	1
22	$y'' + 5y' - 6y^2 + 6y = 0$	5	5	2
23	$y'' - 3y' - 2y^3 + 2y = 0$	4	4	2
24	$y'' - 7y'/2 - 45y(y^9 - 1)/16 = 0$	4	4	1
25	$y'' + y' + 2y^8 = 0$	4	4	1
26	$y'' + y' + 2x^3y^8 = 0$	5	5	0
27	$x^4y'' + y^8 = 0$	4	4	1
28	$x^4y'' - x(x^2 + 2y)y' + 4y^2 = 0$	4	4	2
29	$x^4y'' - x^2(x + y')y' + 4y^2 = 0$	4	4	1
30	$x^4y'' + (xy' - y)^3 = 0$	1	0	8
31	$y'' + yy' - y^3 + y = 0$	4	4	1
32	$y'' + (y + 3)y' - y^3 + y^2 + 2y = 0$	4	4	2
33	$x^4y'' + (x^4y + 3x^3)y' - x^4y^3 + x^3y^2 + 2x^2y = 0$	4	4	1
34	$y'' + 2yy' + xy' + y = 0$	5	5	0
35	$y'' + 2yy' + x(y' + y^2) - 1 = 0$	5	5	0
36	$y'' + 3yy' + y^3 + yx - 1 = 0$	1	0	8
37	$y'' + (3y + x)y' + y^3 + xy^2 = 0$	1	0	8
38	$y'' - 3yy' - 3y^2 - 4y - 2 = 0$	4	4	1
39	$y'' - (3y + x)y' + y^3 + xy^2 + x^2y + x^4 = 0$	1	0	8
40	$y'' - 2yy' = 0$	4	4	2
41	$y'' + yy' + 2y^3 = 0$	4	4	2
42	$y'' + x^2y' + x^3 = 0$	1	0	8
43	$y'' + y'^2 + 2y = 0$	4	4	1
44	$y'' + y'^2 + 2y' + 3y = 0$	4	4	1

TABLE 3. continued

	ODE	k	l	$\dim(\mathcal{G})$
45	$y'' + y'^2 + 2y' + 3y^3 = 0$	4	4	1
46	$y'' + y'^2 + 2x^3y = 0$	5	5	0
47	$y'' + y'^2 + 2 = 0$	1	0	8
48	$y'' + yy'^2 + 2y = 0$	4	4	1
49	$y'' + x^2y'^2 + y' = 0$	4	4	1
50	$4x^2y'' - x^4y'^2 + 4y = 0$	4	4	1

defining system (evaluated at $z = z_0$) of a nearby ODE (in general $B^{(1)} \neq B(\mathbf{b})$ for any $\mathbf{b}$). To project back to the class of symmetry defining systems C we use a Newton iteration.

Rather than carrying out this Newton iteration in matrix spaces, as we will show in a forthcoming work, it is more convenient to work in the space of parameters $\mathbf{b}$ versus singular values. Specifically let σ be a singular value of a matrix A, then σ^2 is an eigenvalue of $\bar{A}A$, where bar denotes complex conjugate transpose. Also $\sigma^2 = \overline{\mathbf{w}}\bar{A}A\mathbf{w}$ where $\mathbf{w}$ satisfying $\overline{\mathbf{w}}\mathbf{w} = 1$ is a normalized eigenvector corresponding to σ^2. In the case where the original ODE is a real analytic function of its (real) Taylor coefficients b_k use of the results of [**5, 32**] shows that $\partial\sigma^2/\partial b_k = \overline{\mathbf{w}}\partial\bar{A}A/\partial b_k\mathbf{w}$. This allows us to compute the derivatives with respect the parameters b_k, which are needed in the Newton iteration at $\mathbf{b} = \mathbf{a}$. With $A = B^{(1)}$, $\sigma = \sigma_{r+1}$, using the direction of steepest descent for the Newton iteration, we obtain a new parameter $\mathbf{a}^{(1)}$. This process when repeated yields a sequence of iterates $\mathbf{a}^{(0)} = \mathbf{a}, \mathbf{a}^{(1)}, \mathbf{a}^{(2)}, \ldots$. If this sequence converges (to some $\widehat{\mathbf{a}}$ say) and the corresponding sequence of singular values $\sigma_{r+1}, \sigma^{(1)}, \sigma^{(2)}, \ldots,$ converges to zero, then a nearby symmetry defining system $B(\widehat{\mathbf{a}})v = 0$ with the desired dimension solution space has been determined.

Any g with its Taylor coefficients at $z = z^0$ of order p or less with values $\widehat{\mathbf{a}}$ yields a nearby ODE $y'' = \widehat{g}(x, y, y')$ in the sense that $\|\mathbf{a} - \widehat{\mathbf{a}}\|$ is small.[1]

Our methods apply to systems of polynomial equations as explored in [**20**]. The analytic backward error analysis above is applied as follows. Approximate coefficients in the input system are replaced by symbolic parameters. The procedure is as above with a Newton refinement process on the parameters to yield convergence to a member of the class, and the exact solution of a nearby system. In some cases, the approximate parameters may satisfy some analytic constraints known a priori and an additional projection back to those constraints is needed.

References

1. V. A. Baikov, R. K. Gazizov, and N. H. Ibragimov, *Approximate symmetries*, Mat. Sb. (N.S.) **136(178)** (1988), no. 4, 435–450, 590 (Russian).
2. G. W. Bluman and S. Kumei, *Symmetries and differential equations*, Springer Verlag, New York, 1989.
3. F. Boulier, D. Lazard, F. Ollivier, and M. Petitot, *Representation for the radical of a finitely generated differential ideal*, Proceedings of the 1995 international symposium on Symbolic

[1]Note the 2-norm on the matrices $B(\mathbf{b})$ yields a condition number with respect to the norm on the Taylor coefficients $\mathbf{a}$, expressing how the closeness of the symmetry defining systems is related to the closeness of the ODE at $\mathbf{b} = \mathbf{a}$.

and algebraic computation table of contents (Montréal, 1995), ACM Press New York, NY, USA, 1995, pp. 158–166.

4. R. L. Bryant, S. S. Chern, R. B. Gardner, H. L. Goldschmidt, and P. A. Griffiths, *Exterior differential systems*, Math. Sci. Res. Inst. Publ., vol. 18, Springer-Verlag, New York, 1991.
5. A. Bunse-Gerstner, R. Byers, V. Mehrmann, and N. K. Nichols, *Numerical computation of an analytic singular value decomposition of a matrix valued function*, Numer. Math. **60** (1991), no. 1, 1–39.
6. F. Tony, Chan, *Rank revealing* QR *factorizations*, Linear Algebra Appl. **88/89** (1989), 67–82.
7. R. M. Corless, P. M. Gianni, B. M. Trager, and S. M. Watt, *The singular value decomposition for polynomial systems*, Proceedings of the 1995 international symposium on Symbolic and algebraic computation table of contents (Montréal, 1995), ACM Press New York, NY, USA, 1995, pp. 195–207.
8. W. I. Fushchich and W. M. Shtelen, *On approximate symmetry and approximate solutions of the non-linear wave equation with a small parameter*, J. Phys. A **22** (1989), L887–L890.
9. R. K. Gazizov, *Lie algebras of approximate symmetries*, Nonlinear Math. Phys. **3**(1996), no. 1-2, 96–101; `www.sm.luth.se/~norbet/home_journal/electroic/3-1_2art9.pdf`.
10. G. Golub and C. V. Loan, *Matrix computations*, 3rd ed., John Hopkins Univ. Press, 1996.
11. D. Hartley and R. Tucker, *A constructive implementation of the Cartan–Kähler theory of exterior differential systems*, J. Symbolic Comput. **12** (1991), 655–667.
12. W. Hereman, *Review of symbolic software for the computation of Lie symmetries of differential equations*, Euromath Bull. **1** (1994) 45–79.
13. E. Kamke, *Differentialgleichungen*, N.Y. Chelsea Publ. Co., 1947.
14. E. Mansfield, *Differential Gröbner bases*, Ph.D. thesis, Univ. of Sydney, 1991.
15. P. J. Olver, *Applications of Lie groups to differential equations*, 2nd ed., Grad. Texts in Math., vol. 107, Springer-Verlag, New York, 1993.
16. ———, *Equivalence, invariants, and symmetry*, Cambridge Univ. Press, 1995.
17. J. F. Pommaret, *Systems of partial differential equations and Lie pseudogroups*, Gordon and Breach Science Publ., Inc. 1978.
18. G. J. Reid, P. Lin, and A. D. Wittkopf, *Differential elimination-completion algorithms for* DAE *and* PDAE, Stud. Appl. Math. **106** (2001), no. 1, 1–45.
19. G. J. Reid, C. Smith, and J. Verschelde, *Geometric completion of differential systems using numeric-symbolic continuation*, SIGSAM Bulletin **36** (2002), no. 2, 1–17.
20. G. J. Reid, J. Tang, and L. Zhi, *A complete symbolic-numeric linear method for camera pose determination*, Proceedings of the 2003 international symposium on Symbolic and algebraic computation (Philadelphia, 2003), ACM Press New York, NY, USA, 2003, pp. 215–223.
21. G. Reid, A. Wittkopf, and A. Boulton, *Reduction of systems of nonlinear partial differential equations to simplified involutive forms*, Eur. J. Appl. Math. **7** (1996), 635–666.
22. C. Rust, G. J. Reid, and A. D. Wittkopf, *Existence and uniqueness theorems for formal power series solutions of analytic differential systems*, Proceedings of the 1999 International Symposium on Symbolic and Algebraic Computation (Vancouver, 1999), ACM, New York, pp. 105–112 (electronic).
23. A. Sedoglavic, *Fast computation of discrete invariants associated to a differential rational mapping*, International Symposium on Symbolic and Algebraic Computation (ISSAC'2002) (Lille), J. Symbolic Comput. **36** (2003), no. 3-4, 473–499.
24. W. M. Seiler, *Analysis and application of the formal theory of partial differential equations*, Ph.D. thesis, Lancaster University, 1994.
25. ———, *Involution and constrained dynamics* II. *The Faddeev–Jackiw approach*, J. Phys. A. **28** (1995), 7315–7331.
26. W. M. Seiler and R. W. Tucker, *Involution and constrained dynamics* I. *The Dirac approach*, J. Phys. A. **28** (1995), 4431–4451.
27. A. J. Sommese, J. Verschelde, and C. W. Wampler, *Numerical decomposition of the solution sets of polynomial systems into irreducible components*, SIAM J. Numer. Anal. **38** (2001), no. 6, 2022–2046.
28. Lloyd N. Trefethen and David Bau(3rd), *Numerical linear algebra*, SIAM, Philadelphia, PA, 1997.
29. J. Tuomela and T. Arponen, *On the numerical solution of involutive ordinary differential systems*, IMA J. Numer. Anal. **20** (2000), 561–599.

30. A. Wittkopf and Gregory J. Reid, *Fast differential elimination in C: The* `CDiffElim` *environment*, Computer Physics Communications **139** (2001), 192-217.
31. T. Wolf, `Crack, LiePDE, ApplySym` *and* `ConLaw`*, Section* 4.3.5 *and computer program on* CD-ROM, (J. Grabmeier, E. Kaltofen, and V. Weispfenning, eds.), Computer Algebra Handbook Springer, 2002, pp. 465–468.
32. K. Wright, *Differential equations for the analytic singular value decomposition of a matrix*, Numer. Math. **63** (1992), no. 2, 283–295; `http://www.cs.ncl.ac.uk/research/pubs/articles/papers/125.pdf`.

Ontario Research Centre for Computer Algebra, c/o Department of Applied Mathematics, University of Western Ontario, London, ON N6A 5B7, Canada.
E-mail address: `eigenvalue57@hotmail.com`

Équipe Calcul Formel, Laboratoire d'Informatique Fondamentale de Lille, Université de Lille I, 59655 Villeneuve d'Acsq, France.
E-mail address: `lemaire@lifl.fr`
URL: `http://www.orcca.on.ca/~lemaire`.

Ontario Research Centre for Computer Algebra, c/o Department of Applied Mathematics, University of Western Ontario, London, ON N6A 5B7, Canada.
E-mail address: `reid@uwo.ca`
URL: `http://www.orcca.on.ca/~reid`.

Ontario Research Centre for Computer Algebra, c/o Department of Applied Mathematics, University of Western ON London, Ontario, N6A 5B7, N6A 5B7, Canada.
E-mail address: `rscott2@uwo.ca`

Mathematics Mechanization Research Center, Institute of Systems Science, Academy of Mathematics and System Sciences, Academia Sinica, China.
URL: `http://www.mmrc.iss.ac.cn/~lzhi`.
E-mail address: `lzhi@mmrc.iss.ac.cn`

Centre de Recherches Mathématiques
CRM Proceedings and Lecture Notes
Volume **39**, 2005

Discrete and Finite Fractional Fourier Transform

K. B. Wolf

ABSTRACT. The fractional Fourier integral transform has been of interest for signal analysis in phase space, and it can be implemented by optical means for parallel processing. Discrete and finite versions of the fractional Fourier transform have been sought because such are data to be sensed and processed in shallow waveguides. We use Lie algebras and q-algebras to set up finite oscillator models, in one and two dimensions, in Cartesian and polar coordinates, whose configuration space consists of a finite number of 'sensor' points, and whose unitary evolution defines corresponding finite fractional Fourier–Kravchuk, Fourier–q–Kravchuk, and Hankel–Hahn transforms. Such models support coherent states and have well-defined contraction limits to their continuous counterparts.

1. Introduction

The Fourier integral transform $\mathcal{F}$ is the bridge between the descriptions of physical systems in position and in momentum. It plays a fundamental rôle in Hamiltonian systems and is used widely in signal and image analysis. The *fractional* Fourier transforms $\mathcal{F}^\alpha$ form a 1-parameter subgroup $\mathcal{C}\left(\begin{smallmatrix}\cos\theta & \sin\theta\\ -\sin\theta & \cos\theta\end{smallmatrix}\right)$, $\theta = \frac{1}{2}\pi\alpha$, of the canonical transforms studied by Moshinsky and Quesne [**29, 30, 34**] in the context of quantum mechanics. This fractional transform can be produced optically [**33**], within the paraxial régime, with lenses, or within a segment of harmonic waveguide. Quite naturally, it is of interest to have a *finite* version of this transform.

Finite Fourier transforms between $\Re^N$ vector spaces of elements $\mathbf{f} = \{f_n\}_{n=1}^N$ and $\widetilde{\mathbf{f}} = \{\widetilde{f}_n\}_{n=1}^N$ are also well known [**40**, Part I],

$$(1.1)\qquad \widetilde{f}_n := \frac{1}{\sqrt{N}}\sum_{m=1}^{N} e^{2\pi i mn/N} f_m, \quad f_n = \frac{1}{\sqrt{N}}\sum_{m=1}^{N} e^{-2\pi i mn/N}\widetilde{f}_m,$$

and can be found numerically by means of the Cooley–Tukey fast Fourier transform (FFT) algorithm. It has been therefore quite natural to search for a fractionalization of this finite transform. Now, both the finite and integral Fourier transforms are fourth roots of unity, $\mathcal{F}^4 = \mathit{1}$; and of course, when we have an rth root of unity, linear combinations with trigonometric coefficient functions of that operator and its

2000 *Mathematics Subject Classification.* 22E45, 39A70, 93C57, 94A11.
Work supported by project DGAPA–UNAM IN112300 *Optica Matemática*.
This is the final form of the paper.

r powers can be used to produce a 'fractionalization' such that $\mathcal{F}^{\alpha_1}\mathcal{F}^{\alpha_2} = \mathcal{F}^{\alpha_1+\alpha_2}$. Several authors have done so [**18, 21, 22, 27, 28, 35–37**], but the result does not approximate the Moshinsky–Quesne continuous oscillator integral transform at all. There are no coherent states, no good energy levels, and no dynamic support for such a theory.

In this contribution I will give an overview of current research of the *Optica Matemática* project where we have introduced a set of finite Fourier-*like* transforms based on the very foundations of the harmonic oscillator, as Moshinsky and Quesne did for the continuous model, but where the configuration space is discrete and finite *ab initio*. The following Sections will give a résumé of finite fractional Fourier–Kravchuk, Fourier-q-Kravchuk, and Hankel–Hahn transforms. As will be seen, the construction is based on very-well-known group-theoretical results from the three- and four-dimensional rotation groups. All finite oscillator models are realized as Hamiltonian systems with unitary evolution, equally-spaced energy levels (bounded from below and from above), and satisfy a Schrödinger *difference* equation [**3, 13**]. Elsewhere, we have also shown that these systems support coherent states [**14, 16**], have a compact phase space, and admit a Wigner function which is covariant with the dynamical group of the system [**4**].

2. One-dimensional finite Fourier–Kravchuk transform

The following axioms define both the common well-known continuous quantum mechanical oscillator, and the finite oscillator model.

(1) There exists an essentially self-adjoint *position* operator Q, whose spectrum $\{q\}$ is the set of positions of the system.
(2) There exists a *Hamiltonian* operator H, self-adjoint and compact, whose spectrum $\{h\}$ is the set of energies of the system.
(3) The two operators satisfy the Hamilton equations with the commutators:

$$[H, Q] =: -iP, \quad [H, P] = iQ. \tag{2.1}$$

The first equation defines the *momentum* operator; the second Hamilton equation contains the oscillator dynamics.

(4) Under commutators, all operators close into an *associative* algebra, *i.e.*, they satisfy the Jacobi identity, which in turn implies

$$\underbrace{[Q, \overbrace{[P, H]}^{-iQ}]}_{=0} + \underbrace{[P, \overbrace{[H, Q]}^{-iP}]}_{=0} + [H, [Q, P]] = 0. \tag{2.2}$$

These axioms contain several oscillator models. First of all we should note that (2.2) gives the only 'condition' on our oscillator structure: the *as-yet undefined* commutator between the position and momentum operators, $[Q, P]$, *must commute* with the Hamiltonian: $[H, [Q, P]] = 0$. Thus we have several choices for the key commutator (cf. Ref. [**39**]), each of which entails a distinct model of the harmonic oscillator:

$$(2.3)\qquad [Q,P] = \begin{cases} i\hbar & \Longrightarrow \text{ordinary quantum harmonic oscillator,} \\ i(H-\tfrac{1}{2}N) & \Longrightarrow \text{Lie algebra u(2) } (N+1)\text{-point finite system,} \\ iF_q(H,N) & \Longrightarrow q\text{-algebra } \mathrm{u}_q(2)\ N+1 \text{ points } x_s = \tfrac{1}{2}[2s]_q \text{ unequally spaced,} \\ \text{other } F\text{'s} & \quad \ldots \text{other models} \ldots \\ \text{in } D \text{ dimensions} & \quad \text{Cartesian: } \mathrm{u}(1)\oplus[\mathrm{su}(2)]^D, \text{ radial: } \mathrm{u}(1)\oplus\mathrm{so}(D). \end{cases}$$

The ordinary quantum harmonic oscillator model is obtained with the Heisenberg–Weyl Lie algebra in the irreducible representation labelled by $\hbar \in \Re^+$ [**41**], and the Hamiltonian is realized by $H = \frac{1}{2}(P^2 + Q^2)$.

On the other hand, the *finite* oscillator model satisfies the axioms with the Lie algebra su(2),

$$(2.4)\qquad [H,P] = iQ, \quad [H,Q] = -iP, \quad [Q,P] = i(H - j - \tfrac{1}{2}),$$

in the unitary irreducible representation labelled by $j := \frac{1}{2}N \in \{0, \frac{1}{2}, 1, \ldots\}$. There, the Casimir operator is a multiple of unity, $J^2 := J_1^2 + J_2^2 + J_3^2 = j(j+1)\hat{1}$, and the number j is used to displace the spectrum of H to $h = \nu + \frac{1}{2}$, $\nu \in \{0, 1, \ldots, N\}$. We use the central operator $E_J := j\hat{1}$, generator of an u(1) algebra, which specifies the representation. In this representation (as is well known), both the spectra $\{q\}$ and $\{h\}$ consist of $N+1$ points.

The identification of physical observables in the finite oscillator with the generators in the algebra and their spectra is:

$$(2.5)\qquad \begin{array}{llll} \text{position:} & \leftrightarrow & Q = J_1, & q \in \{-j, -j+1, \ldots, j\}, \\ \text{- momentum:} & \leftrightarrow - & P = J_2, & p \in \{-j, -j+1, \ldots, j\}, \\ \text{Hamiltonian:} & \leftrightarrow & H = J_3 + j + \frac{1}{2}, & h \in \{\frac{1}{2}, \frac{3}{2}, \ldots, 2j + \frac{1}{2}\}, \\ \text{mode number:} & & H - \frac{1}{2} = J_3 + j, & \nu \in \{0, 1, \ldots, N\}. \end{array}$$

The first two commutators in (2.4) are common to all models; the third is the *nonstandard* commutator which characterizes the finite oscillator model.

The ket states of definite position and of definite energy are defined by

$$(2.6)\qquad \begin{array}{ll} E_J|N;q\rangle_1 = \frac{1}{2}N|N;q\rangle_1, & E_J|N;\nu\rangle_H = \frac{1}{2}N|N;\nu\rangle_H, \\ Q|N;q\rangle_1 = q|N;q\rangle_1, & H|N;\nu\rangle_H = (\nu + \frac{1}{2})|N;\nu\rangle_H. \end{array}$$

The *wavefunctions* of the 1-dim finite oscillator system are the *overlaps* between the position and energy eigenbases. Within su(2), these are nothing else but the Wigner little-d functions for the rotation of $\frac{1}{2}\pi$, which relates one axis to the other:

$$(2.7)\qquad \Phi_\nu^{(N)}(q) := {}_1\langle N; q \mid N; \nu\rangle_H = {}_H\langle N; q + j|e^{+i\pi/2J_2}|N;\nu\rangle_H$$

$$(2.8)\qquad = d^j_{q,\nu-j}(-\tfrac{1}{2}\pi) = d^j_{\nu-j,q}(\tfrac{1}{2}\pi) = (-1)^{\nu-j-q}\, d^j_{j-\nu,-q}(\tfrac{1}{2}\pi)$$

$$(2.9)\qquad = \frac{(-1)^\nu}{2^j}\sqrt{\binom{2j}{\nu}\binom{2j}{j+q}}\, K_\nu(j+q; \tfrac{1}{2}, 2j).$$

Here we note the fact that the Wigner d-functions are given in terms of the symmetric *Kravchuk* polynomials,

$$K_\nu(x;\tfrac{1}{2},N) := {}_2F_1(-\nu,-\tfrac{1}{2}N-x;-N;2), \tag{2.10}$$

of degree ν in x; they are among the families of 'discrete' polynomials, and are orthogonal with respect to the binomial distribution on the one-dimensional, finite and equally-spaced spectrum of positions.

Plotting the Kravchuk functions (2.8) (see Refs. [**14, 16**]) we can see that the lower-energy states closely resemble the well-known Hermite functions of the ordinary quantum oscillator on the finite interval $q \in [-j,j]$. By the middle of the energy multiplet, $\nu \approx \frac{1}{2}N$, the wavefunction zig-zags every two points with an amplitude which increases somewhat at the interval ends. Finally, near to the top state $\nu \approx N$, they repeat the $\nu \approx 0$-functions but with alternating signs—due to the factor $(-1)^q$ in (2.9). Beyond the highest energy state $\Phi_N^{(N)}(q)$ the wavefield cannot propagate in the shallow optical waveguide—or vacuum breaks with pair creation.

The $N+1$ Kravchuk functions (2.9) satisfy no differential equation, but a *difference* one, which relates the values of the polynomial on three neighboring points (not necessarily integer), separated by one unit. The Schrödinger difference equation for the finite oscillator wavefunctions $\Phi_\nu^{(N)}(q)$ is

$$H^{(N)}(q)\Phi_\nu^{(N)}(q) = (\nu+\tfrac{1}{2})\Phi_\nu^{(N)}(q), \quad \nu = 0,1,\ldots,2j =: N, \tag{2.11}$$

with the Hamiltonian difference operator

$$H^{(N)}(q) := -\tfrac{1}{2}[\alpha^j(q)e^{-\partial_q} - (2j+1) + \alpha^j(-q)e^{\partial_q}], \tag{2.12}$$

$$\alpha^j(m) := \sqrt{(j+m+1)(j-m)} = \alpha^j(-m-1), \tag{2.13}$$

with the unit-shift operators $e^{a\partial_q}f(q) = f(q+a)$, for $a = \pm 1$.

We note that the Hamiltonian (2.12) does not separate into a sum of 'kinetic' plus 'potential energy' terms, and it would be difficult to 'guess' its form out of the well-known continuous form $\frac{1}{2}(\partial_q^2 + q^2)$. Yet, as we know from ordinary quantum mechanics, the ground state $\Psi_0(x)$ of a system with a potential $V(x)$ and energy $E_0 > -\infty$, has no zeros; thus, the Schrödinger equation

$$(-\tfrac{1}{2}\nabla^2 + V(q) - E_0)\Psi_0(q) = 0 \implies V(q) - E_0 = \tfrac{1}{2}\nabla^2\Psi_0(q)/\Psi_0(q). \tag{2.14}$$

In our discrete system, an *equivalent* potential may be defined following (2.14), as [**6**]

$$V(q) - E_0 := \frac{\psi(q+1) - 2\psi(q) + \psi(q-1)}{2\psi(q)}. \tag{2.15}$$

In the finite Kravchuk oscillator, the set of positions is finite. Yet, the wavefunctions (2.9) can be analytically continued in q everywhere except for branchpoint zeros at $q = \pm(j+1)$, which are due to the square root of the binomial distribution. Thus the second difference in Eq. (2.15) is defined for real values of q in the interval $|q| \leq j+1$,

$$2\big(V(q) - E_0\big) + 1 = \frac{\sqrt{(j+q)(j+q+1)} + \sqrt{(j-q)(j-q+1)}}{\sqrt{(j+1)^2 - q^2}}. \tag{2.16}$$

The time evolution of the finite oscillator is generated by the Hamiltonian H. For the common quantum oscillator this the canonical transform of Moshinsky and Quesne mentioned above, which is the phase $\exp(i\frac{1}{4}\pi\alpha)$ times the fractional Fourier transform $\mathcal{F}^\alpha$ of power $\alpha = 2t/\pi$. There is a straightforward extension to the finite oscillator, where we define the fractional Fourier–*Kravchuk* operator with the summation kernel of the *fractional Fourier–Kravchuk transform* to be

$$\mathcal{K}^\alpha := e^{i1/4\pi\alpha} \exp(-i\tfrac{1}{2}\pi\alpha H), \tag{2.17}$$

$$K^{(N,\alpha)}_{q,q'} := {}_1\langle N, q|\mathcal{K}^\alpha|N, q'\rangle_1 = i^{q'-q} d^{N/2}_{q,q'}(\tfrac{1}{2}\pi\alpha). \tag{2.18}$$

As was shown in Ref. [**11**], the contraction limit $N \to \infty$ returns the ordinary fractional Fourier kernel.

3. Finite q-oscillator

The theory of q-algebras has been a disappointment, in a sense, because they have not provided the plethora of applications initially expected from their extra parameter. However, as shown in Ref. [**2**], a q-algebra *is* solution to the oscillator axioms (2.3) and moreover, serves for signal analysis on a non-equally-spaced set of points. [Note: in this section, position will be denoted by x_s, $s|_{-j}^{j}$, while q will be the quantum deformation number.]

A finite q-oscillator is being proposed in Ref. [**6**],

$$[J_2, J_3] = iJ_1, \quad [J_3, J_1] = iJ_2, \quad [J_1, J_2] = i\tfrac{1}{2}[2J_3]_q = i\frac{\sinh \kappa J_3}{2\sinh \kappa/2}, \qquad \kappa = -\ln q. \tag{3.1}$$

on the basis of a good assignment of the position, momentum and energy operators, which takes the place of Eqs. (2.5). This is the *nonstandard* basis established by Atakishiyev and Winternitz in Ref. [**5, 15**], for the q-algebra $\mathrm{su}_q(2)$: The first two commutators have the structure of the oscillator Hamilton equations (2.1), while the third one involves the q-number, which distinguishes the q-algebras from a Lie algebra ($q = 1$). The identifications are the following:

$$\begin{array}{llll} \text{position:} & Q \leftrightarrow \widetilde{J}_1 := q^{1/4J_3} J_1 q^{1/4J_3}, & x_s = \tfrac{1}{2}[2s]_q, \\ \text{-momentum:} & -P \leftrightarrow \widetilde{J}_2 := q^{1/4J_3} J_2 q^{1/4J_3}, & p_s = \tfrac{1}{2}[2s]_q, \\ \text{mode number as before:} & H - \tfrac{1}{2} = J_3 + j, & \nu \in \{0, 1, \ldots, N\}. \end{array} \tag{3.2}$$

We note that the q-number gives anti-symmetric wavefunctions under $q \leftrightarrow q^{-1}$: the ground state of a $q < 1$ oscillator becomes the top state of its $q^{-1} > 1$ partner on the *same* set of points

$$x_s = \frac{1}{2}[2s]_q = \frac{\sinh s\kappa}{2\sinh \kappa/2} = \frac{1}{2}[2s]_{q^{-1}}, \quad q := e^{-\kappa}, \quad s \in \{-j, -j+1, \ldots, j\}. \tag{3.3}$$

As before, the overlap between the position and the mode basis yields the wavefunctions of the system: a finite, complete, orthonormal set of functions $\Psi_\nu^{(j|q)}(x_s)$, on the non-equally-spaced set of points (3.3). These involve *dual q-Kravchuk* functions [**6, 38**].

In the model of finite optics that we have developed in Cuernavaca over the past few years, the 'quantum' model proposed above is the first departure from equally-spaced sets of sensors. The retina of mammals also has discrete sensor *areas* each covered by a single neuron, and they tend to crowd towards the fovea. The distribution of sensor points (3.3) follows closely this pattern: dense at the center, sparse at the edges, as the hyperbolic sine of $s\kappa$, with s an integer symmetric and bounded by the representation parameter j, and the 'q'-parameter $\kappa = -\ln q$. We have observed that the range of q on which the lowest-state wavefunction is a 'credible' discrete Gaussian is above $q \approx 0.7$, including our previous results on the finite Kravchuk oscillator for $q = 1$ [**14, 16**]. It may well be that the 'best' analytic eigenfunctions for retina-like sensor distributions is the basis of dual q-Kravchuk functions [**6**].

Contractions of the finite sensor set to the continuum have been made for the Lie algebras, Schrödinger equations and wavefunctions in the case of equally-spaced sensors, in Refs. [**11**]. We had thought that one could produce a 'continuous' q-oscillator based on q-phase space with a proper q-Hamiltonian. In fact, the model shows that q-phase space is a *spheroïd*—a q-deformation of the spherical phase space of a standard discrete system [**4, 14**]. Yet, and although the contraction of the q-algebra yields the Biedenharn–Macfarlane q-oscillator algebra [**19, 26**], we have found a paradoxical situation: the bounded spectrum of q-position Q in (2.5) becomes unbounded as the number of points is made to grow with $N \to \infty$. In particular the extreme points $x_{\pm j}$ race exponentially ahead of the rest; this discrete set can never become dense, therefore. We do not recall any such 'forbidden' contraction in other areas of physics.

4. Two-dimensional oscillator models

Finite signal analysis is performed on a line of sensor points, but image processing requires a two-dimensional array of pixels. The simplest generalization of the one-dimensional oscillator of the algebra $\mathrm{u}(2) = \mathrm{u}(1) \otimes \mathrm{su}(2)$ is as a direct product of two orthogonal Cartesian coordinates, $\mathrm{u}(1) \otimes \mathrm{su}(2)_x \mathrm{su}(2)_y = \mathrm{u}(1) \otimes \mathrm{so}(4)$ [**10**], where the number of sensors in the x- and in the y-directions is $N+1$, as determined by the eigenvalue of the central $\mathrm{u}(1)$ generator. This finite 2-dim oscillator model applies thus to optical harmonic waveguides whose fields are sensed on a square pixellated screen.

The continuous oscillator model in quantum mechanics is separable in Cartesian, polar, and elliptic coordinate systems. The process of separation of variables under the ægis of a Lie algebra is well known for partial differential equations [**17, 23–25**]; for *difference* equations, the maximal abelian subalgebras and subalgebra chains of so(4) are relevant. The Gel'fand–Tsetlin chain

$$\mathrm{so}(4) \supset \mathrm{so}(3) \supset \mathrm{so}(2) \tag{4.1}$$

serves to separate the $(N+1)^2$-point 2-dim finite oscillator in polar coordinates of radius and angle. This requires a 'vectorized' set of axioms for the finite oscillator that include angular momentum. In [**9**], we proposed:

1. There exists a **position** subalgebra of generators $\mathbf{Q} := \mathrm{span}\{Q_x, Q_y\}$, plus those obtained from their commutators (which need not vanish), whose spectra are the position coordinates of the system.

2. There exists one **Hamiltonian** operator, again self-adjoint and compact, which satisfies the vector Hamilton-Lie equations (2.1), $[H, \mathbf{Q}] := -i\mathbf{P}$ and $[H, \mathbf{P}] = i\mathbf{Q}$, so there also exists a **momentum** subalgebra of algebraic basis $\mathbf{P} := \mathrm{span}\{P_x, P_y\}$, plus their commutators.

3. There exists an **angular momentum** subalgebra so(2) whose generator is $M = M_{x,y}$ under which $\mathbf{Q}$ and $\mathbf{P}$ transform as vectors, and commuting with H,

$$\begin{aligned} &[M_{\alpha,\beta}, Q_\gamma] = -i(\delta_{\beta,\gamma}Q_\alpha - \delta_{\alpha,\gamma}Q_\beta), \\ &[M_{\alpha,\beta}, P_\gamma] = -i(\delta_{\beta,\gamma}P_\alpha - \delta_{\alpha,\gamma}P_\beta), \end{aligned} \qquad [M_{\alpha,\beta}, H] = 0. \tag{4.2}$$

4. Under Lie brackets, all operators close into a **Lie algebra** (we have not yet extended this to q-algebras, but see Ref. [**2**]).

The Lie algebra $\mathrm{u}(1)\otimes\mathrm{so}(4)$ that we propose thus requires the following further *nonstandard* commutators between the generators

$$[Q_x, P_x] = iJ = [Q_y, P_y], \tag{4.3}$$

$$[Q_x, Q_y] = iM = [P_x, P_y], \tag{4.4}$$

where the Hamiltonian H is J plus $2j+1$ (for ground energy 1),

$$H = J + 2E_J + \hat{1}, \tag{4.5}$$

where $E_J = j\hat{1} \in \mathrm{u}(1)$ in each totally symmetric irreducible representation space of so(4) characterized by $j \in \{0, \frac{1}{2}, 1, \dots\}$. While the Lie bracket (4.3) is familiar from the one-dimensional finite oscillator model, the new finite radial oscillator model has the commutator (4.4), which states that Q_x and Q_y, or P_x and P_y, cannot be simultaneously diagonalized; an so(3)-complementarity principle holds with angular momentum.

For the so(4) dynamical algebra, we proposed the following identification between the generators $\Lambda_{\alpha,\beta}$ and physical quantities of position, momentum, shifted mode (J), and angular momentum [**9**]:

$$\begin{array}{|c|c|c|} \hline \Lambda_{1,2} & \Lambda_{1,3} & \Lambda_{1,4} \\ \hline & \Lambda_{2,3} & \Lambda_{2,4} \\ \cline{2-3} & & \Lambda_{3,4} \\ \cline{3-3} \end{array} = \begin{array}{|c|c|c|} \hline J & P_x & P_y \\ \hline & Q_x & Q_y \\ \cline{2-3} & & M \\ \cline{3-3} \end{array} = \begin{array}{|c|c|c|} \hline S_3 & -S_2 & S_1 \\ \hline & R_1 & R_2 \\ \cline{2-3} & & R_3 \\ \cline{3-3} \end{array}$$

In the last pattern we indicate the so(3) Gel'fand–Tsetlin subalgebra of position by $\{R_1, R_2, R_3\} = \{Q_x, Q_y, M\}$, whose Casimir operator $\mathbf{R}^2$ has the eigenvalues of *'squared' radius*, (shifted) mode, and angular momentum,

$$R^2 := \mathbf{Q}^2 + M^2, \quad \rho(\rho+1), \quad 0 \le \rho \le N = 2j, \tag{4.6}$$

$$J + j, \quad \nu, \quad 0 \le \nu \le N, \tag{4.7}$$

$$M, \quad m, \quad |m| \le \rho \le N. \tag{4.8}$$

From the overlap between the eigenstates classified by the chains $\mathrm{so}(4) \supset \mathrm{so}(2)_J \oplus \mathrm{so}(2)_M$ and $\mathrm{so}(4) \supset \mathrm{so}(3)_R \supset \mathrm{so}(2)_M$, yields the finite oscillator radial wavefunctions,

$$(-1)^\rho \Phi^{(N)}_{\nu,m}(\rho) := {}_{\mathrm{RM}}\langle N; \rho, m \mid N; \nu, m\rangle_{\mathrm{JM}} = C^{N/2,N/2,\rho}_{(m+\nu)/2,(m-\nu)/2,m} \tag{4.9}$$

$$= \frac{N!}{m!}\sqrt{\frac{(2\rho+1)}{(N-\rho)!(N+\rho+1)!}\frac{(\rho+m)!}{(\rho-m)!}\frac{(j+m_b)!(j-m_b)!}{(j+m_a)!(j-m_a)!}} \times (-1)^{j-m_a} R_{j-m_a}\big(\rho(\rho+1); m, -m, N\big), \tag{4.10}$$

where $C^{N/2,N/2,\rho}_{(m+\nu)/2,(m-\nu)/2,m}$ are the su(2) *Clebsch–Gordan* coefficients, and

$$R_n\big(\rho(\rho+1);m,m',N\big) := {}_3F_2\left(\begin{matrix}-n,-\rho,\rho+m+m'+1\\ m+1,-N\end{matrix}\,\middle|\,1\right) \tag{4.11}$$

are the dual *Hahn* polynomials.

Out of these states of definite angular momentum m and radius ρ, we obtain the radius-and-angle states through the discrete $(2\rho+1)$-dim Fourier transform (1.1),

$$\Phi^{(N)}_\nu(\rho,\theta_k) := \sum_{m=-\rho}^{\rho} e^{im\theta_k}\Phi^{(N)}_{\nu,m}(\rho), \qquad \begin{matrix}\theta_k = 2\pi k/N,\\ k\in\{-\rho,-\rho+1,\ldots,\rho\}.\end{matrix} \tag{4.12}$$

The surface element in these coordinates is $\Delta\sigma := \rho\Delta\rho\Delta\theta_k$; with $\Delta\rho = 1$ and $\Delta\theta_k = \theta_k - \theta_{k-1} = 2\pi/(2\rho+1)$; the 'density of sensor points' or pixels per unit are is thus $\pi\rho/(\rho+\frac{1}{2}) \approx \pi$. In the contraction limit $N \to \infty$ [**12**], these are unitary approximants to the well-known radial harmonic oscillator wavefunctions with Laguerre functions.

The generator of Fourier transforms is again the mode number operator, and its exponential is the radial analogue of Eqs. (2.17)–(2.18),

$$\mathcal{K}^\alpha := e^{i1/2\pi\alpha}\exp(-i\tfrac{1}{2}\pi\alpha H), \tag{4.13}$$

$$K^{(N,\alpha)}_{q,q'} = d^{N,0}_{\rho,m,\rho'}(\tfrac{1}{2}\pi\alpha). \tag{4.14}$$

These are *fractional Hankel–Hahn transforms*, which are the finite-N unitary approximations to the fractional Hankel integral transforms, on a sensor grid of $N+1$ concentric circles of integer radius ρ, where on each circle are $\rho+1$ equidistant points. We conjecture that an appropriate q-algebra so$_q(4)$ can be built so that the $N+1$ radii have a sine-hyperbolic spacing, which is closer to the observed neuron density in the retina around the fovea.

5. Outlook

We are currently looking at various discrete and finite models that can be realized with micro-optical devices studded with sensors and processed in shallow planar, rectangular and circular cylindrical waveguides. Concretely, we are interested in the general case of discrete *elliptic* coordinates, related to a non-subalgebra reduction of so(4), and also in non-equally spaced 2-dim q-arrangements.

In every case there is a phase space picture with a Wigner function over the (space of coadjoint orbits of the) algebra [**1**]. Symplecticity of the continuous models corresponds to unitarity in their discrete counterparts. The information on square and on round pixellated screens, can be unitarily mapped from one to the other, as well as linearly Fourier-transformed or rotated. And we uphold harmonic motion, which is reflected in the existence and the dynamics of coherent states in these finite Hamiltonian systems.

The understanding of models whose energy eigenfunctions are Wigner or Bargmann d's [**7, 8**], and Clebsch–Gordans, has value in pure as well as applied mathematics. Particularly, measurement theory for discrete systems, such as uncertainty relations [**20,31,32**] for spot widths, requires further development. Other topics that we can touch are, as mentioned above, the separation of discrete variables and the factorization of their dynamical (Schrödinger) difference equations, and concrete applications for the theory of discrete polynomials.

References

1. S. T. Ali, N. M. Atakishiyev, S. M. Chumakov, and K. B. Wolf, *The Wigner function for general Lie groups and the wavelet transform*, Ann. Henri Poincaré **1** (2000), no. 4, 685–714.
2. M. Arık, N. M. Atakishiyev, and K. B. Wolf, *Quantum algebraic structures compatible with the harmonic oscillator Newton equation*, J. Phys. A **32** (1999), L371–L376.
3. R. Askey, N. M. Atakishiyev, and S. K. Suslov, *Fourier transformations for difference analogs of the harmonic oscillator*, Proceedings of the XV Workshop on High Energy Physics and Field Theory (Protvino, 1992) (A. P. Samokhin and G. L. Rcheulishvili, eds.), IHEP, Serpukhov, 1995, pp. 140–144.
4. N. M. Atakishiyev, S. M. Chumakov, and K. B. Wolf, *Wigner distribution function for finite systems*, J. Math. Phys. **39** (1998), 6247–6261.
5. N. M. Atakishivev and A. U. Klimyk, *Diagonalization of operators and one-parameter families of nonstandard bases for representations of* $\mathrm{su}_q(2)$, J. Phys. A **35** (2002), 5267–5278.
6. N. M. Atakishiyev, A. U. Klimyk, and K. B. Wolf, *Finite q-oscillator*, J. Phys. A **37** (2004), no. 21, 5569–5587.
7. N. M. Atakishiyev, E. I. Jafarov, Sh. M. Nagiyev, and K. B. Wolf, *Meixner oscillators*, Rev. Mexicana Fís. **44** (1998), 235–244.
8. N. M. Atakishiyev, Sh. M. Nagiyev, L. E. Vicent, and K. B. Wolf, *Covariant discretization of axis-symmetric linear optical systems*, J. Opt. Soc. Amer. A **17** (2000), 2301–2314.
9. N. M. Atakishiyev, G. S. Pogosyan, L. E. Vicent, and K. B. Wolf, *Finite two-dimensionsal oscillator. II. The radial model*, J. Phys. A **34** (2001), 9399–9415.
10. ———, *Finite two-dimensionsal oscillator. I. The Cartesian model*, J. Phys. A **34** (2001), 9381–9398.
11. N. M. Atakishiyev, G. S. Pogosyan, and K. B. Wolf, *Contraction of the finite one-dimensional oscillator*, Internat. J. Modern Phys. A **18** (2003), 317–327.
12. ———, *Contraction of the finite radial oscillator*, Internat. J. Modern Phys. A **18** (2003), 329–341.
13. N. M. Atakishiyev and S. K. Suslov, *Difference analogs of the harmonic oscillator*, Theoret. and Math. Phys. **85** (1991), 1055–1062.
14. N. M. Atakishiyev, L. E. Vicent, and K. B. Wolf, *Continuous vs. discrete fractional Fourier transforms*, J. Comput. Appl. Math. **107** (1999), 73–95.
15. N. M. Atakishiyev and P. Winternitz, *Bases for representations of quantum algebras*, J. Phys. A **33** (2000), no. 30, 5303–5313.
16. N. M. Atakishiyev and K. B. Wolf, *Fractional Fourier–Kravchuk transform*, J. Opt. Soc. Amer. A **14** (1997), 1467–1477.
17. L. Barker, *Continuum quantum systems as limits of discrete quantum systems. II. State functions*, J. Phys. A **34** (2001), 4673–4682.
18. L. Barker, Ç. Çandan, T. Hakioğlu, A. Kutay, and H. M. Ozaktas, *The discrete harmonic oscillator, Harper's equation, and the discrete fractional Fourier transform*, J. Phys. A **33** (2000), 2209–2222.
19. L. C. Biedenharn, *The quantum group* $\mathrm{SU}_q(2)$ *and a q-analogue of the boson operators*, J. Phys. A **22** (1989), L873–L878.
20. G. W. Forbes and M. A. Alonso, *Measures of spread for periodic distributions and the associated uncertainty relations*, Amer. J. Phys. **69** (2001), 340–347.
21. F. A. Grünbaum, *The eigenvectors of the discrete Fourier transform: A version of the Hermite functions*, J. Math. Anal. Appl. **88** (1982), 355–363.
22. T. Hakioğlu, *Finite-dimensional Schwinger basis, deformed symmetries, Wigner function, and an algebraic approach to quantum phase*, J. Phys. A **31** (1998), 6975–6994.
23. A. A. Izmest'ev, G. S. Pogosyan, A. N. Sissakian, and P. Winternitz, *Contractions of Lie algebras and separation of variables*, J. Phys. A **29** (1996), 5940–5962.
24. ———, A. N. Sissakian, and P. Winternitz, *Contractions of Lie algebras and separation of variables. The n-dimensional sphere*, J. Math. Phys. **40** (1999), 1549–1573.
25. ———, *Contractions of Lie algebras and the separation of variables. Interbasis expansions*, J. Phys. A **34** (2001) 521–554.
26. A. J. Macfarlane, *On q-analogues of the quantum harmonic oscillator and the quantum group* $\mathrm{SU}(2)_q$, J. Phys. A **22** (1989), 4581–4588.

27. J. H. McClellan and T. W. Parks, *Eigenvalue and eigenvector decomposition of the discrete Fourier transform*, IEEE Trans. Audio Electroacoust. **AU 20** (1972), no. 1, 66–74.
28. M. L. Mehta, *Eigenvalues and eigenvectors of the finite Fourier transform*, J. Math. Phys. **28** (1987), 781–785.
29. M. Moshinsky and C. Quesne, *Oscillator systems*, Proceedings of the 15th Solvay Conference in Nuclei, Gordon and Breach, New York, 1974, pp. 239-242.
30. ———, *Linear canonical transformations and their unitary representation*, J. Math. Phys. **12** (1971), 1772–1780.
31. T. Opatrný, *Mean value and uncertainty of optical phase—A simple mechanical analogy*, J. Phys. A **27** (1994), 7201–7208.
32. ———, *Number-phase uncertainty relations*, J. Phys. A **28** (1995), 6961–6975.
33. H. M. Ozaktas, Z. Zalevsky, and M. Alper Kurtay, *The fractional Fourier transform, with applications in optics and signal processing*, Wiley, Chichester, 2001.
34. C. Quesne and M. Moshinsky, *Canonical transformations and matrix elements*, J. Math. Phys. **12** (1971), 1780–1783.
35. T. S. Santhanam, *Finite-space quantum mechanics and Krawtchuk functions*, Proceedings of the Workshop on Special Functions and Differential Equations (K. Srinivasa Rao, R. Jagannathan, J. Van der Jeugt, and G. Van der Berghe (eds.), Allied, Publ., Delhi, 1998, pp. ???–???.
36. T. S. Santhanam and A. R. Tekumalla, *Quantum mechanics in finite dimensions*, Found. Phys. **6** (1976), 583–589.
37. J. Tolar and G. Chadzitaskos, *Quantization on Z_M and coherent states over $Z_M \times Z_M$*, J. Phys. A **30** (1997), 2509–2517.
38. N. Ja. Vilenkin and A. U. Klimyk, *Representation of Lie groups and special functions*, Kluwer Academic Publ., vol. 1, Dordrecht, 1991.
39. E. P. Wigner, *Do the equations of motion determine the quantum mechanical commutation relations?*, Physical Rev. (2) **77** (1950), 711–712.
40. K. B. Wolf, *Integral transforms in science and engineering*, Plenum Press, New York, 1979.
41. ———, *The Heisenberg–Weyl group in quantum mechanics*, Group Theory and its Applications (E. M. Loebl, ed.), vol. III, Academic Press, New York, 1975.

Centro de Ciencias Físicas, Universidad Nacional Autónoma de México, Apartado Postal 48–3, Cuernavaca 62251, Morelos, México.

E-mail address: bwolf@fis.unam.mx

Centre de Recherches Mathématiques
CRM Proceedings and Lecture Notes
Volume **39**, 2005

Some Nanotube-Like Systems and Their Discrete Equations

W. J. Zakrzewski

Abstract. We present here some results, obtained in collaboration with Drs L. Brizhik and A. Eremko from the Bogoliubov Institute in Kiev, Ukraine and B. Hartmann and B. Piette from Durham, on the properties of some one and two dimensional discrete (lattice) systems involving an electron field interacting with the deformations of this lattice. We show that, in the static case, the resultant equations, in one dimension, can be solved in a closed (and local) form. This is also true for the systems involving hexagonal lattices but not for the rectangular ones. We present a solution on the hexagonal lattice and discuss some of its properties.

1. Introduction

In this talk we discuss some of our recent work performed on systems involving an electron field (or sometimes other, more general, exciton field) defined on a one- or two-dimensional lattice which interacts with the deformations of this lattice. Such systems were first studied in the case of one dimensional lattices. They have lead to the idea of Davydov solitons which has received a lot of attention over the last few years (see a review by Scott [**6**]).

The Hamiltonians describing such systems consist of three terms; the purely electron part, the phonon part describing the deformations of the lattice and the interaction terms whose strength is controlled by a coupling constant g. The interaction brings nonlinearity into play and so g controls its strength. Clearly, when the coupling is weak (i.e. g is small) we have effectively free systems while for large values of g the system is very nonlinear resulting in highly localised "polaron states". This last phenomenon follows from the observation that in the continuum limit the description of the system reduces to the nonlinear Schrödinger equation with an attractive interaction.

Unfortunately, the equations that govern the dynamics of such systems are too complicated to be solved analytically. In fact, explicit solutions are not known even

---

2000 *Mathematics Subject Classification.* Primary: 39A12, 39A99; Secondary: 65N22.

I would like to thank Pavel Winternitz for his invitation to present this talk at the Workshop. I would also like to thank Drs. L. Brizhik, A. Eremko, B. Hartmann, and B. Piette for their collaboration with me on the subjects of this talk.

This is the final form of the paper.

in one dimension. Most of the known results have been obtained numerically and, until recently, very little was known about systems in dimensions higher than one.

Recently we have performed some studies of such systems in 2 dimensions. In this talk we will report first on these studies performed in the case of two dimensional rectangular lattices [**1, 2**]. Later, we will describe also our recent studies in which we have looked at systems based on hexagonal lattices [**4**].

In this talk we will concentrate on the discrete aspects of the work (i.e. on the discrete equations that govern the dynamics of these systems), Hence we will say very little about the applications and various properties of the derived solutions. The interested reader can find them in [**1, 2, 4**].

In the next section we present the general equations which describe our systems when, first, we consider rectangular lattices. The following section specialises them to an isotropic case—which is then shown to result in an interesting **discrete** equation. The next section repeats the discussion for the hexagonal case when, again, we have a very interesting equation, which this time has an explicit and localised solution. In the following section we show that the continuum limit of the equations from the previous sections leads to a nonlinear Schrödinger equation with an extra term—a modified nonlinear Schrödinger equation. The last section presents some conclusions and suggestions for further work.

2. Equations

The simplest system involving a quadratic lattice with lattice spacing $a = 1$ is described by the Fröhlich Hamiltonian which is given by a sum of Hamiltonians describing electron, phonons and electron-phonon interactions:

$$\widehat{H} = \widehat{H}_e + \widehat{H}_{\rm ph} + \widehat{H}_{\rm int}$$

i.e., in the site representation,

$$\widehat{H}_e = \sum_{i,j} [\widetilde{E}_0 A^+_{i,j} A_{i,j} - j_x (A^+_{i,j} A_{i+1,j} + A^+_{i+1,j} A_{i,j}) - j_y (A^+_{i,j} A_{i,j+1} + A^+_{i,j+1} A_{i,j})],$$

$$\widehat{H}_{\rm ph} = \frac{1}{2} \sum_{i,j} \Bigg(\frac{\widehat{p}^2_{i,j}}{M} + \frac{\widehat{q}^2_{i,j}}{M} + k_x [(\widehat{u}_{i,j} - \widehat{u}_{i+1,j})^2 + (\widehat{v}_{i,j} - \widehat{v}_{i+1,j})^2] \\ + k_y [(\widehat{u}_{i,j} - \widehat{u}_{i,j+1})^2 + (\widehat{v}_{i,j} - \widehat{v}_{i,j+1})^2] \Bigg),$$

$$\widehat{H}_{\rm int} = \sum_{i,j} A^+_{i,j} A_{i,j} [\chi_x (\widehat{u}_{i+1,j} - \widehat{u}_{i-1,j}) + \chi_y (\widehat{v}_{i,j+1} - \widehat{v}_{i,j-1})].$$

Here $A^+_{i,j}(A_{i,j})$ are the creation (annihilation) operators of the electron on the site (i, j), $\widehat{u}_{i,j}$, $\widehat{v}_{i,j}$ and $\widehat{p}_{i,j}$, $\widehat{q}_{i,j}$ describe the longitudinal and transverse components of the vector operator of molecule displacements and their conjugated momenta.

Moreover, $\widetilde{E}_0 - 2j_x - 2j_y$ is the bottom of the electron energy band, j_x, j_y are the strengths of the exchange interaction energies, while χ_x, χ_y are the electron-phonon coupling constants in the x and y directions of the lattice. Finally, k_x, k_y are the corresponding elasticity coefficients.

Next, one performs a standard semi-classical analysis [**6**] which leads to an effective classical Hamiltonian H, with φ_{ij}—the probability amplitude for the electron, and u_{ij}, v_{ij}—the variables describing molecular displacements, from their positions of equilibrium, in the x and y directions.

This way we find that H is given by

$$H = \sum_{m,n} (\widetilde{E}_0 + W)\varphi_{m,n} - j_x\varphi^*_{m,n}(\varphi_{m+1,n} + \varphi_{m-1,n}) - j_y\varphi^*_{m,n}(\varphi_{m,n+1} + \varphi_{m,n-1}) + |\varphi_{m,n}|^2[b\chi_x(u_{m+1,n} - u_{m-1,n}) + a\chi_y(v_{m,n+1} - v_{m,n-1})],$$

where

$$W = \frac{1}{2}\sum_{m,n}\left(\frac{p^2_{m,n}}{M} + \frac{q^2_{m,n}}{M} + k_x[(u_{m,n} - u_{m+1,n})^2 + (v_{m,n} - v_{m+1,n})^2] + k_y[(u_{m,n} - u_{m,n+1})^2 + (v_{m,n} - v_{m,n+1})^2]\right).$$

Moreover, we also have a constraint: the electron wave function has to satisfy the normalization condition:

$$\sum_{m,n} |\varphi_{m,n}|^2 = 1.$$

We note [**6**] that in the $D = 1$ case this condition is less important for one extra electron but it is essential in the many-electron problems. However, this condition leads to nontrivial effects in the 2D lattice even for just one extra electron.

Finally, to derive the equations of motion, it is convenient to introduce dimensionless units:

$$\tau = \frac{j_x t}{\hbar}, \quad U = C_x\frac{u}{b}, \quad V = C_x\frac{v}{b}, \quad E_s = \frac{b^2 M j_x}{\hbar^2}, \quad E_0 = \frac{\widetilde{E}_0}{j_x},$$
$$C_x = \frac{\chi_x b^2}{j_x}, \quad C_y = \frac{\chi_y ab}{j_x}, \quad K_x = \frac{k_x\hbar^2}{Mj_x^2}, \quad K_y = \frac{k_y\hbar^2}{Mj_x^2}, \quad g = \frac{2C_x^2}{K_x E_s}$$

and to define the anisotropy parameters

$$A_j = \frac{j_y}{j_x}, \quad A_c = \frac{C_y}{C_x}, \quad A_k = \frac{K_y}{K_x}.$$

The equations of motion then become:

$$i\frac{d\varphi_{m,n}}{d\tau} = (E_0 + W)\varphi_{m,n} - (\varphi_{m+1,n} + \varphi_{m-1,n}) - A_j(\varphi_{m,n+1} + \varphi_{m,n-1}) + [(U_{m+1,n} - U_{m-1,n}) + A_c(V_{m,n+1} - V_{m,n-1})]\varphi_{m,n}, \tag{2.1}$$

$$\frac{d^2U_{m,n}}{d\tau^2} = -\frac{K_x}{C_x}\Big[(2U_{m,n} - U_{m+1,n} - U_{m-1,n}) + A_k(2U_{m,n} - U_{m,n+1} - U_{m,n-1}) - \frac{g}{2}(|\varphi_{m+1,n}|^2 - |\varphi_{m-1,n}|^2)\Big], \tag{2.2}$$

$$\frac{d^2V_{m,n}}{d\tau^2} = -\frac{K_x}{C_x}\Big[(2V_{m,n} - V_{m+1,n} - V_{m-1,n}) + A_k(2V_{m,n} - V_{m,n+1} - V_{m,n-1}) - \frac{gA_c}{2}(|\varphi_{m,n+1}|^2 - |\varphi_{m,n-1}|^2)\Big]. \tag{2.3}$$

Here

$$W = \frac{1}{2} E_s \sum_{m,n} \left(P_{m,n}^2 + Q_{m,n}^2 + \frac{K_x}{C_x} \left([(U_{m,n} - U_{m+1,n})^2 + (V_{m,n} - V_{m+1,n})^2] \right.\right.$$

$$\left.\left. + A_k [(U_{m,n} - U_{m,n+1})^2 + (V_{m,n} - V_{m,n+1})^2] \right) \right),$$

$$P_{m,n} = \frac{dU_{m,n}}{d\tau}, \quad Q_{m,n} = \frac{dV_{m,n}}{d\tau}.$$

3. Isotropic case

In this talk we restrict our attention to the isotropic case; thus we can set $A_j = A_k = A_c = 1$. The anisotropic cases are not that different; the interested reader can consult [**1**].

Let us look at the stationary solutions of (2.1)–(2.3). Thus we assume that $U_{ij} \neq U_{ij}(t)$ and $V_{ij} \neq V_{ij}(t)$.

Note that the equation for φ involves the combination

$$Z_{mn} = (U_{m+1,n} - U_{m-1,n}) + (V_{m,n+1} - V_{m,n-1}).$$

However, the U (and V) fields satisfy

$$4U_{m,n} - U_{m+1,n} - U_{m-1,n} - U_{m,n+1} - U_{m,n-1} = \frac{g}{2}(|\varphi_{ma+1,n}|^2 - |\varphi_{m-1,n}|^2)$$

and so we see that they lead to a **very interesting discrete equation**

$$\Delta(1) Z_{mn} = -\alpha \Delta(2) |\varphi_{m,n}|^2, \tag{3.1}$$

where $\Delta(1)$ is the 5 point Laplacian, while $\Delta(2)$ is the same operator but involving only every second point on the lattice.

Note that, had we considered the same problem in one dimension, we would have had only one field, say U, and all the fields would have had only one subscript. Then, following the same steps as above we would have ended up with

$$P_{i+1} + P_{i-1} - 2P_i = \alpha(2R_i - R_{i+2} - R_{i-2}). \tag{3.2}$$

Its continuum limit ($a = 1 =$ lattice spacing $\to 0$) would have given us:

$$P_{i+1} = P_i + a\frac{\partial P_i}{\partial x} + \frac{1}{2}a^2 \frac{\partial^2 P}{\partial x^2} + \cdots$$

and so would have lead to

$$\frac{\partial^2 P}{\partial x^2} = -4\alpha \frac{\partial^2 R}{\partial x^2}.$$

Of course, this equation is easy to solve. Clearly, its solution is

$$P = -4\alpha R + \mu x + \nu.$$

In the discrete case it is easy to check that a solution of (3.2) is given by

$$P_n = -\alpha(R_{n+1} + R_{n-1} + 2R_n),$$

where we have neglected all the boundary terms.

So what can we say about the system in 2 dimensions, i.e. Eq. (3.1)

Clearly, in the continuum limit we have

$$\left(\frac{\partial^2}{\partial x^2} + \frac{\partial^2}{\partial y^2} \right) Z = -4\alpha \left(\frac{\partial^2}{\partial x^2} + \frac{\partial^2}{\partial y^2} \right) R.$$

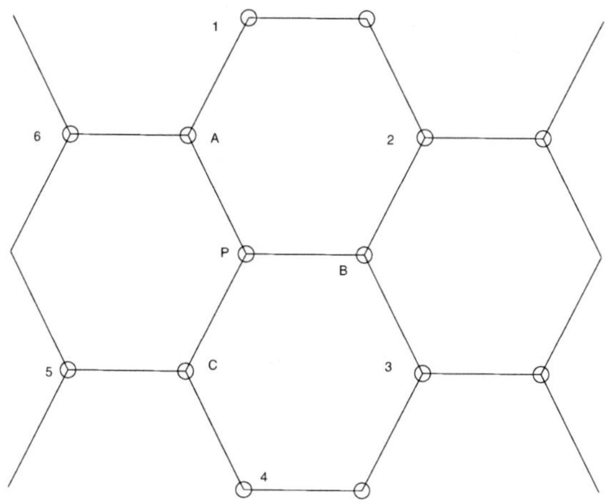

FIGURE 1. This is our hexagonal lattice.

Then, its solution is

$$Z = -4\alpha R + \text{solutions} \quad \text{of} \quad \nabla^2 Z = 0.$$

But what about the discrete case (3.1)? What are the solutions in this case? Unfortunately, we have not been able to find them in an analytical form. The problem is with the "corners", i.e. terms $i \pm 1$, $j \pm 1$.

4. Hexagonal lattice

Next we look at the similar equations defined on a hexagonal lattice given below.

The system is very similar and the equations are similar (for more details see [**4**]).

The discrete Schrödinger equation for the ψ field at the point P (see our figure) thus becomes:

$$\text{(4.1)} \quad i\hbar\frac{\partial\psi_P}{\partial t} = (E+W)\psi_P - 2j_x(\psi_A + \psi_B + \psi_C) + \psi_P\left[\frac{c_x}{3}(u_A + u_C - 2u_B) + \frac{c_x}{\sqrt{3}}(v_A - v_C)\right],$$

while the equations for the displacement fields u_P and v_P are given by:

$$\frac{d^2u_P}{dt^2} = k_x(3u_P - u_A - u_B - u_C) + \frac{c_x}{3M}(2|\psi_B|^2 - |\psi_A|^2 - |\psi_C|^2)$$

and

$$\frac{d^2v_P}{dt^2} = k_x(3v_P - v_A - v_B - v_C) - \frac{c_x}{\sqrt{3}M}(|\psi_A|^2 - |\psi_C|^2).$$

Again, restricting our attention to stationary fields we define

$$\Delta(1)U_P = U_A + U_B + U_C - 3U_P$$

and introduce

$$Z_P = U_A + U_C - 2U_B + \sqrt{3}(V_A - V_C)$$

and find that, again, we have an **interesting equation**

$$\Delta(1)Z_P = -\tilde{g}\Delta(2)|\psi_P|^2, \tag{4.2}$$

where

$$\Delta(2)K_P = (6K_P - K_1 - K_2 - K_3 - K_4 - K_5 - K_6)$$

and where $K_P = |\psi_P|^2$.

Note that $\Delta(1)$ involves a four-point Laplacian while $\Delta(2)$ is a corresponding seven-point Laplacian.

However, this time we have been able to find a **solution** of this equation, namely

$$Z_P = -\tilde{g}(|\psi_A|^2 + |\psi_B|^2 + |\psi_C|^2 + 3|\psi_P|^2). \tag{4.3}$$

What about other solutions?

What about solutions of $\Delta(2)K_P = 0$? and/or $\Delta(1)Z_P = 0$?

Incidentally, using our solution given above, i.e. (4.3), we find that the stationary solution of our system is described by the following form of the Discrete Nonlinear Schrödinger Equation:

$$i\frac{\partial\psi_P}{\partial t} - 2\Delta(1)\psi_P - \tilde{g}\psi_P(\Delta(1)|\psi_P|^2 + 6|\psi_P|^2) = 0.$$

5. Continuum limit—square lattice

Finally, let us look at the continuum limit of our equations. Let us present our discussion in the case of a square lattice.

To do this we define: $\varphi(x,y) = \varphi_{m,n}$ where $x = m$ and $y = n$.

Then we perform the expansions of all the fields and find:

$$\varphi_{m+1,n} = \varphi_{m,n} + \frac{\partial\varphi(x,y)}{\partial x} + \frac{1}{2}\frac{\partial^2\varphi(x,y)}{\partial x^2} + \frac{1}{6}\frac{\partial^3\varphi(x,y)}{\partial x^3} + \frac{1}{24}\frac{\partial^4\varphi(x,y)}{\partial x^4} + \cdots$$

$$U_{m\pm1,n} = U_{m,n} \pm \frac{\partial U}{\partial x} + \frac{1}{2}\frac{\partial^2 U}{\partial x^2} \pm \frac{1}{6}\frac{\partial^3 U}{\partial x^3} + \frac{1}{24}\frac{\partial^4 U}{\partial x^4} + \cdots$$

$$V_{m,n\pm1} = V_{m,n} \pm \frac{\partial V}{\partial y} + \frac{1}{2}\frac{\partial^2 V}{\partial x^2} \pm \frac{1}{6}\frac{\partial^3 V}{\partial x^3} + \frac{1}{24}\frac{\partial^4 V}{\partial x^4} + \cdots$$

thus giving us

$$\frac{\partial^2 U}{\partial x^2} = -g\left(\frac{\partial|\varphi|^2}{\partial x} + \frac{1}{6}\frac{\partial^3}{\partial x^3}|\varphi|^2\right), \quad \frac{\partial^2 V}{\partial y^2} = -g\left(\frac{\partial|\varphi|^2}{\partial y} + \frac{1}{6}\frac{\partial^3}{\partial y^3}|\varphi|^2\right).$$

Integrating these equations (and assuming that the constants of integrations, which could be the functions of other variables, all vanish) we get

$$\frac{\partial U}{\partial x} = -g\left(|\varphi|^2 + \frac{1}{6}\frac{\partial^2}{\partial x^2}|\varphi|^2\right), \quad \frac{\partial V}{\partial y} = -g\left(|\varphi|^2 + \frac{1}{6}\frac{\partial^2}{\partial y^2}|\varphi|^2\right)$$

and so we find that our system is described by

$$i\frac{d\varphi}{d\tau} + \Delta\varphi + 2g\left(|\varphi|^2 + \frac{1}{12}\Delta|\varphi|^2\right)\varphi = 0.$$

i.e. the **Nonlinear Schrödinger Equation** for the φ field (with an extra term).

In fact it is easy to see that the same is true in the hexagonal case, only the coefficients of the various terms are different).

This equation has been studied in some detail in [**3**]. There it has been shown that in one dimension an extended (solitonic) solution exists for a range of g and that the extra term is not that important. In two dimensions the extra term is needed for the solution to exist.

6. Conclusions and general comments

In this talk we have presented some aspects of the work described in more detail in [**1, 4**].

In fact, as shown in these papers, our equations admit solitonic-like solutions for the Schrödinger field. Thus we have demonstrated the existence of solitons in a class of discrete models. Their existence is related to the existence of solitons in a **modified** nonlinear Schrödinger model. However, our work has lead to many interesting questions like: can we find all solutions of (3.2) or (4.2)? What are the solutions of $\nabla(1)Z = 0$ on a hexagonal lattice? Or of $\nabla(2)Z = 0$? Furthermore, why can we find explicit and localised solutions in some cases and not in the others?

We have not discussed the applications of our equations. Thus we have not said anything about nanotubes where, as is easy to see (see, e.g. [**5**]), similar equations are relevant. Finally, what about more general lattices? (so far we have looked only at very simple or hexagonal lattices).

Clearly, a lot of work still has to be done before we have a reasonable understanding of our (and related) systems.

References

1. L. Brizhik, A. Eremko, B. Piette, and W. J. Zakrzewski, *Electron self-trapping in a discrete two-dimensional Lattice*, Phys. D **159** (2001), 71–90.
2. ———, *Spontaneous localisation of electrons in two-dimesional lattices within the adiabatic approximation*, J. Math. Phys. **44** (2003) no. 9, 3689–3697.
3. ———, *Static solutions of a D-dimensional modified nonlinear Schrödinger equation*, Nonlinearity **16** (2003), 1481–1497.
4. B. Hartmann and W. J. Zakrzewski, *Hexagonal lattices and nanotubes*, Preprint, Durham, 2003.
5. R. Saito, G. Dresselhaus, and M. S. Dresselhaus, *Physical properties of carbon nanotubes*, Imperial College Press, London, 1999.
6. A. C. Scott, *Davydov's soliton*, Phys. Rep. **217** (1992), 1–67.

DEPARTMENT OF MATHEMATICAL SCIENCES, UNIVERSITY OF DURHAM, DURHAM DH1 3LE, UK.

E-mail address: W.J.Zakrzewski@durham.ac.uk

Centre de Recherches Mathématiques
CRM Proceedings and Lecture Notes
Volume **39**, 2005

Explicit Multipoint Rational Interpolation Padé Table for Exponential and Power Functions

A. Zhedanov

Abstract. We construct explicit Padé table $Q_m(x;n)/P_n(x;m)$, for the Cauchy–Jacobi interpolation problem for arbitrary geometric progression on either uniform or exponential grid of interpolation points. Both $Q_m(x;n)$ and $P_n(x;m)$ polynomials are expressed in terms of Krawtchouk or q-Krawtchouk polynomials. This allow us to write down explicitly recurrence relations for $P_n(x;m)$, $Q_m(x;n)$ and to derive rational functions which are biorthogonal with respect to binomial distribution. We construct an explicit Thiele-type continued fractions for the exponential and power functions.

1. Introduction

The Cauchy–Jacobi interpolation problem (CJIP) for the sequence a_j of complex numbers can be formulated as follows [**1, 11**]. Given two nonnegative integers n, m, choose a system of (distinct) points $x_j, j = 0, 1, \dots, n+m$ on the complex plane. We are seeking polynomials $Q_m(x;n)$, $P_n(x;m)$ of degrees m and n correspondingly such that

$$a_j = \frac{Q_m(x_j;n)}{P_n(x_j;m)}, \quad j = 0, 1, \dots, n+m \tag{1.1}$$

(in our notation we stress, e.g. that polynomial $Q_m(x;n)$, being degree m in x, depends on n as a parameter). The CJIP is called *normal* if polynomials $Q_m(x;n)$, $P_n(x;m)$ exist for all values of $m, n = 0, 1, \dots$ and polynomials $Q_m(x;n)$, $P_n(x;m)$ have no common zeroes. This means, in particularly, that polynomials $Q_m(x;n)$, $P_n(x;m)$ have no roots, coinciding with interpolatory points, i.e.

$$Q_m(x_j;n) \neq 0, \quad P_n(x_j;m) \neq 0, \qquad j = 0, 1, \dots, n+m. \tag{1.2}$$

In a special case when there exists analytic function $f(z)$ of complex variable such that $f(x_j) = a_j$ the corresponding CJIP is called multipoint Padé approximation problem [**1**]. The standard Padé approximation problem consists in finding polynomials $Q_m(x;n)$, $P_n(x;m)$ such that for given function $f(x)$ (which is assumed to

---

2000 *Mathematics Subject Classification.* Primary: 41A21; Secondary: 41A05, 41A20, 33C45, 33D45.

This is the final form of the paper.

be analytical near $x = 0$) we have the condition

$$f(x) - \frac{Q_m(x;n)}{P_n(x;m)} = O(x^{n+m+1}). \tag{1.3}$$

It is clear that the ordinary Padé problem can be obtained by the limiting process $x_j \to 0$ for all j. In this respect, CJIP can be considered as a generalization of the ordinary Padé problem.

Both standard and multipoint Padé approximation problems are important from practical point of view (see, e.g. [**1**]). Therefore it is interesting to investigate the cases when Padé approximates $Q_m(x;n)$, $P_n(x;m)$ are known explicitly for all $m, n = 0, 1, 2, \ldots$. In this case we say that the Padé table is known.

For the standard Padé problem explicit Padé table is known for several functions. The simplest and very important example is exponential function, whose Padé approximants were calculated by Padé himself (see, e.g. [**1**]). In this case we have

$$Q_m(x;n) = {}_1F_1\left(\begin{matrix}-m, \\ -m-n\end{matrix}; x\right), \quad P_n(x;m) = {}_1F_1\left(\begin{matrix}-n, \\ -m-n\end{matrix}; -x\right), \tag{1.4}$$

where ${}_1F_1$ denotes standard hypergeometric function. Zeroes and poles of the standard Padé approximation problem (i.e. zeroes of polynomials $Q_m(x;n)$, $P_n(x;m)$) for exponential function were intensively studied in many papers (see, e.g. [**2, 13**]).

In 1981 A. Iserles [**6**] (see also more detailed analysis of this problem in [**10**]) constructed explicit multipoint Padé table (MPT) for the exponential function $\exp(x)$ on the uniform grid. As far as we know this is the only known example of explicit multipoint Padé table for nontrivial transcendent functions.

In this paper we derive explicit MPT for the arbitrary geometric progression

$$a_j = rq^j, j = 0, 1, 2, \ldots \tag{1.5}$$

with arbitrary complex $q \neq 0, 1$ and r. We obtain explicit Padé table for two cases of interpolation grid:

(i) arbitrary arithmetic progression (uniform grid) and
(ii) arbitrary geometric progression (exponential grid).

In the case of uniform interpolation grid we arrive at Iserles' result. However,our method of derivation is more simple and moreover we derive explicitly biorthogonal rational functions corresponding to this problem. These functions are expressed in terms of the ordinary Krawtchouk polynomials. For the case of the exponential grid we obtain new result: both numerator and denominator polynomials $P_n(x;m)$, $Q_m(x;n)$ are expressed in terms of q-Krawtchouk polynomials. We also construct a new representation of the exponential and power functions in terms of the Thiele continued fraction.

2. Padé interpolation table for the exponential function on the uniform grid

We choose the uniform grid of interpolation points

$$x_j = hj + \beta, \quad j = 0, 1, \ldots \tag{2.1}$$

where $h \neq 0$ and β are arbitrary real parameters.

We are seeking solution of the CJIP

$$rq^j = \frac{Q_m(x_j;n)}{P_n(x_j;m)}, \quad j = 0, 1, \dots, n+m \tag{2.2}$$

i.e. we would like to construct polynomials $Q_m(x;n)$, $P_n(x;m)$ of degrees m and n satisfying $n+m+1$ conditions (2.2). As was shown in [**5**], the CJIP (2.2) for the exponential function is normal for any grid x_j. So, polynomials Q_m, P_n do exist, have no common zeroes, and these polynomials are unique.

Instead of derivation of the polynomials Q_m, P_n we give the main result and then check that conditions (2.2) hold.

Theorem 1. *The polynomials $Q_m(x;n)$, $P_n(x;m)$ providing solution of the* CJIP *for the exponential function on the uniform grid* (2.1) *are given by*

$$\begin{aligned} Q_m(x;n) &= r\,{}_2F_1\left({-m, (\beta-x)/h \atop -m-n}; 1-q\right), \\ P_n(x;m) &= {}_2F_1\left({-n, (\beta-x)/h \atop -m-n}; 1-1/q\right), \end{aligned} \tag{2.3}$$

where ${}_2F_1$ is standard Gauss hypergeometric function.

Remark. In our case the bottom parameter $-n-m$ is a negative integer. We adopt standard rule [**3**] that in this case only a finite number of terms is taken in definition of hypergeometric function

$${}_2F_1\left({-n, b \atop -N}; z\right) \equiv \sum_{k=0}^{n} \frac{(-n)_k (b)_k}{k!(-N)_k} z^k, \tag{2.4}$$

where $(b)_k = b(b+1)\dots(b+k-1)$ is standard Pochhammer symbol. The obvious necessary condition for such sum to exist is $n \le N$.

Proof. It is clear from (2.3) that $Q_m(x;n)$ is a polynomials of degree m in x and $P_n(x;m)$ is a polynomial of degree n in x. In order to verify conditions (2.2), we use Pfaff identity for the Gauss hypergeometric function [**3**]

$${}_2F_1\left({a, b \atop c}; z\right) = (1-z)^{-a}\,{}_2F_1\left({a, c-b \atop c}; \frac{z}{z-1}\right). \tag{2.5}$$

This formula is valid under condition that c is not negative integer. In our case, nevertheless, $c = -n-m$, hence validity of Pfaff formula should be checked additionally. It is well known that the Pfaff transformation formula (2.5) for $c = -N$, where $N = 1, 2, \dots$ is still valid identically for all z if and only if $a = -s = 0, -1, -2, \dots, -N$, where s is a nonnegative integer such that $s < N$.

Now we apply the Pfaff transformation formula to our case choosing $a = -s = 0, -1, \dots, -n-m$, $b = -n$, $c = -n-m$ and $z = 1-q$. Then relation (2.5) is equivalent to (2.2).

Uniqueness and normality of the CJIP for the exponential function was already proven in [**5**]. Thus theorem is proven. □

Formulas (2.3) were obtained by Iserles (see also [**10**]) using a more difficult (but equivalent) approach.

Note that polynomials $P_n(x;m)$, $Q_m(x;n)$ can be expressed in terms of the Krawtchouk polynomials. Recall that (non-monic) Krawtchouk polynomials

$K_n(x,p,N)$ depend on two real parameters: p and $N = 1,2,\dots$. They are defined as [9]

$$K_n(x,p,N) = {}_2F_1\left({-n,-x \atop -N}; 1/p\right). \tag{2.6}$$

We see, that

$$Q_m(x;n) = K_m(x, 1/(1-q), n+m), \quad P_n(x;m) = K_n(x, 1/(1-1/q), n+m). \tag{2.7}$$

For standard Krawtchouk polynomials there is a restriction $0 < p < 1$. Only under this restriction the Krawtchouk polynomials are orthogonal on a finite set of real integer points $0, 1, \dots, N$ and, moreover, all zeroes of the Krawtchouk polynomials are real and lie inside the interval $[0, N]$. Recall (see, e.g. [9]) that standard Krawtchouk polynomials are orthogonal

$$\sum_{s=0}^{N} w_s K_n(s,p,N) K_m(s,p,N) = 0, \quad \text{if} \quad n \neq m \tag{2.8}$$

with respect to the binomial distribution

$$w_s = \binom{N}{s} p^s (1-p)^{N-s}. \tag{2.9}$$

In our case the polynomials $P_n(x;m)$, $Q_m(x;n)$ coincide with "true" Krawtchouk polynomials if and only if $q < 0$. This corresponds to a sign-varied geometric progression $a_j = r(-1)^j |q|^j$. Curiously enough that "degenerated" geometric progression with $q = -1$ (consisting only from two values r and $-r$) means $p = \frac{1}{2}$ which corresponds to so-called symmetric Krawtchouk polynomials [9] having the property

$$K_n(-x, 1/2, N) = (-1)^n K_n(x, 1/2, N). \tag{2.10}$$

For the ordinary exponent function $f(x) = \exp(\omega x)$ one has $q = e^{\omega}$; hence for the most important practical case of real values of ω the parameter p lies beyond the interval $(0, 1)$. This leads to some unusual properties of corresponding Krawtchouk polynomials. In particular, their zeroes are no more real (apart from possible one equal to zero). Moreover, orthogonality weights w_s are no more positive as is clear from (2.9). Nevertheless, relation with Krawtchouk polynomials is very useful in establishing further explicit formulas for multipoint Padé approximations.

3. Recurrence relations

In what follows we will assume that shift parameter $\beta = 0$. Indeed, as is seen from (2.3), nonzero β leads merely to shift of argument of polynomials $Q_m(x;n)$, $P_n(x;m)$. Thus for interpolation points we will have $a_j = hj$, $j = 0, 1, \dots$. Moreover, we will assume $r = 1$ because rescaling of this parameter leads merely to the same rescaling of numerator polynomials $Q_m(x;n)$. Clearly, polynomials $P_n(x;m)$ and $Q_m(x;n)$ can be multiplied by the same nonzero multiplier. It is convenient to choose denominator polynomials $P_n(x;m)$ in monic form $P_n(x;m) = x^n + O(x^{n-1})$, i.e.

$$P_n(x;m) = \kappa_{nm}\, {}_2F_1\left({-n, -x/h \atop -m-n}; 1-q^{-1}\right) \tag{3.1}$$

where

$$\kappa_{nm} = (-1)^n h^n (m+1)_n (1-q^{-1})^{-n}. \tag{3.2}$$

Using known relations for Krawtchouk polynomials (see, e.g. [**9**]) we can establish recurrence relations for interpolation polynomials $P_n(x;m)$ (recurrence relations for polynomials $Q_m(x;n)$ are quite similar and obtained by trivial change of parameters).

We start from two such relations (we denote $p = (1-1/q)^{-1}$)

$$P_n(x;m+1) - P_n(x;m) + pnhP_{n-1}(x;m+1) = 0 \tag{3.3}$$

and

$$(x - h(n+m))P_n(x;m-1) - P_{n+1}(x;m-1) + hm(1-p)P_n(x;m) = 0 \tag{3.4}$$

which can be easily verified (in fact, they correspond to so-called Christoffel and Geronimus transformations [**14**], [**17**] for the Krawtchouk polynomials).

Starting from these relations we can derive recurrence relation for interpolation polynomials with fixed value of m:

$$\begin{aligned}(x - hn - ph(n+m+1))P_n(x;m) - P_{n+1}(x;m)&\\ + pnh(x - h(n+m))P_{n-1}(x;m) &= 0.\end{aligned} \tag{3.5}$$

It is interesting to note that recurrence relation (3.5) belongs to the class of so-called R_I type recurrence relations for polynomials leading to some generalized orthogonality properties [**7**]. Note also that for $m = 0$ polynomials $P_n(x;0)$ coincide with Lagrange interpolation polynomials for the sequence q^{-j}, $j = 0, 1, \ldots$.

Diagonal ($m = n$) and subdiagonal ($m = n \pm 1$) cases are of great importance in theory and practice of interpolation. These cases correspond to polynomials which will be denoted as $A_n(x) = P_n(x;n)$, $B_n(x) = P_n(x;n-1)$ and $C_n(x) = P_n(x;n+1)$. We have two pairs of interrelations between them:

$$\begin{aligned}(1-p)A_n(x) - p(x-(2n-1)h)A_{n-1}(x) + (2p-1)B_n(x) &= 0,\\ B_{n+1}(x) - (x-2nh)B_n(x) + (p-n+2pn)hA_n(x) &= 0\end{aligned} \tag{3.6}$$

and

$$\begin{aligned}(1-p)(n+1)C_n(x) - pn(x-2nh)C_{n-1}(x)&\\ + (n(2p-1)+p-1)A_n(x) &= 0,\\ A_n(x) - (x-(2n-1)h)A_{n-1} + n(2p-1)hC_{n-1}(x) &= 0.\end{aligned} \tag{3.7}$$

From (3.6) we can obtain recurrence relation involving polynomials $B_n(x)$ only:

$$\begin{aligned}\left(\frac{1-p}{h\sigma_n}\right)(B_{n+1}(x) - (x-2hn)B_n(x)) - (2p-1)B_n(x)&\\ - \frac{p(x-h(2n-1))}{h\sigma_{n-1}}(B_n(x) - (x-h(2n-2))B_{n-1}(x))&\end{aligned} \tag{3.8}$$

where $\sigma_n = p + n(2p-1)$. Note this recurrence relation belongs to so-called R – II-type relations for polynomials [**7**]

$$P_{n+1}(x) - (\alpha_n x + \beta_n)P_n - \gamma_n(x - a_{2n-2})(x - a_{2n-1})P_{n-1}(x) = 0 \tag{3.9}$$

with some parameters $\alpha_n, \beta_n, \gamma_n, a_n$. In our case

$$\alpha_n = 1 + \frac{p\sigma_n}{(1-p)\sigma_{n-1}}, \quad \beta_n = -2nh + \frac{(2p-1)h\sigma_n}{1-p} - \frac{ph(2n-1)\sigma_n}{(1-p)\sigma_{n-1}}, \quad \gamma_n = 1 - \alpha_n. \tag{3.10}$$

Polynomials satisfying relations of the R – II-type have a remarkable generalized orthogonality property [**7**]. In turn, it can be shown [**18**] that this orthogonality leads to some biorthogonality between two sets of rational functions of type $[n/n]$ with prescribed denominators.

4. Location of poles and zeroes of interpolants of the exponential function

In this section we prove a theorem concerning location of poles and zeroes of the interpolant $Q_m(x;n)/P_n(x;m)$ for all possible values of n and m. Clearly, we need only to locate zeroes of polynomials $Q_m(x,n)$ and $P_n(x;m)$.

Our key tool will be Henrici theorem proposed in [**4**]:

Theorem 2. *Assume that a sequence ψ_n satisfies* 3*-term recurrence relation*

$$\psi_{n+1} = (\beta_n + z_{n+1})\psi_n - \varepsilon_n z_n \psi_{n-1} \tag{4.1}$$

with initial conditions

$$\psi_0 = 1, \quad \psi_1 = \beta_0 + z_1 \tag{4.2}$$

where β_n, $n = 0, 1, \dots$ and ε_n, $n = 1, 2, \dots$ are positive real numbers satisfying condition

$$\alpha \equiv \inf_{n\geq 0} (\beta_n - \varepsilon_n) > 0 \tag{4.3}$$

(we put $\varepsilon_0 = 0$ in (4.3)).

Let $\mathcal{P}_\alpha$ be a parabolic region in a complex plane

$$\mathcal{P}_\alpha : |z| < Re(z) + 2\alpha$$

such that all parameters z_n in the recurrence relation (4.1) *belong to the region $\mathcal{P}_\alpha$. Then all possible zeroes of the sequence ψ_n, $n = 1, 2, \dots$ lie outside the parabolic region $\mathcal{P}_\alpha$.*

For the proof of this theorem see [**4**].

In our case we put for simplicity $h = 1$, $\omega > 0$, fix parameter m and define $\psi_n = (-1)^n P_n(-x;m)$ Then from recurrence relation (3.5) we see that all conditions of the above theorem are fulfilled with

$$\varepsilon_n = pn, \quad \beta_n = pn + (m+1)(p-1), \quad z_n = x+n+m, \quad \alpha = (m+1)(p-1). \tag{4.4}$$

(Recall that $p = 1/(1-\exp(-\omega))$ and hence all parameters ε_n, β_n and α are positive, as required). Note that if $z \in \mathcal{P}_\alpha$ then obviously $z_n \in \mathcal{P}_\alpha$, $n = 1, 2, \dots$. Thus all zeroes of sequence $\psi_n = (-1)^n P_n(-x;m)$ lie outside the parabolic region $\mathcal{P}_\alpha$. Hence we arrive at

PROOF. Let interpolation points be $x_s = 0, 1, \dots, n+m$. Then for the denominator interpolants $P_n(x;m)$ of the exponential function $e^{\omega x}$ all zeroes (for fixed m) lie outside the parabola

$$\eta^2 = 4\alpha_m(\alpha_m - \xi), \tag{4.5}$$

in complex plane $x = \xi + i\eta$, where

$$\alpha_m = \frac{m+1}{e^{\omega} - 1}. \qquad \square$$

It is interesting to consider when zeroes of polynomials $P_n(x;m)$ lie outside of the interpolation interval $[0, n+m]$. Obviously, for this it is sufficient that the parabola vertex $x = \alpha_m$ is located to the right from the end of the interval: $\alpha_m > m+n$ which implies the condition

$$e^{\omega} < 1 + \frac{m+1}{m+n}.$$

In particular, for diagonal case $m = n$ we have $e^{\omega} < 3/2 + 1/n$. Hence if $\exp(\omega) < 5/2$ then poles of all diagonal interpolants lie outside the interpolation interval.

Consider now location of zeroes of the polynomials $Q_m(x;n)$ (i.e. zeroes of the interpolants). Polynomials $Q_m(x;n)$ are related with $P_m(x;n)$ by

$$Q_n(x,m) = \kappa_{nm} P_n(n+m-x, m), \tag{4.6}$$

with some nonzero normalization constant κ_{nm}. Thus location of zeroes of numerator polynomials $Q_m(x;n)$ is obtained from location of zeroes of the denominator polynomials $P_m(x;n)$ by simple reflection $z \to n+m-z$ in the complex plane z with respect to middle-point of the interpolation interval $[0, n+m]$. In particular, it is clear, that if all zeroes of the polynomials $P_m(x;n)$ lie outside the interpolation interval $[0, n+m]$, then the same is true for zeroes of the polynomials $Q_n(x;m)$.

We thus see, that if $e^{\omega} < \frac{3}{2}$ then all zeroes and poles of the diagonal Padé interpolants $Q_n(x;n)/P_n(x;n)$ of the exponent function $e^{\omega x}$ lie outside the interpolation interval $[0, 2n]$.

5. Continued fractions corresponding to interpolation of exponential function

In this section we construct Thiele-type and R − II-type continued fractions for the exponential function. First, recall the Thiele algorithm for rational interpolation [**1, 8, 12**]. Let $f(x)$ be a function and $x_0, x_1, \ldots$—corresponding interpolation points. Let us construct a set of functions $f_0(x)$ by the procedure

$$f_0(x) = f(x), \quad f_1(x) = \frac{x - x_0}{f_0(x) - f_0(x_0)} \tag{5.1}$$

and then recursively

$$f_{n+1}(x) = \frac{x - x_n}{f_n(x) - f_n(x_n)}, \quad n = 1, 2, \ldots. \tag{5.2}$$

Then it is clear from this definition that the function $f(x)$ can be expanded into the continued fraction (so-called Thiele-type continued fraction [**8**])

$$f(x) = \eta_0 + \cfrac{x - x_0}{\eta_1 + \cfrac{x - x_1}{\eta_2 + \cfrac{x - x_2}{\eta_3 + \cdots}}} \tag{5.3}$$

where quantities $\eta_i = f_i(x_i)$, $i = 0, 1, \ldots$ are found from already constructed functions $f_n(x)$ by algorithm (5.2). Let $U_n(x)/T_n(x)$ be the nth approximate to Thiele continued fraction with $U_n(x)$, $T_n(x)$ being n-th numerator and denominator. They satisfy three-term recurrence relations

(5.4)
$$U_{n+1}(x) = \eta_{n+1}U_n(x)+(x-x_n)U_{n-1}(x), \quad T_{n+1}(x) = \eta_{n+1}T_n(x)+(x-x_n)T_{n-1}(x)$$

with initial conditions

$$U_0 = \eta_0, \quad T_0 = 1, \quad U_1(x) = x - x_0 + \eta_0\eta_1, \quad T_1 = \eta_1. \tag{5.5}$$

From the Thiele algorithm it follows that $U_n(x)$ and $T_n(x)$ form a special sequence of multipoint Padé approximants to function $f(x)$ at interpolation points x_i, $i = 0, 1, \ldots, n$. More precisely, (see, e.g. [**8**]) we have

$$\begin{aligned} &Q_n(x;n) = U_{2n}(x), \quad Q_{n+1}(x;n) = U_{2n+1}(x), \\ &P_n(x;n) = T_{2n}(x), \quad P_n(x;n+1) = T_{2n+1}(x), \quad n = 0, 1, \ldots. \end{aligned} \tag{5.6}$$

Thus denominator polynomials $T_{2n}(x)$ have degree n in x, and polynomials $T_{2n+1}(x)$ have degree n as well.

The Thiele algorithm is thus a simple tool for obtaining these special multipoint Padé approximants. Nevertheless, so far, no explicit Thiele continued fractions for transcendent functions were known.

Now we can present the Thiele fraction for the exponential function, because all members of the Padé table are known explicitly. The only problem is calculation of the quantities η_n. But they are easily found from recurrence relations (3.7) (recall that both $Q_m(x)$ and $P_n(x)$ are defined up to an arbitrary common nonzero factor; we can choose this freedom to reduce recurrence relations (3.7) to the form (5.4)). More exactly, in our case we have $x_n = n$ and

$$T_{2n}(x) = \mu_n A_n(x), \quad T_{2n+1}(x) = \nu_n C_n(x),$$

where $A_n(x) = P_n(x;n)$ and $C_n(x) = P_n(x;n+1)$ satisfy recurrence relations (3.7). The coefficients μ_n, ν_n are easily calculated and we thus have

Theorem 3. *The continued fraction* (5.3) *with* $x_i = ih$, $i = 0, 1, \ldots$ *and*

$$\eta_0 = 1,\ \eta_{2n} = (-1)^n e^{\omega(n-1)h}(1+e^{\omega h}),\ \eta_{2n+1} = (-1)^n e^{-\omega hn}(n+1+ne^{-\omega h})\frac{h}{e^{\omega h}-1}$$

represents expansion of the exponential function:

$$e^{\omega x} = 1 + \cfrac{x}{\eta_1 + \cfrac{x-h}{\eta_2 + \cfrac{x-2h}{\eta_3 + \ldots}}}$$

It is instructive to note that for $h \to 0$ we have

$$\eta_{2n} = 2(-1)^n, \quad \eta_{2n+1} = (-1)^n(2n+1)/\omega$$

and in this limit we arrive at the classical Euler continued fraction for the exponential function [**8**]

$$e^{\omega x} = 1 + \cfrac{x}{\omega^{-1} + \cfrac{x}{-2 + \cfrac{x}{-3\omega^{-1} + \ldots}}}$$

Analogously one can find continued fraction of R – II-type starting from recurrence relation (3.9) (for description of such continued fractions and their relation with (3.9) see [7]):

$$e^{\omega x} = \cfrac{1}{\alpha_0 x + \beta_0 + \cfrac{\gamma_1 x(x-h)}{\alpha_1 x + \beta_1 + \cfrac{\gamma_2 (x-2h)(x-3h)}{\alpha_2 x + \beta_2 + \dots}}} \tag{5.7}$$

6. Biorthogonal rational functions related with interpolation problem

It is well known that the Cauchy–Jacobi interpolation problem implies some orthogonality relation for the denominator polynomials $P_n(x;m)$ [10, 11].

Indeed, consider a set of conditions for CJIP for some function $Ff(x)$

$$f(x)P_n(x;m) - Q_m(x;n) = 0, \quad x = x_0, \dots, x_{n+m}. \tag{6.1}$$

Multiply both sides of (6.1) by x^j and apply operator of finite difference $\mathcal{D}^{n+m}$ on the set of points $x_0, \dots x_{n+m}$. Recall that the first-order finite difference operator is defined as

$$\mathcal{D}f(x) = \frac{f(x) - f(x_0)}{x - x_0}$$

and then by induction

$$\mathcal{D}^n f(x) = \frac{\mathcal{D}^{n-1} f(x) - \mathcal{D}^{n-1} f(x_n)}{x - x_n}.$$

For any polynomial $\pi(x)$ of degree N we have

$$\mathcal{D}^n f(x) = 0, \quad \text{if n} > N. \tag{6.2}$$

There are two useful explicit forms for divided difference operator. The first one is Hermite formula

$$\mathcal{D}^n f(x) = (2\pi i)^{-1} \int_\Gamma \frac{f(\zeta)\, d\zeta}{(\zeta - x_0)\cdots(\zeta - x_n)}, \tag{6.3}$$

where the contour Γ should include interpolation points $x_0, \dots, x_n$. This formula is valid if the function $f(x)$ has no singularities inside of the contour Γ.

The second formula yields the value of the function $\mathcal{D}^n f(x)$ at the point $x = x_n$:

$$\mathcal{D}^n f(x_n) = \sum_{k=0}^{n} \varepsilon_k(n) f(x_k) \tag{6.4}$$

where

$$\varepsilon_k(n) = \frac{1}{(x_k - x_0)(x_k - x_1)\cdots(x_k - x_{k-1})(x_k - x_{k+1})\cdots(x_k - x_n)}. \tag{6.5}$$

Using property (6.2) we can conclude that

$$\mathcal{D}^{n+m}\{f(x)x^j P_n(x;m)\} = \mathcal{D}^{n+m}\{x^j Q_m(x;n)\} = 0, \quad \text{if} \quad 0 \le j < n. \tag{6.6}$$

From (6.6) one can obtain orthogonality property for the polynomials $P_n(x;m)$. It can be presented in different forms. Using (6.5) we have

$$\sum_{s=0}^{n+m} w_s(n+m) x_s^j P_n(x_s;m) = 0, \quad j = 0, 1, \dots, n-1 \tag{6.7}$$

where the "discrete weights" $w_s(n)$ are defined as

$$w_s(n) = f(x_s)\varepsilon_s(n). \tag{6.8}$$

If the function $f(x)$ has no singularities in finite part of the complex plane, then we have also

$$\int_\Gamma \frac{f(\zeta)\zeta^j P_n(\zeta;m)\,d\zeta}{(\zeta - x_0)\cdots(\zeta - x_{n+m})} = 0, \quad j = 0,1,\dots,n-1. \tag{6.9}$$

Orthogonality relations (6.7), (6.9) resemble those for the ordinary orthogonal polynomials, the main difference is that the weights $w_s(n+m)$ depend explicitly on parameter n (degree of polynomial $P_n(x)$).

For the case of the uniform grid $x_k = k$, $k = 0,1,\dots$ we have obviously from (6.5)

$$\varepsilon_k(n) = (-1)^{n+k}\frac{1}{k!(n-k)!}. \tag{6.10}$$

Thus for the exponential function $f(x) = \exp(\omega x)$ we get for the weights

$$w_s(n) = (-1)^{s+n} e^{\omega s}\frac{1}{s!(n-s)!}, \tag{6.11}$$

i.e. discrete weight function coincides with (sign-changed) binomial distribution.

Nevertheless, as shown in [**15, 16, 18**], if $m = n-1$ (subdiagonal case) it is possible to construct a pair of rational functions $U_n(x)$ and $V_n(x)$ which are *biorthogonal* to one another:

$$\int_\Gamma \frac{U_n(\zeta)V_m(\zeta)f(\zeta)\,d\zeta}{\zeta - x_0} = h_n\delta_{nm} \tag{6.12}$$

where

$$\begin{aligned} U_n(x) &= \frac{P_n(x;n-1)}{(x-x_1)(x-x_3)\cdots(x-x_{2n-1})}, \\ V_n(x) &= \frac{\widetilde{P}_n(x;n-1)}{(x-x_2)(x-x_4)\cdots(x-x_{2n})}. \end{aligned} \tag{6.13}$$

The polynomials $P_n(x;n-1)$, $n = 1,2,\dots$ are denominator polynomials of CJIP for the function $f(x)$ on the set of interpolation points $X_{2n} = \{x_0, x_1, \dots, x_{2n-1}\}$:

$$f(x) - \frac{Q_{n-1}(x;n)}{P_n(x;n-1)} = 0, \quad x \in X_{2n}. \tag{6.14}$$

The polynomials $\widetilde{P}_n(x;n-1)$ are denominator interpolation polynomials of the CJIP for the same function $f(x)$ but on a slightly "surgered" grid $\widetilde{X}_{2n} = \{x_0, x_1, \dots, x_{2n-2}, x_{2n}\}$ (i.e. the grid $\widetilde{X}_{2n}$ is obtained from the grid X_{2n} by replacing the final interpolation point x_{2n} by the point x_{2n}):

$$f(x) - \frac{\widetilde{Q}_{n-1}(x;n)}{\widetilde{P}_n(x;n-1)} = 0, \quad x \in \widetilde{X}_{2n}. \tag{6.15}$$

In what follows we will assume that polynomials $P_n(x;n-1)$ and $\widetilde{P}_n(x;n-1)$ are monic. It is not difficult to show (we will not derive this formula here) that polynomials $\widetilde{P}_n(x;n-1)$ are obtained from $P_n(x;n-1)$ by the formula

$$\widetilde{P}_n(x;n-1) = \frac{\mu_n P_{n+1}(x;n) + (1-\mu_n)(x - x_{2n})P_n(x;n-1)}{x - x_{2n-1}}, \tag{6.16}$$

where parameters μ_n are determined as

$$\mu_n = \frac{(x_{2n} - x_{2n-1})P_n(x_{2n-1}; n-1)}{P_{n+1}(x_{2n-1}; n) + (x_{2n} - x_{2n-1})P_n(x_{2n-1}; n-1)}. \tag{6.17}$$

Thus for the exponential function we have a pair of rational functions satisfying biorthogonality relation

$$\int_\Gamma e^{\omega\zeta} U_n(\zeta) V_k(\zeta) \zeta^{-1}\, d\zeta = h_n \delta_{nk} \tag{6.18}$$

with some coefficients h_n. We can present this relation in the form of finite-term sum:

$$\sum_{s=0}^{2n} \operatorname{res}\big(U_n(x)V_k(x)\big)\big|_{x=s} = 0, \quad n \neq k. \tag{6.19}$$

There is one more biorthogonality relation which can be easily obtained from general formula (6.7):

PROOF. For any grid x_s let $P_n(x, m)$ be denominators of CJ interpolation problem. Fix positive integer m and define rational functions

$$W_n(x; m) = \frac{P_n(x; m+1)}{(x - x_0)\cdots(x - x_{n+m+1})} \tag{6.20}$$

of order $[n/(n+m+2)]$. Then biorthogonality relation

$$\int_\Gamma f(\zeta) P_{n_1}(\zeta; m) W_{n_2}(\zeta; m)\, d\zeta = 0, \quad \text{if} \quad n_1 \neq n_2 \tag{6.21}$$

holds, where contour Γ contains interpolation points $x_0, x_1, \ldots, x_{n_0+m+1}$ with $n_0 = \max(n_1, n_2)$. ☐

Biorthogonality relation (6.21) can be rewritten in terms of finite sum:

$$\sum_{s=0}^{n_0} e^{\omega s} P_{n_1}(s; m) \operatorname{res}(W_{n_2}(x; m))\big|_{x=s} = 0, \quad n_1 \neq n_2. \tag{6.22}$$

We can replace n_0 in (6.22) with any $N > n_0$, hence one can rewrite this relation in the form

$$\sum_{s=0}^{\infty} e^{\omega s} P_{n_1}(s; m) \operatorname{res}\big(W_{n_2}(x; m)\big)\big|_{x=s} = 0, \quad n_1 \neq n_2. \tag{6.23}$$

The reason is that for $s > n_0$ one has $\operatorname{res}\big(W_{n_2}(x; m)\big)\big|_{x=s} = 0$.

Note that in the special case $m = 0$ we have biorthogonality relation for polynomials $P_n(x; 0)$ appearing in the ordinary Lagrange interpolation problem. This question is studied in [**19**].

7. Multipoint interpolation of the power function on the exponential grid

In this section we announce a new result concerning multipoint interpolation of the power function $f(x) = x^\alpha$ with arbitrary real parameter α on the exponential grid $x_s = q^{-s}$ with arbitrary positive parameter q (of course, we assume that $q \neq 0, 1$). It is convenient to introduce the parameter $p = q^{-\alpha}$.

Theorem 4. *Assume that the interpolation grid x_s is a geometric progression $x_s = q^{-s}$, $s = 0, 1, \ldots$. Then solution of the Cauchy–Jacobi interpolation problem*

$$a_s = q^{-\alpha s} = p^s = \frac{Q_m(q^{-s}; n)}{P_n(q^{-s}; m)}, \quad s = 0, 1, \ldots, n+m \tag{7.1}$$

is given explicitly by the polynomials

$$\begin{aligned} P_n(x; m) &= {}_3\Phi_2\left(\begin{matrix} q^{-n}, x, q^{-m}/p \\ q^{-n-m}, 0 \end{matrix}; q, q\right), \\ Q_m(x; n) &= {}_3\Phi_2\left(\begin{matrix} q^{-m}, x, pq^{-n} \\ q^{-n-m}, 0 \end{matrix}; q, q\right). \end{aligned} \tag{7.2}$$

We adopt standard notation for basic hypergeometric function [**9**]:

$${}_3\Phi_2\left(\begin{matrix} q^{-n}, b, c \\ d, e \end{matrix}; q, z\right) = \sum_{k=0}^{n} \frac{(q^{-n}; q)_k (b; q)_k (c; q)_k}{(q; q)_k (d; q)_k (e; q)_k} z^k,$$

where $(a; q)_k = (1-a)(1-aq)\cdots(1-aq^{k-1})$ is q-shifted factorial (Pochhammer q-symbol).

Note that the polynomials

$$K_n(x; \tau, N; q) = {}_3\Phi_2\left(\begin{matrix} q^{-n}, x, -\tau q^n \\ q^{-N}, 0 \end{matrix}; q, q\right) \tag{7.3}$$

are called *q-Krawtchouk polynomials* [**9**], where $N = 1, 2, \ldots$ is a positive integer and p is an arbitrary parameter. Comparing (7.2) with (7.3) we notice that both $P_n(x; m)$ and $Q_m(x; n)$ interpolation polynomials are expressed in terms of q-Krawtchouk polynomials with $N = n + m$ and $\tau = -q^{-N}/p$ for $P_n(x; m)$ and $\tau = -q^{-N} p$ for $Q_m(x; n)$. Thus our result can be considered as exact q-analogue of the multipoint interpolation of the exponential function, where the ordinary Krawtchouk polynomials arise.

As in the previous section, we can construct a pair of biorthogonal functions:

$$\sum_{s=0}^{\infty} q^{-\alpha s} P_{n_1}(q^{-s}; m) \operatorname{res}\big(W_{n_2}(x; m)\big)\big|_{x=q^{-s}} = 0, \quad n_1 \neq n_2.. \tag{7.4}$$

We give also explicit expression for the Thiele continued fraction of the power function:

$$x^\alpha = 1 + \cfrac{x-1}{\eta_1 + \cfrac{x - q^{-1}}{\eta_2 + \cfrac{x - q^{-2}}{\eta_3 + \cdots}}}, \tag{7.5}$$

where

$$\begin{aligned} \eta_{2n+2} &= \frac{p^{2n}(p^2-1)(q/p; q)_n}{(pq^2; q)_n (1-pq)}, \\ \eta_{2n+1} &= \frac{(pq^2; q)_n (p - q^{-1})((p+1)q^{n+1} - 1 - qp)}{(qp)^{2n}(q/p; q)_n (1 - pq^{n+1})(p-1)}, \quad n = 0, 1, \ldots \end{aligned} \tag{7.6}$$

and $p = q^{-\alpha}$.

In the special case $p = q^{-1/2}$ (i.e. for the simplest radical function $f(x) = \sqrt{x}$) we have drastic simplification of the coefficients:

$$\eta_n = (1 + q^{1/2}) q^{-n/2}. \tag{7.7}$$

This result follows from the fact that for $f(x) = x^{1/2}$ and for any positive interpolation sequence x_k, $k = 0, 1, 2, \dots$ we have from the Thiele algorithm (5.2): $f_n(x) = x^{1/2} + x_{n-1}^{1/2}$, i.e. in this case $f_n(x)$ coincides with $f_0(x)$ to within an additive constant. Hence $\eta_n = f_n(x_n) = x_n^{1/2} + x_{n-1}^{1/2}$ which coincides with (7.7).

Note that we can combine our results in the following form. Consider an arbitrary geometric progression $a_s = p^s$, $s = 0, 1, \dots$. Then the Cauchy–Jacobi interpolation problem for the sequence a_s admits explicit solutions for two types of interpolation grid:

(i) arbitrary arithmetic progression $x_s = sh - s_0$ (in this case we obtain the ordinary Krawtchouk polynomials), and

(ii) arbitrary geometric progression $x_s = q^s$ (in this case we obtain q-Krawtchouk polynomials).

References

1. G. A. Baker and P. Graves-Morris, *Padé approximants. I–II*, Encyclopedia Math. Appl. **13**, **14**, Addison-Wesley Publishing Co., Reading, Mass., 1981.
2. K. Driver and N. Temme, *Zero and pole distribution of diagonal Padé approximants to the exponential function*, Quaestiones Math. **22** (1999), no. 1, 7–17.
3. A. Erdélyi, W. Magnus, F. Oberhettinger, and F. Tricomi, *Higher transcendental functions. I–II*, McGraw-Hill Book Company, Inc., New York-Toronto-London, 1953.
4. P. Henrici, *Note on a theorem of Saff and Varga*, Padé and Rational Approximation, Proc. Internat. Sympos. (Univ. South Florida, Tampa, 1976), Academic Press, New York, 1977, pp. 157–161.
5. A. Iserles, *On the generalized Padé approximations to the exponential function*, SIAM J. Numer. Anal. **16** (1979), no. 4, 631–636.
6. ———, *Rational interpolation to* $\exp(-x)$ *with application to certain stiff systems*, SIAM J. Numer. Anal. **18** (1981), no. 1, 1–12.
7. M. Ismail and D. Masson, *Generalized orthogonality and continued fractions*, J. Approx. Theory **83** (1995), no. 1, 1–40.
8. W. B. Jones and W. J. Thron, *Continued fractions analytic theory and applications*, Encyclopedia Math. Appl. **11**, Addison-Wesley Publishing Co., Reading, Mass., 1980.
9. R. Koekoek and R. Swarttouw, *The Askey-scheme of hypergeometric orthogonal polynomials and its q-analogue*, Report no. 98–17, Delft University of Technology, 1998.
10. A. Magnus *Asymptotic estimates in complex rational approximation*, Report 2002, `http://www.math.ucl.ac.be/ magnus/num3/m3xxx00.pdf`.
11. J. Meinguet, *On the solubility of the Cauchy interpolation problem*, 1970 Approximation Theory, Proc. Sympos. (Lancaster, 1969), Academic Press, London, pp. 137–163.
12. L. M. Milne-Thomson, *The calculus of finite differences*, Macmillan and Co., Ltd., London, 1951.
13. E. B. Saff and R. S. Varga, *On the zeros and poles of Paé approximants to* e^z. III, Numer. Math. **30** (1978), 241–266.
14. V. Spiridonov and A. Zhedanov, *Discrete-time Volterra chain and classical orthogonal polynomials*, J. Phys. A **30** (1997), no. 24, 8727–8737.
15. ———, *Generalized eigenvalue problem and a new family of rational functions biorthogonal on elliptic grids*, Special Functions 2000: Current Perspective and Future Directions (Tempe, AZ), NATO Sci. Ser. II Math. Phys. Chem., vol. 30, Kluwer Acad. Publ., Dordrecht, 2001, pp. 365–388.
16. ———, *To the theory of biorthogonal rational functions*, New Developments in the Research of Integrable Systems that are Continuous, Discrete and Ultradiscrete (Japanese) (Kyoto, 2002), Sūrikaisekikenkyūsho Kōkyūroku, no. 1302, 2003, pp. 172–192.
17. A. Zhedanov, *Rational spectral transformations and orthogonal polynomials*, J. Comput. Appl. Math. **85**, (1997), no. 1, 67–86.
18. ———, *Biorthogonal rational functions and the generalized eigenvalue problem*, J. Approx. Theory **101** (1999), no. 2, 303–329.

19. ———, *Biorthogonality of the Lagrange interpolants*, J. Comput. Appl. Math. **172** (2004), 1–6.

Donetsk Institute for Physics and Technology, Donetsk 83114, Ukraine.
E-mail address: zhedanov@kinetic.ac.donetsk.ua

Titles in This Series

39 **P. Winternitz, D. Gomez-Ullate, A. Iserles, D. Levi, P. J. Olver, R. Quispel, and P. Tempesta, Editors,** Group theory and numerical analysis, 2005

38 **Jacques Hurtubise and Eyal Markman, Editors,** Algebraic structures and moduli spaces, 2004

37 **P. Tempesta, P. Winternitz, J. Harnad, W. Miller, Jr., G. Pogosyan, and M. Rodriguez, Editors,** Superintegrability in classical and quantum systems, 2004

36 **Hershy Kisilevsky and Eyal Z. Goren, Editors,** Number theory, 2004

35 **H. E. A. Eddy Campbell and David L. Wehlau, Editors,** Invariant theory in all characteristics, 2004

34 **P. Winternitz, J. Harnad, C. S. Lam, and J. Patera, Editors,** Symmetry in physics, 2004

33 **André D. Bandrauk, Michel C. Delfour, and Claude Le Bris, Editors,** Quantum control: Mathematical and numerical challenges, 2003

32 **Vadim B. Kuznetsov, Editor,** The Kowalevski property, 2002

31 **John Harnad and Alexander Its, Editors,** Isomonodromic deformations and applications in physics, 2002

30 **John McKay and Abdellah Sebbar, Editors,** Proceedings on moonshine and related topics, 2001

29 **Alan Coley, Decio Levi, Robert Milson, Colin Rogers, and Pavel Winternitz, Editors,** Bäcklund and Darboux transformations. The geometry of solitons, 2001

28 **J. C. Taylor, Editor,** Topics in probability and Lie groups: Boundary theory, 2001

27 **I. M. Sigal and C. Sulem, Editors,** Nonlinear dynamics and renormalization group, 2001

26 **J. Harnad, G. Sabidussi, and P. Winternitz, Editors,** Integrable systems: From classical to quantum, 2000

25 **Decio Levi and Orlando Ragnisco, Editors,** SIDE III—Symmetries and integrability of difference equations, 2000

24 **B. Brent Gordon, James D. Lewis, Stefan Müller-Stach, Shuji Saito, and Noriko Yui, Editors,** The arithmetic and geometry of algebraic cycles, 2000

23 **Pierre Hansen and Odile Marcotte, Editors,** Graph colouring and applications, 1999

22 **Jan Felipe van Diejen and Luc Vinet, Editors,** Algebraic methods and q-special functions, 1999

21 **Michel Fortin, Editor,** Plates and shells, 1999

20 **Katie Coughlin, Editor,** Semi-analytic methods for the Navier-Stokes equations, 1999

19 **Rajiv Gupta and Kenneth S. Williams, Editors,** Number theory, 1999

18 **Serge Dubuc and Gilles Deslauriers, Editors,** Spline functions and the theory of wavelets, 1999

17 **Olga Kharlampovich, Editor,** Summer school in group theory in Banff, 1996, 1998

16 **Alain Vincent, Editor,** Numerical methods in fluid mechanics, 1998

15 **François Lalonde, Editor,** Geometry, topology, and dynamics, 1998

14 **John Harnad and Alex Kasman, Editors,** The bispectral problem, 1998

13 **Michel C. Delfour, Editor,** Boundaries, interfaces, and transitions, 1998

12 **Peter C. Greiner, Victor Ivrii, Luis A. Seco, and Catherine Sulem, Editors,** Partial differential equations and their applications, 1997

11 **Luc Vinet, Editor,** Advances in mathematical sciences: CRM's 25 years, 1997

10 **Donald E. Knuth,** Stable marriage and its relation to other combinatorial problems: An introduction to the mathematical analysis of algorithms, 1997

9 **D. Levi, L. Vinet, and P. Winternitz, Editors,** Symmetries and integrability of difference equations, 1996

TITLES IN THIS SERIES

8 **J. Feldman, R. Froese, and L. M. Rosen, Editors,** Mathematical quantum theory II: Schrödinger operators, 1995

7 **J. Feldman, R. Froese, and L. M. Rosen, Editors,** Mathematical quantum theory I: Field theory and many-body theory, 1994

6 **Guido Mislin, Editor,** The Hilton Symposium 1993: Topics in topology and group theory, 1994

5 **D. A. Dawson, Editor,** Measure-valued processes, stochastic partial differential equations, and interacting systems, 1994

4 **Hershy Kisilevsky and M. Ram Murty, Editors,** Elliptic curves and related topics, 1994

3 **Rémi Vaillancourt and Andrei L. Smirnov, Editors,** Asymptotic methods in mechanics, 1993

2 **Philip D. Loewen,** Optimal control via nonsmooth analysis, 1993

1 **M. Ram Murty, Editor,** Theta functions: From the classical to the modern, 1993